Wind Energy: Science and Engineering

Wind Energy: Science and Engineering

Edited by Benjamin Wayne

SYRAWOOD
PUBLISHING HOUSE

New York

Published by Syrawood Publishing House,
750 Third Avenue, 9th Floor,
New York, NY 10017, USA
www.syrawoodpublishinghouse.com

Wind Energy: Science and Engineering
Edited by Benjamin Wayne

International Standard Book Number: 978-1-68286-466-1 (Hardback)

Cataloging-in-publication Data

Wind energy : science and engineering / edited by Benjamin Wayne.
    p. cm.
Includes bibliographical references and index.
ISBN 978-1-68286-466-1
1. Wind power. 2. Wind energy conversion systems. 3. Wind turbines--Automatic control. I. Wayne, Benjamin.
TJ820 .W56 2017
621.45--dc23

Printed in the United States of America.

TABLE OF CONTENTS

PREFACE

The ever growing need for energy has caused serious harm to the environment and nature. The use of fossil fuel in energy generation has led to severe pollution. Therefore, alternative energy is the need of the hour. Wind energy is one of the most reliable and cost-effective forms of alternative or green energy. This book explores all the important aspects of wind energy in the present day scenario. It discusses the science behind its conception and engineering behind its application. The various studies that are constantly contributing towards advancing technologies and evolution of the field of wind energy are examined in detail in the text. The extensive content of this book provides the readers with a thorough understanding of the subject. It will serve as a reference to a broad spectrum of readers.

This book is a comprehensive compilation of works of different researchers from varied parts of the world. It includes valuable experiences of the researchers with the sole objective of providing the readers (learners) with a proper knowledge of the concerned field. This book will be beneficial in evoking inspiration and enhancing the knowledge of the interested readers.

In the end, I would like to extend my heartiest thanks to the authors who worked with great determination on their chapters. I also appreciate the publisher's support in the course of the book. I would also like to deeply acknowledge my family who stood by me as a source of inspiration during the project.

Editor

The Measuring System for Estimation of Power of Wind Flow Generated by Train Movement and Its Experimental Testing

Oleksandr Mokin, Borys Mokin, Vadym Bazalytskyy

Department of Renewable Energy and Transport Electrical Systems and Complexes, Vinnytsia National Technical University, Vinnytsia, Ukraine
Email: abmokin@gmail.com

Abstract

The measuring system for estimation of power of wind flow generated by the train movement has been created. The advantages of the proposed system are the cheapness and simple design. With its simplicity of design and easy build-up of channels, designed measuring system can be used for a wide range of technical problems. This paper describes the design process, validation and conducting the first field test of this measuring system.

Keywords

Power Generation, Wind Turbine, Wind Flow, Sensor, Measurement

1. Introduction

Anyone staying on the platform of the station during the freight train passing could fully realize that the speed of wind flow generated by the moving trains can reach significant values and in the case with high-speed trains the wind flow may reach the speed of storm winds. Today the energy of the wind flow is dissipated in the atmosphere without any benefit while it can be used to generate electricity by using wind power plants located close to the railway tracks.

It is known [1] that power P_{wf} (W) of the wind flow with a density $\rho \left(\text{kg}/\text{m}^3 \right)$ received by the wind turbine with vertical axis of rotation and axial sectional area of the wind wheel $S_0 \left(\text{m}^2 \right)$ and power factor ε, is proportional to that area $S_0 \left(\text{m}^2 \right)$ and the cube of the speed v_{wf} (m/s) of the wind flow, that is:

$$P_{wf} = \varepsilon \frac{S_0}{2} \rho v_{wf}^3 \qquad (1)$$

In paper [2], the authors of this paper had made a quantitative assessment of the power of the wind flow generated by the train during movement with using the data presented in paper [3], which states that the sensors for measuring speed of the wind flow generated by the train movement were installed on the wall along the railway track. Obviously, the speed of the wind flow fixed by the sensors on the walls is not adequate to the speed of wind flow generated by the same train in the free space.

In order to make quantitative assessment of the speed of wind flow generated by train, in Vinnytsia National Technical University a measuring system that allows making quantitative assessment of the speed of wind flow in the field conditions has been created. The basic structural element of the measuring system is a digital sensor. In this paper, the development of this measuring system and its digital sensors has been considered and the research of their applicability for the field conditions has been made.

2. Characteristics of the Developed Digital Sensor

Block diagram of the digital sensor, developed by the authors, is shown in **Figure 1**.

Schematic representation of the measuring system and its digital sensors placed in the frame which is clear for wind is shown in **Figure 2**, where the wind speed sensors 1 - 12 are miniature wind wheels with Hall effect sensors.

These sensors are based on the ordinary computer fans. Chip FTC S276 [4] are placed on the rotor and its work is based on the principle of Hall effect. This chip crosses the magnetic reels, placed on the stator during the rotation of the rotor. The result of the Hall effect which is manifested in the potential difference is amplified and supplied to a control system of electronic keys which opens the corresponding one. Since one circuit was sufficient for our experiment, we used only one of the outputs of the chip.

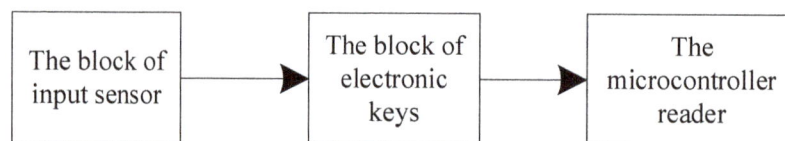

Figure 1. Block diagram of the digital sensor of speed of the wind flow generated by moving train.

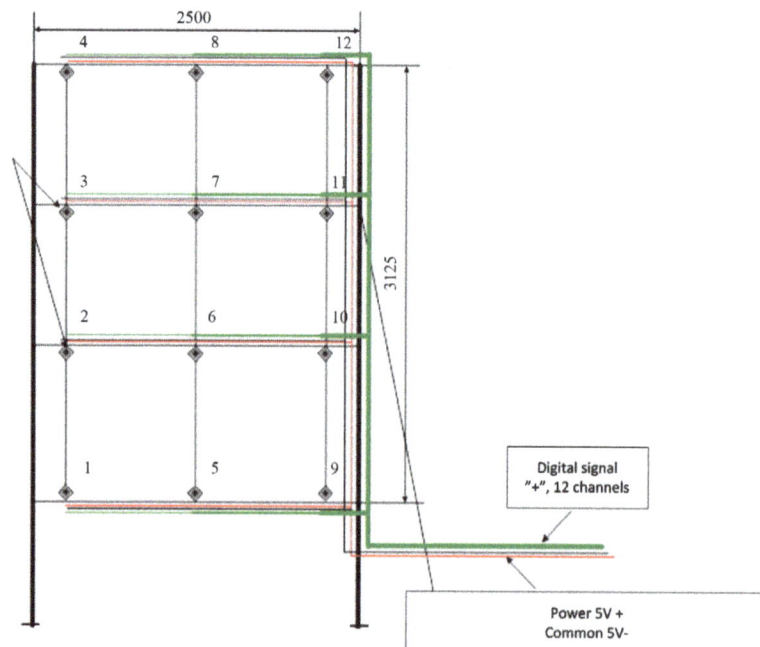

Figure 2. Schematic representation of the placement points measuring sensors for studies of wind flow generated by the train movement.

Sensor works as follows: when a sensor chip with Hall effect sensor passes through the magnetic field of the stator winding, there appears a difference of potentials which then enters an amplifier and commands to open one of the keys. This ensures operation of the digital sensor which upon rotation generates two discrete pulses per revolution in accordance with the number of coils corresponding polarity. This pulse sequence is applied to the digital inputs of the hardware platform based on Arduino Nano microcontroller ATmega 328 with digital inputs D2-D13 on which signals from speed sensors 1 - 12 are applied. And through the serial port COM3 the data is transmitted to a computer where a program designed for reading data from serial port, collects and records this data every second (**Figure 3**).

Each sensor calibration was performed in a laboratory wind tunnel using a reference anemometer.

For each speed sensor 1 - 12 by using the method of least squares [5] there was determined the dependence of the rotational speed from the wind speed (**Figure 4**).

Channel 3 has somewhat different characteristics because it uses sensor with the other ratio of wind speed to the number of turns. But it does not affect the result for each sensor due to using different transformation ratio.

3. Description of Conditions and Locations of the Experiment and the Characteristics of the Measured Values

After the calibration of all the measuring sensors and having determined the conversion factor for each sensor there was performed the experimental verification of their efficiency.

To carry out an experiment on measuring the speed of the wind flow generated by moving train using the designed digital sensor, there had been chosen a segment of a railroad tracks not far from Vinnytsia on the closed railway crossing in the village of Parpurivtsi. This location was chosen with the permission of the railway administration since it is convenient for mounting the measuring frame with the input sensors, there is the service road and the speed of the trains is relatively high. As the safe distance from the railway track according to the regulatory documents is 3 meters, the experimental measuring system with sensors was placed exactly at this distance perpendicularly to the railway track that is perpendicular to the axis of the wind flow moving parallel to the train

The sensors were placed in three columns at 3, 4, 5 meters from the railway track with four sensors in each

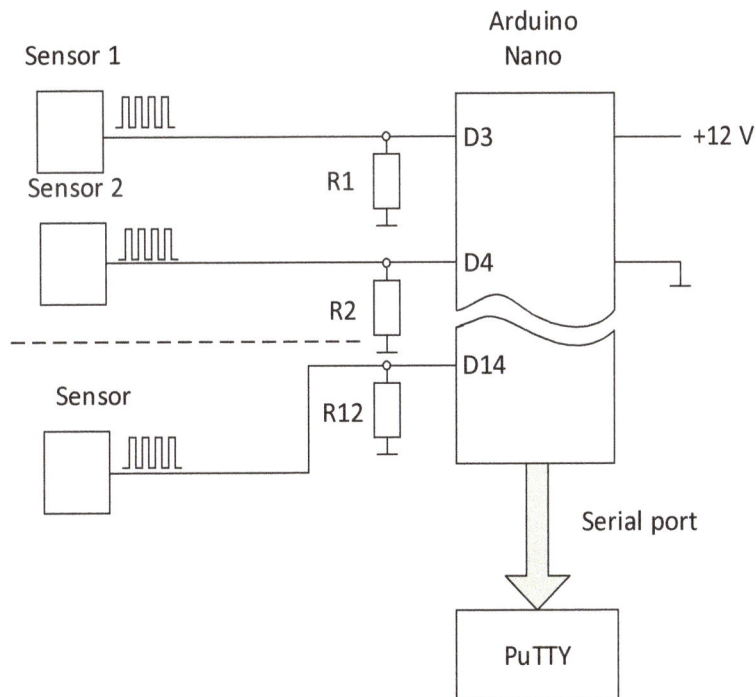

Figure 3. Schematic diagram of the sensors that used to determine the speed characteristics of the wind flow.

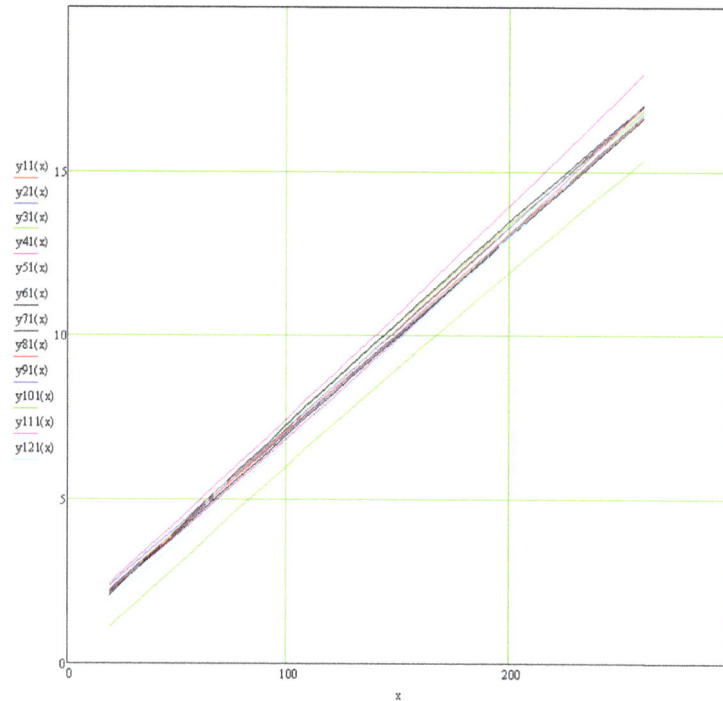

Figure 4. Graphs of the speed of wind flow as a function of the number of pulses for each sensor 1 - 12.

according to the scheme shown in **Figure 2**.

The data was being recorded on a computer simultaneously for each of the sensors with time interval of 1 second as a pulse burst, each of which contained a number of pulses proportional to the angular velocity of the sensor's rotor.

The experimental data were fixed when freight train with locomotive VL80k and the mixed composition of rail cars with the number of 56 of them was passing by. The speed of the train was 66 kilometers per hour and the time of passing by the measuring system was 45 seconds.

Figure 6 represents the graphs that show the speed change measured by each sensor in the coordinates: x-axis—seconds, vertical axis—meters per second.

From **Figure 5** clearly seen point in time $\left(M_t = 43\,\text{s}\right)$ at which the train reaches the measuring system as well as the fact that the wind flow increases several times in comparison to the values of natural wind speed. An interesting fact is that after passing of the train $\left(M_t = 88\,\text{s}\right)$, for some more time which is about 35 seconds (up to $M_t^* = 88\,\text{s}$) for the train with the speed of 66 kilometers per hour, there remains the excess of speed of the wind flow over its value in the unperturbed state.

4. Processing of the Experimental Results

In order to determine the power of the wind flow which crosses the area of the frame of measuring system there had been used an expression (1) which for each sensor would look like

$$P_i = \varepsilon \cdot \rho \cdot \frac{S \cdot V_i^3}{2}, \quad i = 1, \cdots, 12, \tag{2}$$

where S—area of the circle with a diameter equal to the diameter of wheel of the wind speed sensor.

Using the speed curves presented in **Figure 5** and cubic splines of Mathcad for the implementation of (2) we get

$$P_i(u) = \text{interp}\left(\text{cspline}\left(M_t, P_i\right), M_t, P_1, u\right) \tag{3}$$

The graphs of the power of the wind flow for each sensor of the measuring system calculated by the expression (3), shown in **Figure 6** in the coordinates: x-axis—seconds, vertical axis—watts.

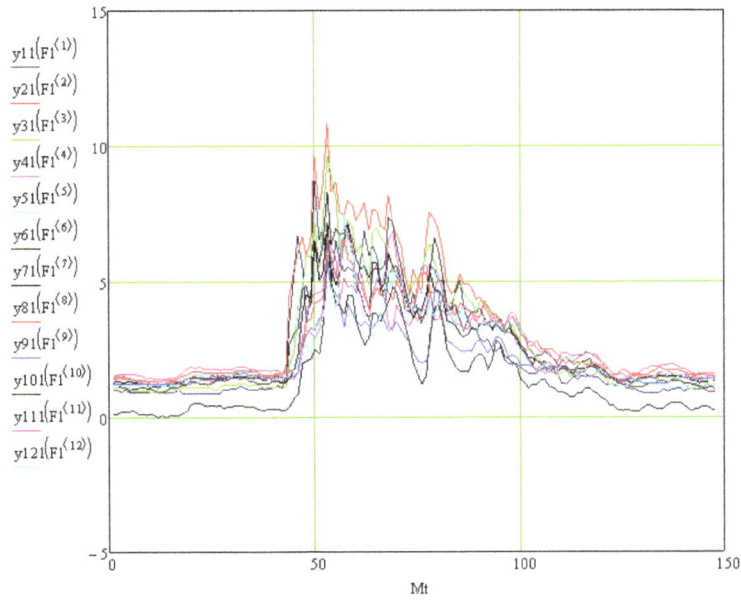

Figure 5. The graphs of speed of the wind flow measured by each sensor.

These graphs show that the values of the wind flow power differ depending on the location of the sensor. For greater clarity, let us show it on a three-dimensional graph that displays the volume of the received capacity for each measuring channel according to the location of sensors in the frame of the measuring system. For this purpose, we integrate each of the dependencies which determines the power of the wind flow for each measuring channel within the time of $M_t = 43$ s to $M_t^* = 121$ s and divide the integration interval $\left(M_t^* - M_t\right)$ and thus find the average power $P_{s_j} = 1, 2, \cdots, 12$ of the wind flow for each measuring channel.

$$Ps_i = \frac{1}{121-43} \int_{43}^{121} P_i(u) \, du, \; i = 1, 2, \cdots, 12 \tag{4}$$

The average power Ps_i, $i = 1, 2, \cdots, 12$ of the wind flow for each measuring channel calculated by the expression (4) can be expressed as matrix-like expression

$$Ps = \begin{pmatrix} 9.924 & 12.263 & 9.276 \\ 22.357 & 12.338 & 18.399 \\ 35.016 & 17.96 & 4.468 \\ 15.495 & 14.71 & 4.511 \end{pmatrix} \tag{5}$$

Figure 7 presents the spatial graphs of average power calculated by the expression (4).

To estimate the total power of the wind flow passing through the sensor frame of the measuring system during perturbation of air masses caused by train movement we calculate the sum of values of power for all measuring channels and find

$$Ps_\Sigma = \sum_{i=1}^{12} Ps_i = 176.715 \, (\text{W}) \tag{6}$$

5. Conclusions

1. The structure and block diagram of the digital sensor of the speed of the wind flow generated by the train movement which measures this parameter of the wind flow with sufficient accuracy for practical purposes has been suggested. Using this digital sensor allowed to design a measuring system applicable to measure the real-time parameters of the wind flow on area of 20 m^2.

2. The field tests of the developed measuring system confirmed its high measuring and computational efficiency and suitability for use in the experiments.

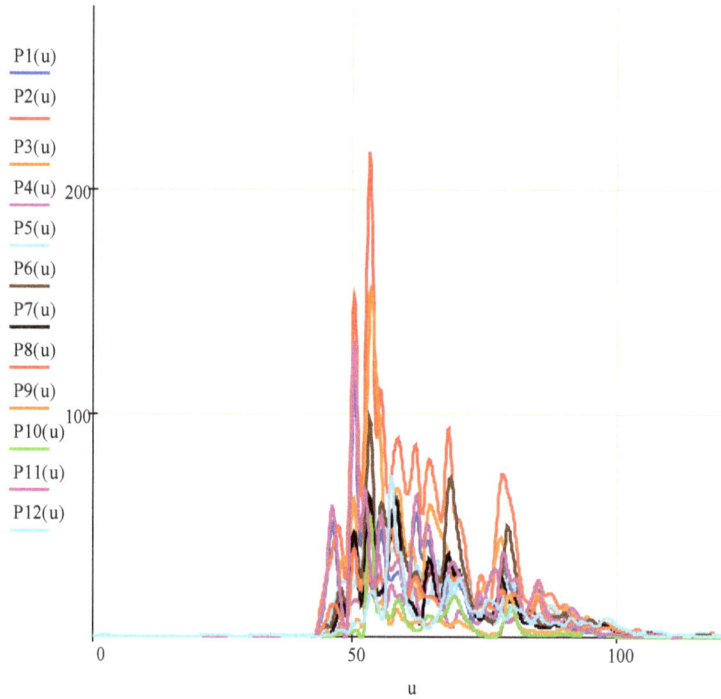

Figure 6. Graphs of power of the wind flow as a function of time for each measuring channel.

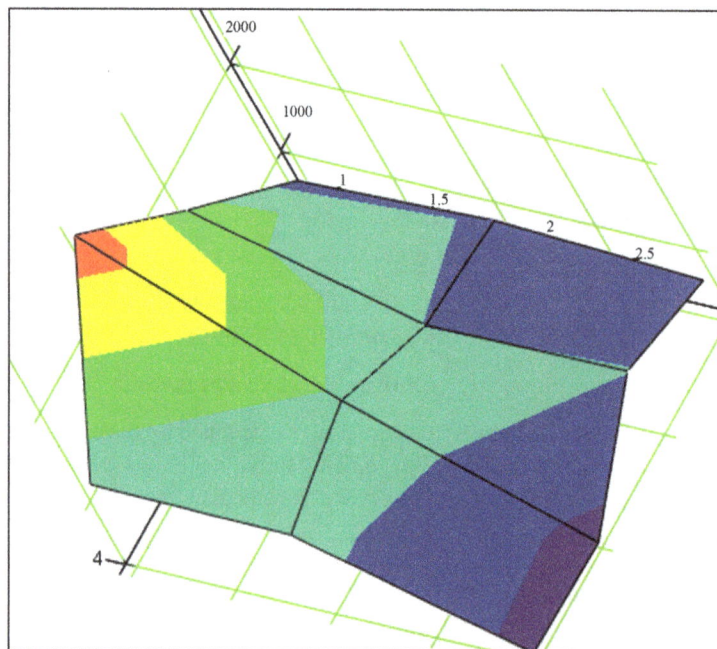

Ps

Figure 7. The surface of the average power of the wind flow measured by measuring system with displaying the fraction of each measuring channel.

3. It had been determined that the powerful wind flow at an authorized three-meter distance from the railway track could be created only by freight trains with different forms of rail cars moving at speeds above 60 kilometers per hour.

References

[1]   Mkhitaryan, N. (1999) Energy Alternative and Renewable Sources. Experience and Prospects. Naukova Dumka, Kiev, 320 p. (in Ukrainian)

[2]   Mokin, B.I., Mokin, O.B. and Bazalytskyy, V.P. (2011) Evaluation of Power, Which Can Be Obtained from the Wind Stream Caused by Traffic Railway Train. Visnyk of Vinnytsia Politechnical Institute, 81-84.

[3]   MacNeill, A. and Holmes, S. (2002) Measurement of the Aerodynamic Pressures Produced by Passing Trains. *Proceedings of the* 2002 *ASME/IEEE Joint Rail Conference*, Washington DC, 23-25 April 2002, 1-8.

[4]   (2003) FTC s276, 2-Phase DC Motor Drive IC. Data Sheets, Felling Technology.

[5]   Mokin, B.I., Mokin, V.B. and Mokin, O.B. (2010) Mathematical Methods of Identification of Dynamical Systems. VNTU, Vinnytsia, 260 p.

2

On the Remapping and Identification of Potential Wind Sites in Nigeria

Zaccheus O. Olaofe

Climate System and Analysis Group, University of Cape Town, Rondebosch, South Africa
Email: zakky201@gmail.com

Abstract

The ERA-Interim reanalysis wind based on the distance-weighted average remapping for studying the wind circulation in Nigeria is presented. The wind flow using this atmospheric model simulation is studied for identification of grid-tie electrification opportunities in different wind locations. A 10-year reanalysis wind speed components at a surface level of the planetary layer at $0.25° \times 0.25°$ spatial resolution is obtained and remapped into a new horizontal wind field at a grid resolution of $0.125° \times 0.125°$ covering longitudinal and latitudinal directions of 3.0 - 15.0°E and 15.0 - 3.0°N, respectively. Using the distance-weighted average technique, the remapped wind field at a new grid resolution of $0.125° \times 0.125°$ is compared at different terrain elevations and approximated close to the actual wind field of the same resolution. To determine the suitability of the prevailing wind for small-scale energy conversion, the magnitude of wind flow across the remapped wind field is studied for a 10-year period. Analysis shows that northern regions of Nigeria have a fair wind potential for a stand-alone application based on the wind flow originated at Gulf of Guinea as well as Chad and Niger. Furthermore, hourly surface wind speed observations from 18 synoptic stations in Nigeria are obtained and compared with the bilinear interpolated wind stations. The reanalysis wind reflects the surface wind observations and proves that the prevailing wind in Nigeria is higher than the reanalysis wind projection obtained from gridded data at resolution of $0.125° \times 0.125°$. The sectorwise wind directions at each synoptic stations for a period of 10 years are presented.

Keywords

ERA-Interim Wind Components, Grid Resolutions, Relative Error, Distance-Weighted Remapping, Bilinear Interpolation, Nigeria

1. Introduction

Wind is produced when different forces of the atmosphere (such as the pressure gradient force, Coriolis force, frictional force and centripetal force) act on a parcel of air in motion from a high pressure to a low pressure region. In addition, the wind is primarily driven by differences in air pressure at two regions or levels, though rotation of the earth does have an influence on the direction of wind flow. The wind flows from a higher pressure to a lower pressure system in an attempt to balance the pressure at two different regions. These variations in air pressure are attributed to temperature differences caused by variations in solar energy received on earth surface. (That is, there is uneven heating of the earth surface by solar radiation which causes temperature differences and hence, atmospheric pressure differences across the surface which produces the wind). The balance of these forces in the vertical and horizontal directions gives rise to the different kinds of winds (gradient wind, thermal wind, geostrophic wind etc.). The frictional force is another atmospheric pressure that opposes the wind flow and causes the air parcel to turn slightly toward lower pressure region. Furthermore, the Coriolis force deflects the direction of wind flow to the right in Northern Hemisphere and to the left in Southern Hemisphere due to earth rotation. Hence, the pressure gradient force (pgf), frictional force, Coriolis force and centripetal force are the main drivers in wind circulation at any considered location. The interaction of these forces and the effects on wind flow are best understood using an atmospheric simulation.

An air parcel which is at initial rest will move from a high pressure to low pressure region due to the pressure gradient force. When the air parcel moves, it is deflected by the Coriolis. As the air parcel gains momentum, this wind deflection increases until the Coriolis force equals the pressure gradient force. This condition is called the geostrophic balance and at this point the air parcels are parallel to the isobars. Therefore, this movement of air parcels parallel to the isobars due to this balance is referred to as geostrophic winds. According to Nilsson and Ivanell [1], the geostrophic wind is found at altitude higher than 1 km, largely driven by air temperature differences, and unaffected by the earth surface.

Several wind studies have been conducted on the analysis of wind speed distributions and power densities using the historical records at a reference or group of synoptic stations at different wind locations in Nigeria. However, no study has been conducted on the regional wind circulation in Nigeria, using a high resolution wind simulation from the climate mode. In the earlier wind study conducted by Aidan et al. [2], the surface wind obtained at 8 synoptic stations in the northern Nigeria where fitted using the Normal, Weibull, Rayleigh and Gamma probability density functions. They concluded their wind study at 80 m hub height to be very satisfactory for utility-scale power generation at four stations: Gusua, Kaduna, Maiduguri and Zaria. As presented by Fadare (2008), the author conducted a statistical wind energy analysis using daily wind records from station mast for 10 years period (1995-2004), obtained from the International Institute of Tropical Agriculture (IITA), Ibadan, Nigeria. The daily, monthly, seasonal, and yearly wind speeds were modeled using Weibull probaility density function. The annual mean wind speed in Ibadan was estimated at 2.75 m/s, while Weibull mean wind speed and the power density were estimated at 2.947 m/s and 15.484 $W \cdot m^{-2}$, respectively. The author classified Ibadan city to be a low wind energy region. Ohunakin [3] in the study conducted a wind energy assessment using a 21 years station observations at Uyo city in Niger-Delta obtained from the Nigerian Meterological Agency (NIMET). The historical speed records (1986-2007) were statistically analyzed using both Weibull and Rayleigh density functions, and results show Uyo station to have mean wind speed of 3.17 m/s and maximum at 3.67 m/s with a corresponding wind power density approximated at 19.91 $W \cdot m^{-2}$ for a year's record. The mean wind speed at this station was not economically viable for a medium nor utility-scale energy application but suitable for small-scale application only as the station falls within class 1 of the international system of wind classification. Furthermore, a number of few wind studies have been conducted on assessments of the wind speed characteristics at different locations in Nigeria by Ojosu et al. [4] [5], Adekoya et al. [6], Anyanwu et al. [7], Agbaka [8], Igbokwe et al. [9], Medugu et al. [10], Ngala et al. [11] and Oriaku et al. [12]. In literature, the wind speed record was modeled using different analytical tools such as: statistical modeling (Weibull and Rayleigh distribution functions); Seasonal Autoregressive Integrated Moving Average Modeling; Linear and Multiple Regression modeling; stochastic simulation; and Artificial Neural Network (ANN).

A recent study on the development of medium-scale wind farm in Katsina-Nigeria was presented by Garba et al. [13]. From the review on the wind studies conducted by different authors in Nigeria; using either the reference, a combination of different meterological observations or monthly mean wind records is not sufficient because most cities and synoptic stations have poor historical wind records and are unsuitable for regional resource assessment in Nigeria. Some of the synoptic observations had missing wind climatologies for different time of days, weeks and month making wind assessment difficult to carry out. This may be attributed to local topographic effects on the wind flow, power supply failure to the data logging systems deployed on meterological masts, missing observation due to data transmission error from the synoptic station to data centre, data logger failure, calibration error or degradation of existing sensors deployed on the various station masts. The effects of boom and other mounting arrangement around a synoptic mast may introduce a large bias in wind sensing as shown in the wind tunnel studies conducted by Pedersen *et al.* [14]. Due to the effect of local topography on the wind flow at different geographical locations in Nigeria, a low quality of historical wind records due to deterioration of weather sensors (such as temperature, pressure, turbulence or cup anemometer mechanisms) as well as insufficient synoptic stations at strategic locations for wind measurements on regional scale in Nigeria, a mesoscale wind modeling covering the landmark is required. The mesoscale modeling has an acceptable accuracy level, very cost-effective and provides long-term wind resource estimates over geographical locations with limited *in situ* measurements or at synoptic stations with no historical wind climates. Using low quality wind observations often leads to a poor guide in the identification and mapping of potential locations for wind atlas development [15]. Furthermore, the solid state wind sensors deployed on synoptic masts have replaced the traditional mechanical cup anemometers because of the following: interference of the station tower, boom and other mounting arrangements obstructing the wind flow, the anemometer design and calibration procedure, response to turbulence characteristics of the wind flow, atmospheric stability measurement needs, deterioration of mechanical parts of the cup anemometer at stations due to harsh weather conditions [16].

The ERA-Interim simulation is a global atmospheric reanalysis dataset produced by the European Centre for Medium-Range Weather Forecasts (ECMWF). The ERA reanalysis dataset provides a multivariate, spatially complete, and coherent records of the global atmospheric circulation. The atmospheric model, data assimilation method, and the *in situ* observations used in producing the ERA-Interim dataset as well as the basic performance evaluation are discussed by Dee *et al.* [17].

Though, ERA-Interimwind datasets have insufficient resolution to allow for direct utilization in wind resource assessment (lower resolution of the model output and the coarse representation of orography over the land-sea surface), however, they can be used to: *provide the boundary conditions to mesoscale modeling capable of deriving wind climatologies at a higher resolution on regional scale; resolve local and regional circulation patterns and the atmospheric boundary layer,* and *can be applied over domains of several hundreds of kilometers squared covered with a grid mesh with a resolutions of a few kilometers* [18] [19]. In summary, the ERA-Interim wind reanalysis dataset provides a global view that encompasses many essential climate variables in a physically consistent framework with only a short time delay. For wind assessment study, it can be used to provide a global view on the wind circulation at desired geographical location. The application of the reanalysis wind has been presented by Buzzi *et al.* [20] where the ECMWF reanalysis data was used to provide boundary conditions for the parallel version of the hydrostatic model BOLAM (Bologna Limited Area Model) for generating the wind statistics over the Mediterranean over a 2 years period at a 10 km grid resolution.

In this study, the ERA-Interim wind speed components are utilized for studying the wind flow into Nigeria, originated at Gulf of Guinea and two neighbouring countries (Chad and Niger). The wind circulation is studied for identification and mapping of wind potential sites in Nigeria. The wind flow from Gulf of Guinea originated from Guinea and Benguela currents, while Chad and Niger flow originated from Mediterranean Sea. The considered geographical coordinates of the proposed wind field covers longitudinal and latitudinal directions of 3.0 - 15.0°E and 15.0 - 3.0°N, respectively as shown in **Figure 1(a)**, **Figure 1(b)**. The 10 meter U and V wind components from the ERA-Interim dataset at a temporal resolution of 00:00:00, 06:00:00, 12:00:00 and 18:00:00 UTC for period of 2005-2014 are obtained on spatial resolutions of 0.125° × 0.125° and 0.25° × 0.25° [21]. Using the wind speed components at a grid resolution of 0.25° × 0.25° (dimension = 2401, number of grid points in the longitudinal direction "n_x" = 49, number of grid points in the longitudinal direction "n_y" = 49), within a geographical coordinates boundary of λ_x = 3 - 15°E, φ_y = 15 - 3°N, a new wind field is desired at a higher resolution of 0.125° × 0.125° (dimension = 9469, n_x = 97, n_y = 97) for the coordinates of λ_x = 3 - 15°E, φ_y = 15 - 3°N. A grid description of the source wind dataset on 0.25° × 0.25° resolution is used to reproduce a wind field on a new grid resolution of 0.125° × 0.125° using the concept of distance-weighted average remapping. To prepare a platform for evaluation of the remapped wind field and determining the accuracy of distance-weighted average technique in remapping from one grid resolution to another, the distance-weighted wind speed components at a new grid resolution of 0.125° × 0.125° (dimension = 9469, n_x = 97, n_y = 97) are compared against the actual wind speed components at the same grid resolution (dimension = 9469, n_x = 97, n_y = 97) using the following tools: 1) the climate data operator and 2) ferret running on a linux-based machine. The estimated daily, monthly and annual wind speed biases show that the gridded wind components on a grid field of 0.25° × 0.25° resolution can be

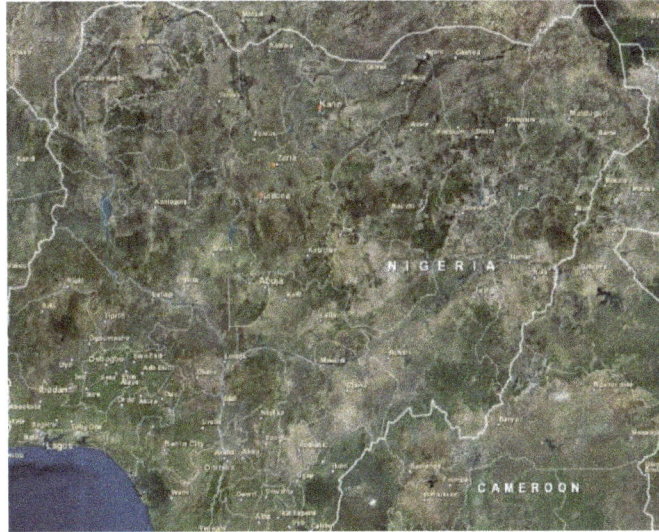

(a)

(b)

Figure 1. (a) Capturing of different locations on Nigeria map for wind flow study; (b) Geographical coordinates mapping of Nigeria land mark.

remapped into a higher resolution for smooth and flat terrain, and may be substituted with gridded dataset on a wind field of 0.125° × 0.125° grid resolution. For complex terrain at different locations in Nigeria, the remapped wind field deviates from the actual wind field of the same resolution. Furthermore, the wind speed chart for 10 years period using the remapped and actual wind fields at 0.125° × 0.125° resolution is produced. To identify wind potential region, the bilinear interpolation on the remapped wind field (0.125° × 0.125°, dimension = 9469, n_x = 97, n_y = 97) was done to obtain the wind speed gridded datasets for 18 synoptic stations within the considered geographical coordinates of the wind field.

According to the World Meteorological Organization (WMO) list containing surface and upper-air weather stations' records for synoptic applications, the Nigerian Meteorological Agency (NIMET) has a number of synoptic stations at coordinates of λ_x = 3 - 15°E, φ_y = 15 - 3°N. Using the geographical coordinates of the considered 18 synoptic stations, the hourly surface wind observations available for these stations in Nigeria are obtained from the National Ocean and Atmospheric Administration (NOAA) database [22]. The hourly surface wind observation for each weather station is available and converted to monthly and annual mean wind observations.

The ERA monthly and annual wind speed for each station are estimated, and compared with the monthly and annual surface wind observation available for each synoptic station. The remapped wind field approximated close to the actual wind field at a smooth and flat terrain while the discrepancies of the remapped wind field from the actual wind field increased with terrain elevations and contours. Comparisons of the wind statistics for 18 synoptic stations using the ERA wind simulation and surface wind are presented. The reanalysis wind reflects the wind flow for identification of potential wind locations in Nigeria. Based on the direction of wind flow from Gulf of Guinea as well as Chad and Niger into Nigeria, the wind sites in northern Nigeria suitable for grid-tie electrification are identified. Section three discussed the methodologies for remapping of gridded dataset and validation of remapped wind field; the wind flow convergence and divergence as well as the wind resource distribution over a 10 m AGL at different geographical locations in Nigeria for identification of wind potential sites in Nigeria is discussed in section four. Conclusion is made on the prospective of wind energy investment in Nigeria.

2. Data Collection

The first step in wind flow study for identification of wind potential wind sites in Nigeria is to identify and map the area of interest. To capture the map of Nigeria as shown in **Figure 1(a)**, a landmark covering longitudinal and latitudinal directions of 3.0 - 15.0°E and 15.0 - 3.0°N is considered. In an attempt to capture the entire landmark of Nigeria, a portion of few cities at Cameroon, Chad and Niger were captured alongside as shown in **Figure 1(b)**. In addition, the considered geographical coordinates are clipped on google earth as x_1 = (lon$_1$ = 3.0°, lat$_1$ = 3.0°), x_2 = (lon$_1$ = 3.0°, lat$_2$ = 15.0°), x_3 = (lon$_2$ = 15.0°, lat$_2$ = 15.0°), and x_4 = (lon$_2$ = 15.0°, lat$_1$ = 3.0°), respectively. The desired wind speed components in the ERA-Interim reanalysis datasets at these geographical coordinates are extracted. The wind speed components (10 meter U and V wind speed components) at temporal resolutions of 00:00:00, 06:00:00, 12:00:00, and 18:00:00 UTC for a period of 2005-2014 were obtained from the European Centre for Medium-Range Weather Forecasts (ECMWF) database [21]. From the wind statistics of the ERA wind speed components, a total of 14, 608 data points for 10 years period was available for this wind flow study. The considered wind speed components on the surface level are on a regular grid resolutions of 0.125° × 0.125° (dimension = 9409, n_x = 97, n_y = 97) and 0.25° × 0.25° (dimension = 2401, n_x = 49, n_y = 49). The wind statistics of these grid resolutions are summarized in **Figure 2(a)**, **Figure 2(b)**, respectively.

In addition, historical wind climatology for global weather stations is freely available on National Ocean and Atmospheric Administration (NOAA) web. The web contains datasets from global automated weather observing systems (AWOS) which are operated and controlled by the Federal Aviation Administration. The AWOS data are archived in the global surface hourly database with record from as early as 1901 [22].

To prepare a platform for evaluation and comparison of the bilinear interpolated stations on the remapped wind field, the historical wind observations at 18 stations on 10 m height AGL were obtained online from NOAA. Quality check on the hourly surface wind observations was made for each synoptic station and **Table 1** summarizes the annual surface wind observations available for each station on a 10 m height above ground level (AGL).

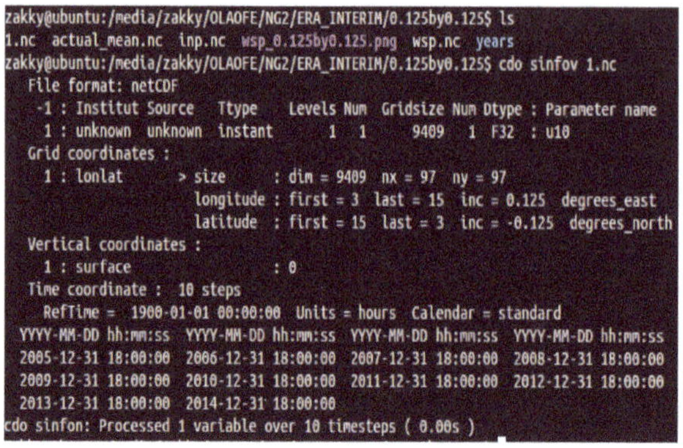

```
zakky@ubuntu:/media/zakky/OLAOFE/NG2/ERA_INTERIM/0.125by0.125$ ls
1.nc  actual_mean.nc  inp.nc  wsp_0.125by0.125.png  wsp.nc  years
zakky@ubuntu:/media/zakky/OLAOFE/NG2/ERA_INTERIM/0.125by0.125$ cdo sinfov 1.nc
   File format: netCDF
    -1 : Institut Source   Ttype    Levels Num  Gridsize Num Dtype : Parameter name
     1 : unknown  unknown  instant       1   1      9409   1  F32  : u10
   Grid coordinates :
     1 : lonlat         > size       : dim = 9409  nx = 97  ny = 97
                          longitude : first = 3  last = 15  inc = 0.125  degrees_east
                          latitude  : first = 15  last = 3  inc = -0.125  degrees_north
   Vertical coordinates :
     1 : surface                   : 0
   Time coordinate :  10 steps
     RefTime =  1900-01-01 00:00:00  Units = hours  Calendar = standard
  YYYY-MM-DD hh:mm:ss  YYYY-MM-DD hh:mm:ss  YYYY-MM-DD hh:mm:ss  YYYY-MM-DD hh:mm:ss
  2005-12-31 18:00:00  2006-12-31 18:00:00  2007-12-31 18:00:00  2008-12-31 18:00:00
  2009-12-31 18:00:00  2010-12-31 18:00:00  2011-12-31 18:00:00  2012-12-31 18:00:00
  2013-12-31 18:00:00  2014-12-31 18:00:00
cdo sinfon: Processed 1 variable over 10 timesteps ( 0.00s )
```

(a)

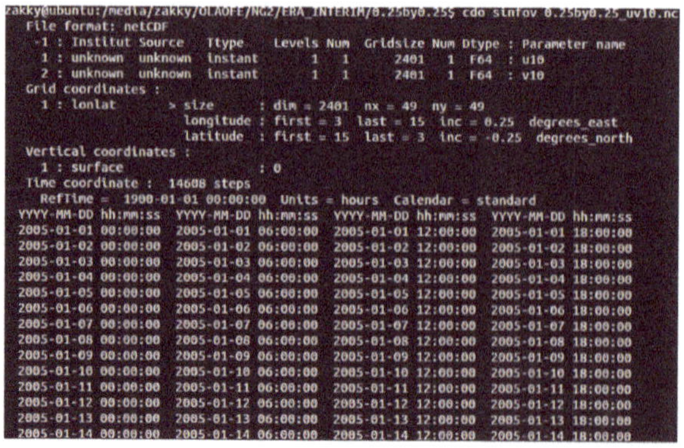

```
zakky@ubuntu:/media/zakky/OLAOFE/NG2/ERA_INTERIM/0.25by0.25$ cdo sinfov 0.25by0.25_uv10.nc
   File format: netCDF
    -1 : Institut Source   Ttype    Levels Num  Gridsize Num Dtype : Parameter name
     1 : unknown  unknown  instant       1   1      2401   1  F64  : u10
     2 : unknown  unknown  instant       1   1      2401   1  F64  : v10
   Grid coordinates :
     1 : lonlat         > size       : dim = 2401  nx = 49  ny = 49
                          longitude : first = 3  last = 15  inc = 0.25  degrees_east
                          latitude  : first = 15  last = 3  inc = -0.25  degrees_north
   Vertical coordinates :
     1 : surface                   : 0
   Time coordinate :  14608 steps
     RefTime =  1900-01-01 00:00:00  Units = hours  Calendar = standard
  YYYY-MM-DD hh:mm:ss  YYYY-MM-DD hh:mm:ss  YYYY-MM-DD hh:mm:ss  YYYY-MM-DD hh:mm:ss
  2005-01-01 00:00:00  2005-01-01 06:00:00  2005-01-01 12:00:00  2005-01-01 18:00:00
  2005-01-02 00:00:00  2005-01-02 06:00:00  2005-01-02 12:00:00  2005-01-02 18:00:00
  2005-01-03 00:00:00  2005-01-03 06:00:00  2005-01-03 12:00:00  2005-01-03 18:00:00
  2005-01-04 00:00:00  2005-01-04 06:00:00  2005-01-04 12:00:00  2005-01-04 18:00:00
  2005-01-05 00:00:00  2005-01-05 06:00:00  2005-01-05 12:00:00  2005-01-05 18:00:00
  2005-01-06 00:00:00  2005-01-06 06:00:00  2005-01-06 12:00:00  2005-01-06 18:00:00
  2005-01-07 00:00:00  2005-01-07 06:00:00  2005-01-07 12:00:00  2005-01-07 18:00:00
  2005-01-08 00:00:00  2005-01-08 06:00:00  2005-01-08 12:00:00  2005-01-08 18:00:00
  2005-01-09 00:00:00  2005-01-09 06:00:00  2005-01-09 12:00:00  2005-01-09 18:00:00
  2005-01-10 00:00:00  2005-01-10 06:00:00  2005-01-10 12:00:00  2005-01-10 18:00:00
  2005-01-11 00:00:00  2005-01-11 06:00:00  2005-01-11 12:00:00  2005-01-11 18:00:00
  2005-01-12 00:00:00  2005-01-12 06:00:00  2005-01-12 12:00:00  2005-01-12 18:00:00
  2005-01-13 00:00:00  2005-01-13 06:00:00  2005-01-13 12:00:00  2005-01-13 18:00:00
  2005-01-14 00:00:00  2005-01-14 06:00:00  2005-01-14 12:00:00  2005-01-14 18:00:00
```

(b)

Figure 2. (a) Wind field information on 0.125° × 0.125° grid resolution; (b) Wind field information on 0.25° × 0.25° grid resolution.

Table 1. Summary of the total hourly surface wind observations available for 18 synoptic stations in Nigeria.

State	2005	2006	2007	2008	2009	2010	2011	2012	2013	2014	Total
Ngu.	-	14	2	35	121	52	6	1	2	1	234
Kat.	5	92	-	120	206	292	822	276	450	1396	3654
Maid.	203	130	1	16	477	603	31	3	170	-	1634
Soko.	367	488	37	161	1823	1515	480	38	266	975	5783
Zar.	1	62	161	442	135	719	509	124	132	406	2690
Gus.	-	1	-	-	-	-	-	-	-	-	1
Pot.	-	83	4	78	44	101	148	128	22	79	687
Kan.	746	1266	295	1681	4330	2377	2689	1368	1980	3056	19,042
Keb.	294	670	361	2011	3518	3144	1870	-	-	1	11,575
Gom	-	-	-	-	-	-	-	21	182	571	774
Kad.	786	919	112	1138	2472	2641	3314	1069	923	396	12,984
Yelwa	1	-	-	453	170	333	816	846	669	290	3578
Bau	220	238	19	134	323	408	278	72	15	-	1707
Makur.	105	38	45	285	280	604	726	527	621	450	3576
Lok	58	-	9	288	286	418	384	924	458	133	2900
Bida	-	-	36	1036	1369	1823	1462	1530	1032	407	8695
Benin	-	433	298	295	446	956	748	225	380	581	4362
Calab	225	144	7	96	1310	1396	1283	732	593	1009	6795

3. Methodology

Two climate data tools running on the linux-based machine (workstation): *climate data operator* for manipulating and analysing climate model output and *ferret* for graphical display of the wind flow, were utilized for this study as shown in **Figures 4-22**. Upon the completion of data collection phase for the proposed wind flow study, the next procedure is to determine the grid information or metadata of the source wind speed components (0.25° × 0.25°, dimension = 2401, n_x = 49, n_y = 49) as presented in **Figure 2(b)**. Thereafter, a new grid description file is written for remapping of the wind field into a new horizontal grid resolution (0.125° × 0.125°, dimension = 9409, n_x = 97, n_y = 97) as summarized in **Figure 3**.

3.1. Remapping of Wind Field

The wind flow originated at Gulf of Guinea as well as two neighbouring countries "Chad and Niger" is studied using the source grid description file produced at a resolution of 0.25° × 0.25° (dimension = 2401, n_x = 49, n_y = 49).

Given a native resolution gridded data on global resolution of 0.125° × 0.125°, the desired horizontal grid associated with reanalysis wind speed components indexed at n_x = 1 and n_y = 1 represents a reference grid point located at longitudinal and latitudinal directions of 3.0°E and 15.0°N, respectively. The longitudinal and latitudinal directions of the desired wind field as a function of their indices (n_x, n_y) can be estimated from the expressions [23]:

$$\lambda_x \left(°E \right) = 3 + \left(\Delta \lambda \right)_n \left(n_x - 1 \right) \quad \text{where} \quad n_x = 1, 2, 3 \cdots 97 \tag{1}$$

$$\varphi_y\left(^\circ\text{N}\right)=15-\left(\Delta\varphi\right)_n\left(n_y-1\right) \quad \text{where} \quad n_y=1,2,3\cdots97 \tag{2}$$

where n_x and n_y are the number of grid points in the longitudinal and latitudinal directions, respectively; $(\Delta\lambda)_n$ is the grid resolution between two points in the longitudinal direction; $(\Delta\varphi)_n$ is the grid resolution between two points in the latitudinal direction.

Note: *3°E and 15°N are the reference grid point of the desire wind field and extends to λ_x (°E) and φ_y (°N) in the longitudinal and latitudinal directions, respectively.*

Using Equations (1)-(2), the longitudinal and latitudinal directions of the desired wind field on a grid indexed $(n_x=97, n_y=97)$, $(\Delta\lambda)_n=0.125°$, $(\Delta\varphi)_n=0.125°$ at reference grid indexed $(n_x=1, n_y=1)$ gives $\lambda_x=(15°\text{E})$, $\varphi_y=(3°\text{N})$. Therefore, a desired wind field on the grid dimension of 9409 $(n_x=97, n_y=97)$, $(\Delta\lambda)_n=0.125°$, $(\Delta\varphi)_n=0.125°$ corresponds to a new wind field on a longitudinal and latitudinal directions of 3.0 - 15.0°E and 15.0 - 3.0°N as shown in **Figure 3**. Hence, a new grid description file is produced for remapping an old wind field on grid dimension of 2401 $(n_x=49, n_y=49)$, $(\Delta\lambda)_n=0.25°$, $(\Delta\varphi)_n=0.25°$ to a new wind field on dimension of 9409 $(n_x=97, n_y=97)$, $(\Delta\lambda)_n=0.125°$ and $(\Delta\varphi)_n=0.125°$. The remapped wind field on a resolution of $0.125° \times 0.125°$ $(n_x=97, n_y=97)$ utilized for the wind flow study for 10 years period is presented in **Figures 4(a)-13(a)**. For comparisons, the wind speed datasets for the actual field on a grid resolution of $0.125° \times 0.125°$ $(n_x=97, n_y=97)$ was obtained as shown in **Figures 4(b)-13(b)**. The annual mean wind flow of the remapped and actual wind fields on the same geographical coordinates of $\lambda_x=(3 - 15°\text{E})$, $\varphi_y=(15 - 3°\text{N})$ are compared. Finally, a 10-year mean wind speed (m/s) chart for the remapped and actual wind fields on a grid resolution of $0.125° \times 0.125°$ are shown in **Figure 14(a)**, **Figure 14(b)**.

Table 2 summarized the minimum (m/s), mean (m/s) and maximum (m/s) wind speeds for the remapped field presented in **Figure 14(a)**, and ranges between 0.00 and 12.40 m/s. From the above mentioned figures, the remapped wind field approximated close to the actual field at smooth and flat terrain and the discrepancies from the actual wind field increased with terrain elevations and contours.

3.2. Validation of Remapped with Actual Wind Fields

In the section 3.1, the gridded datasets of the wind field on a resolution of $0.25° \times 0.25°$ was transformed to a new gridded wind field on a resolution of $0.125° \times 0.125°$. In determining if the distance-weighted remapping of the field underestimated or overestimated the wind originated at Gulf of Guinea as well as at Chad and Niger into Nigeria, the wind speed values of the remapped field is compared with the actual wind field of the same grid resolution, covering the same longitudinal and latitudinal directions of $\lambda_x=3 - 15°\text{E}$, $\varphi_y=15 - 3°\text{N}$.

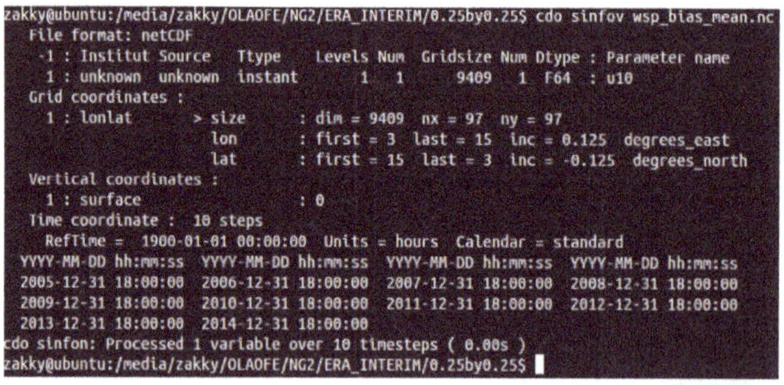

Figure 3. Remapped wind field information on $0.125° \times 0.125°$ grid resolution.

Table 2. Summary of the minimum (m/s), mean (m/s) and maximum (m/s) wind speeds of the remapped field.

WSP	2005	2006	2007	2008	2009	2010	2011	2012	2013	2014	Average
Min	0.001	0.001	0.001	0.001	0.002	0.000	0.001	0.001	0.001	0.001	0.001
Mean	2.69	2.66	2.76	2.67	2.59	2.64	2.71	2.62	2.57	2.60	2.65
Max	10.94	11.20	12.10	11.25	11.24	12.40	11.68	11.91	10.52	10.66	11.39

Note: Min denotes minimum, Max denotes maximum and WSP denotes wind speed.

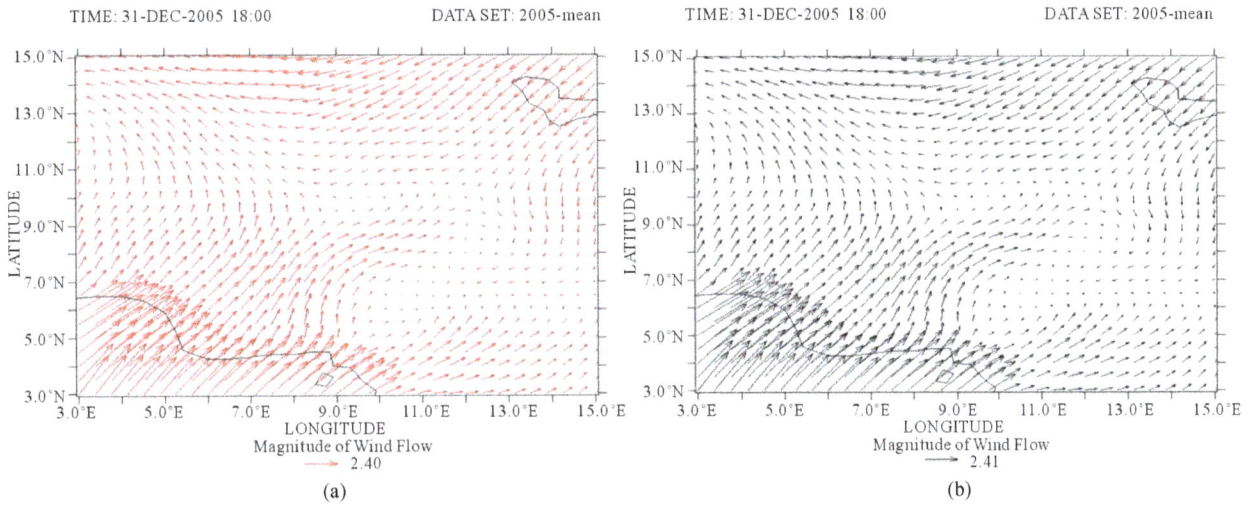

Figure 4. (a) Remapped wind field on 0.125° × 0.125°; (b) Actual wind field on 0.125° × 0.125° grid resolution.

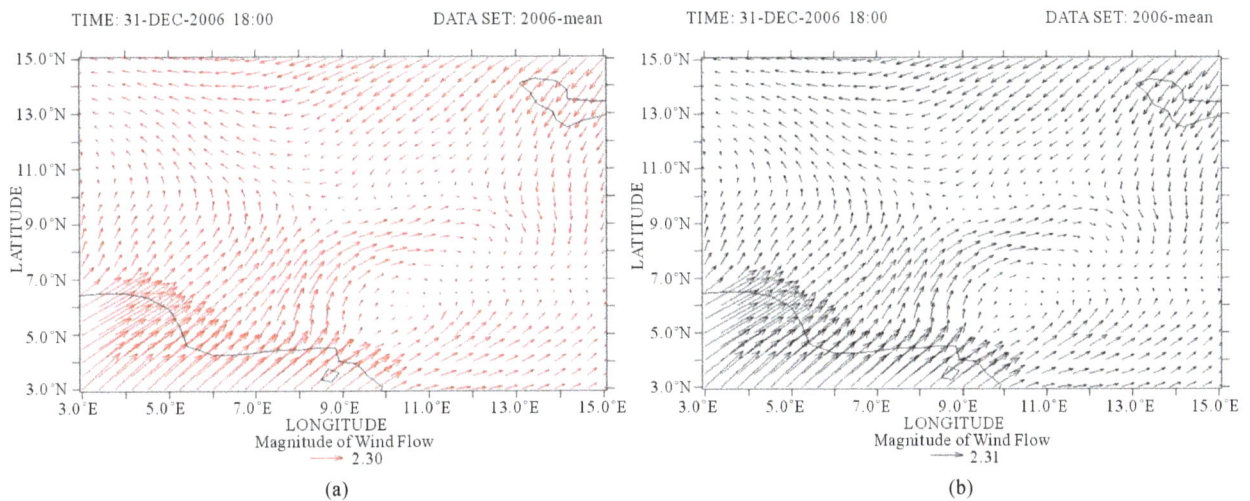

Figure 5. (a) Remapped wind field on 0.125° × 0.125°; (b) Actual wind field on 0.125° × 0.125° grid resolution.

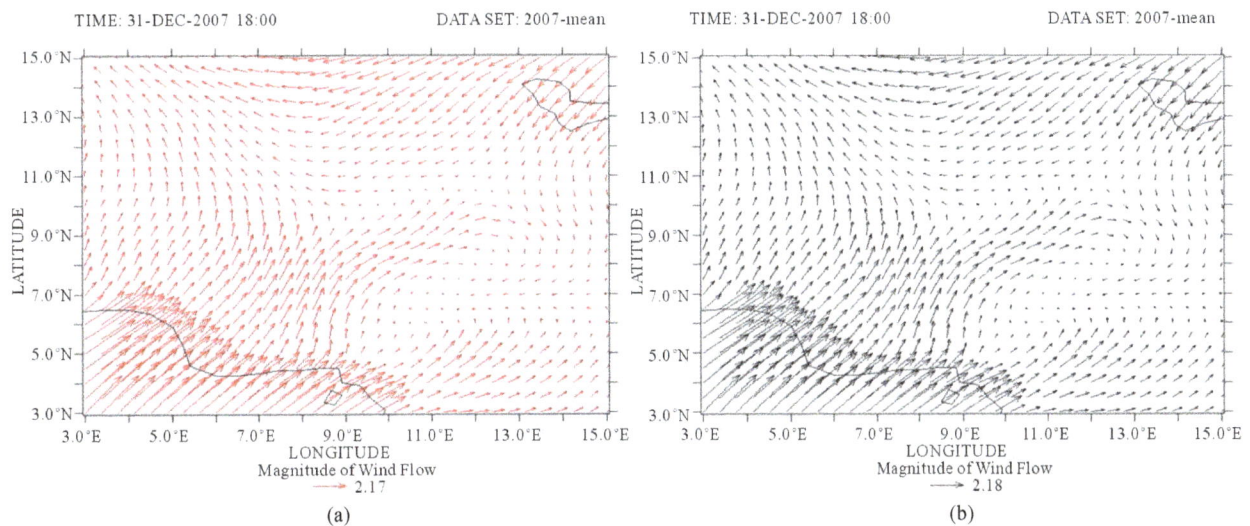

Figure 6. (a) Remapped wind field on 0.125° × 0.125°; (b) Actual wind field on 0.125° × 0.125° grid resolution.

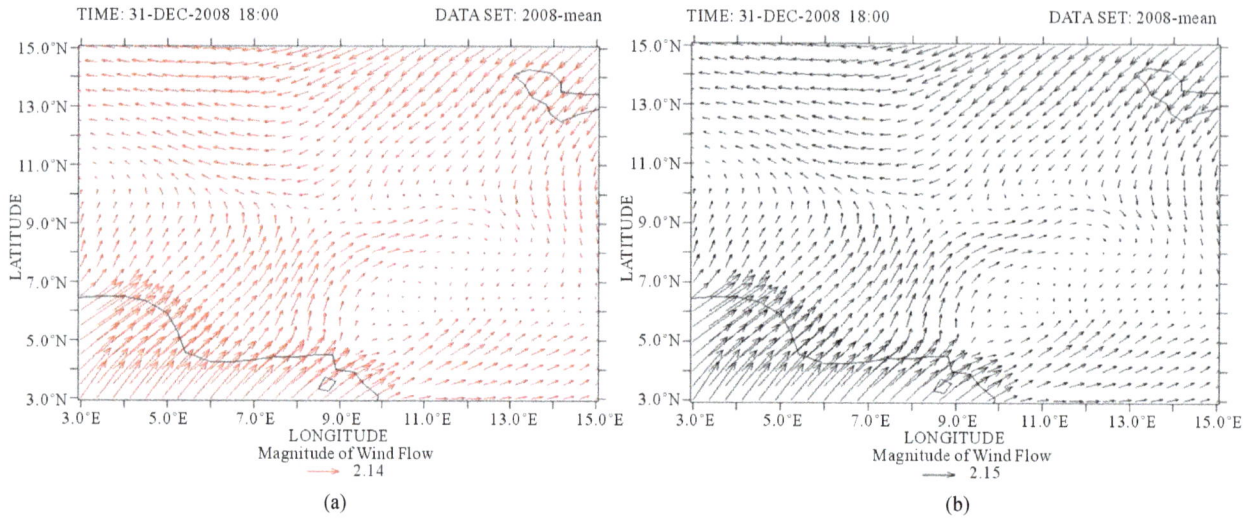

Figure 7. (a) Remapped wind field on 0.125° × 0.125°; (b) Actual wind field on 0.125° × 0.125° grid resolution.

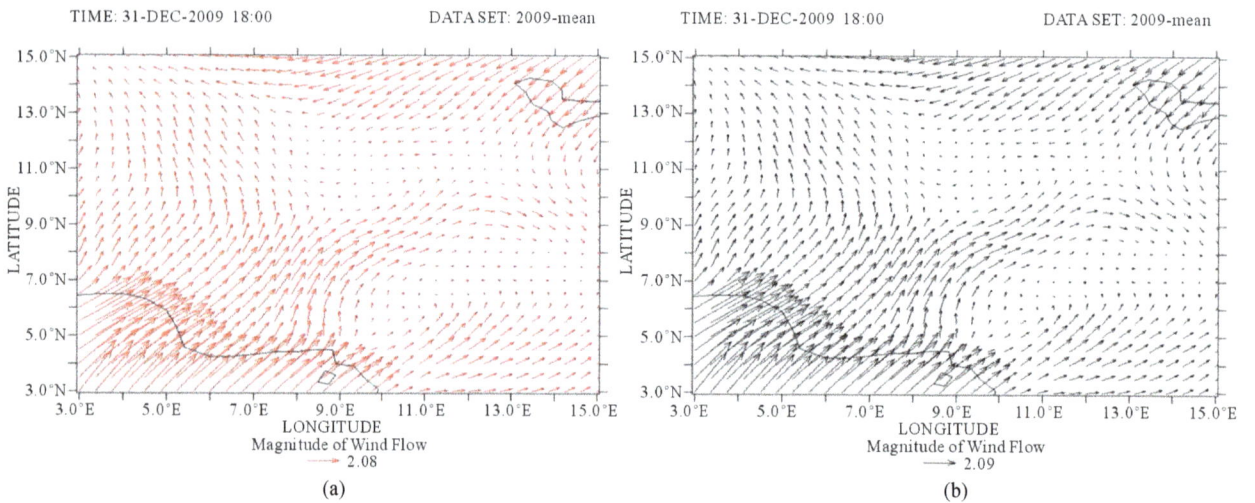

Figure 8. (a) Remapped wind field on 0.125° × 0.125°; (b) Actual wind field on 0.125° × 0.125° grid resolution.

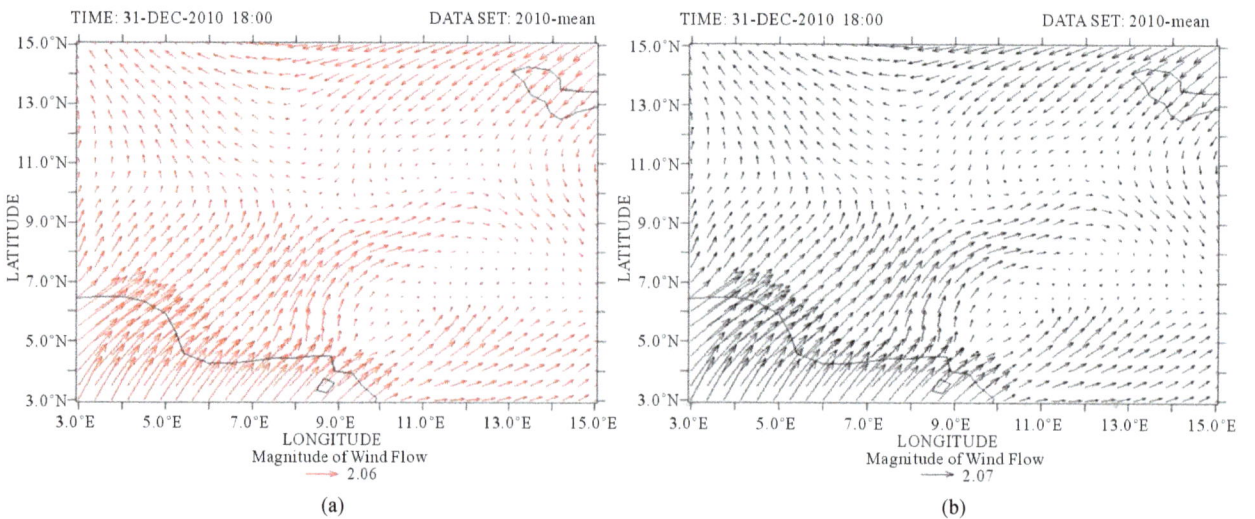

Figure 9. (a) Remapped wind field on 0.125° × 0.125°; (b) Actual wind field on 0.125° × 0.125° grid resolution.

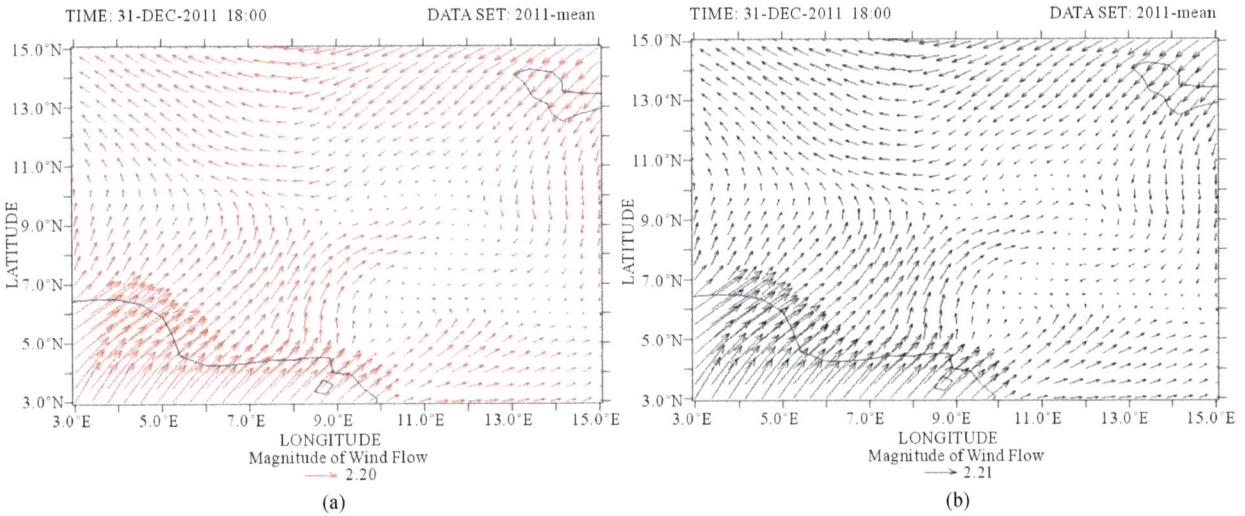

Figure 10. (a) Remapped wind field on 0.125° × 0.125°; (b) Actual wind field on 0.125° × 0.125° grid resolution.

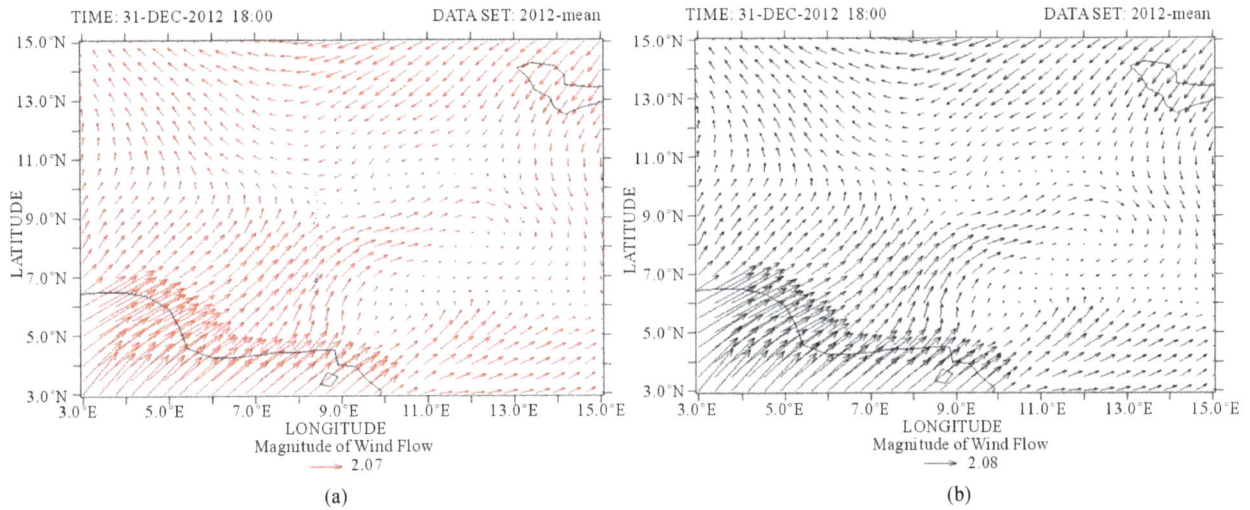

Figure 11. (a) Remapped wind field on 0.125° × 0.125°; (b) Actual wind field on 0.125° × 0.125° grid resolution.

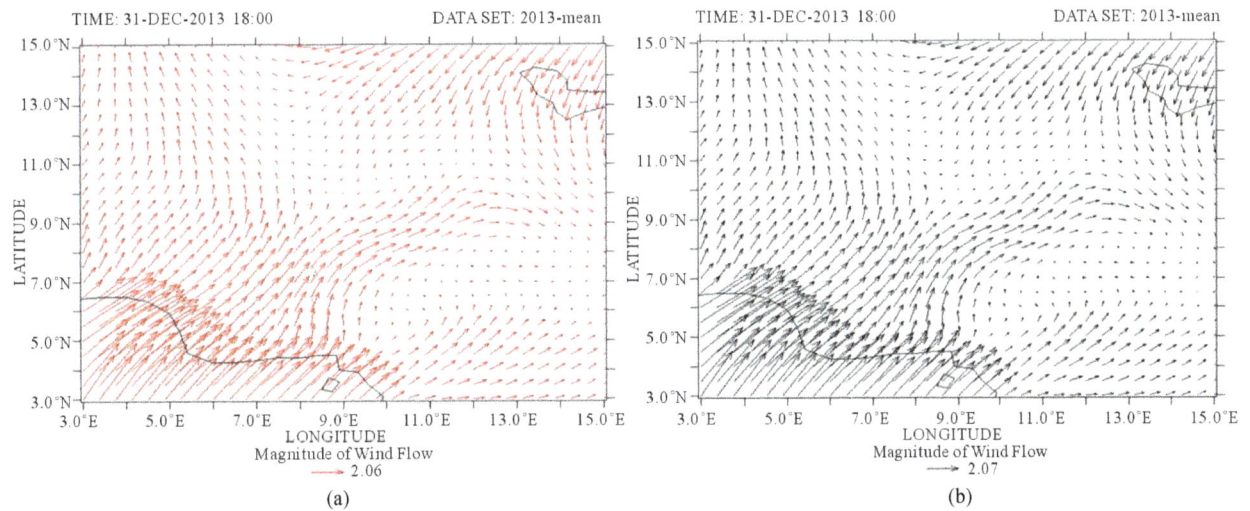

Figure 12. (a) Remapped wind field on 0.125° × 0.125°; (b) Actual wind field on 0.125° × 0.125° grid resolution.

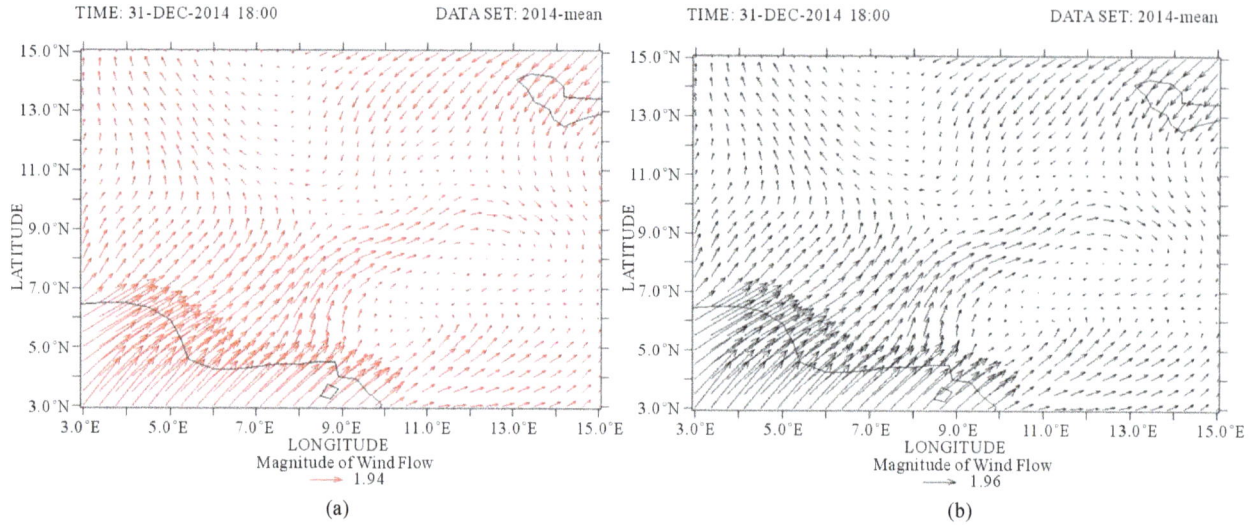

Figure 13. (a) Remapped wind field on 0.125° × 0.125°; (b) Actual wind field on 0.125° × 0.125° grid resolution.

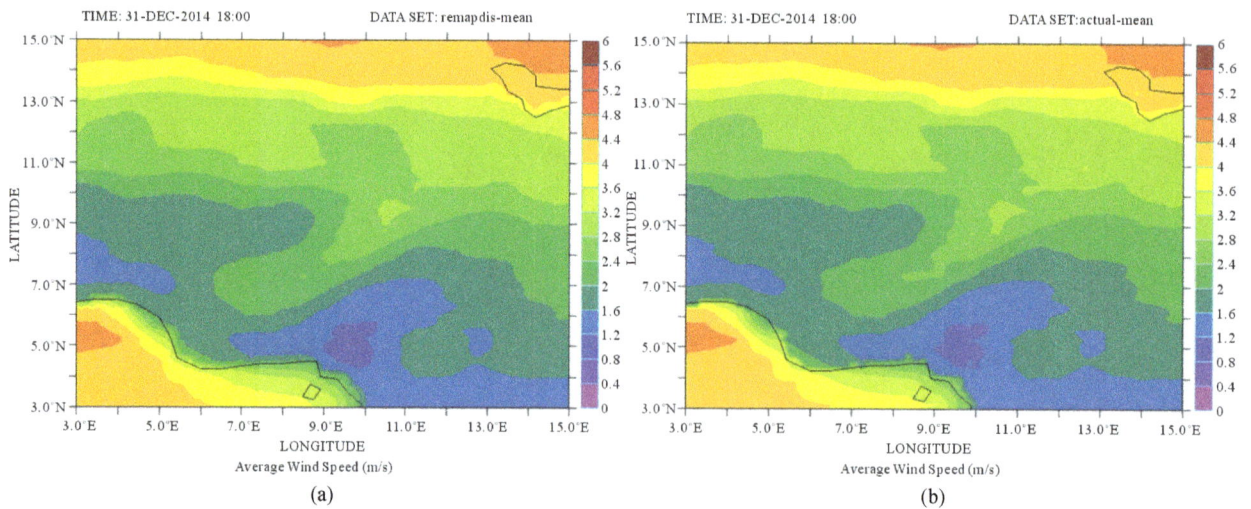

Figure 14. A 10-year wind speed chart of Nigeria on: (a) remapped and (b) actual wind fields at 0.125° × 0.125° grid resolution.

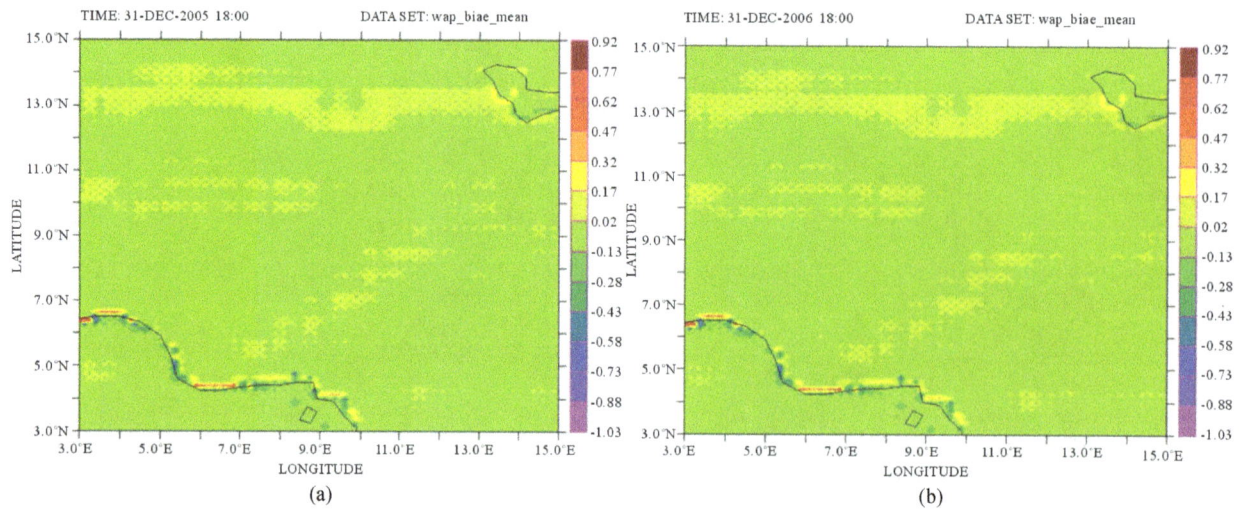

Figure 15. Annual mean wind speed biases (m/s) for (a) 2005 (b) 2006.

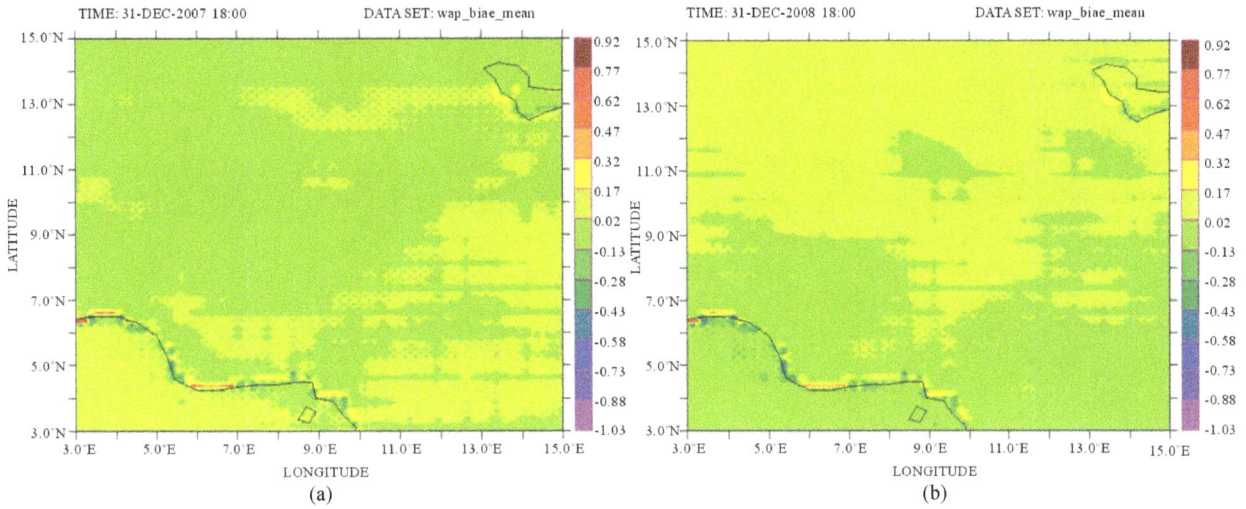

Figure 16. Annual mean wind speed biases (m/s) for (a) 2007 (b) 2008.

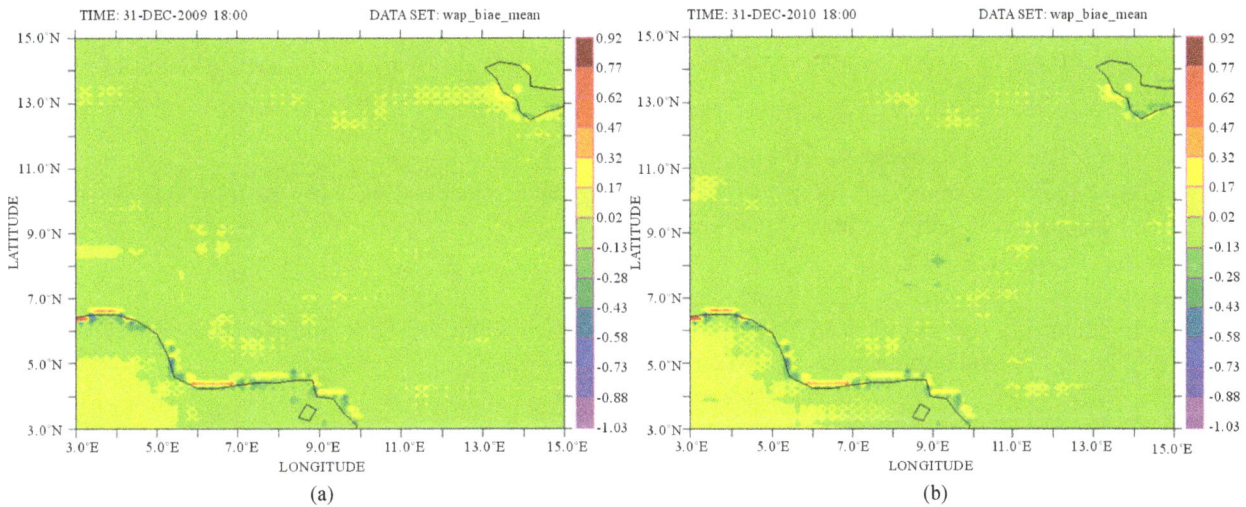

Figure 17. Annual mean wind speed biases (m/s) for (a) 2009 (b) 2010.

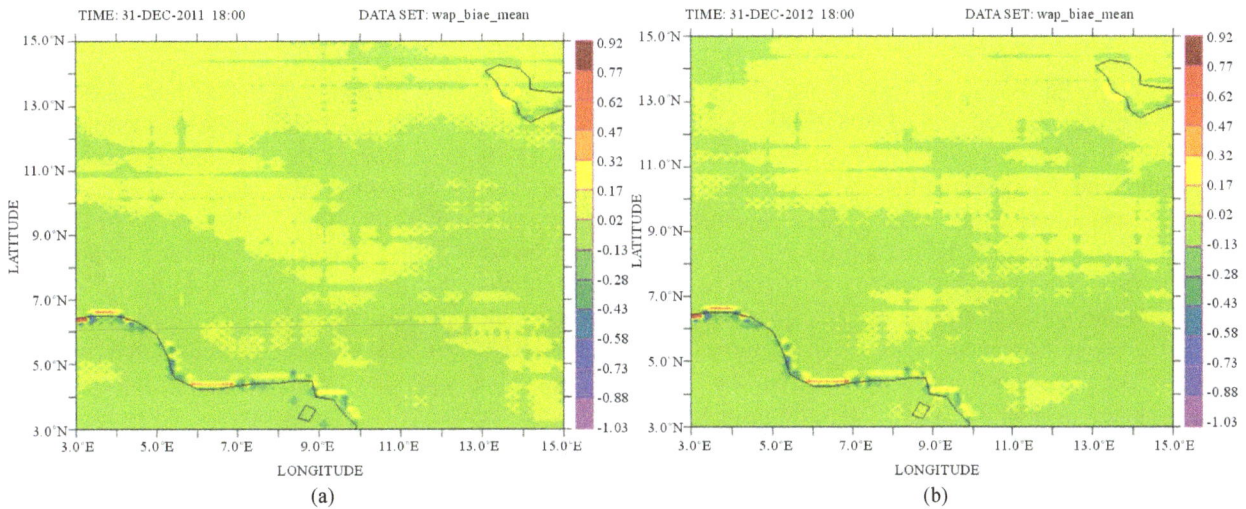

Figure 18. Annual mean wind speed biases (m/s) for (a) 2011 (b) 2012.

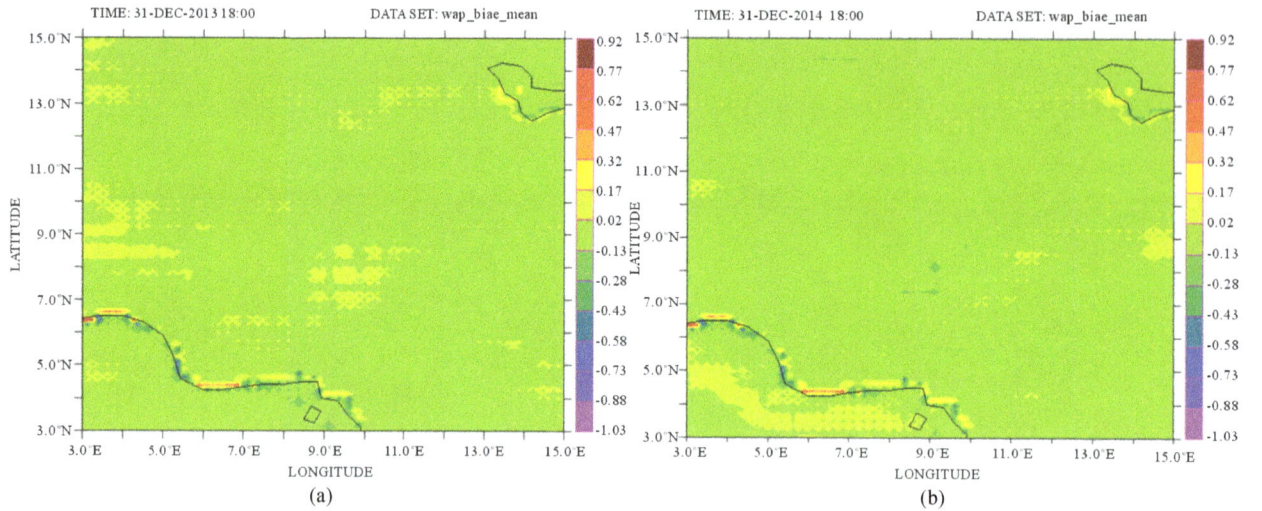

Figure 19. Annual mean wind speed biases (m/s) for (a) 2013 (b) 2014.

Given an estimated distance-weighted gridded wind speed value k_n and the actual wind speed value k at a $0.125° \times 0.125°$ grid resolution, the absolute error of the two gridded wind speed values is defined by the expression:

$$\Delta x_a = |k_n - k| \qquad (3)$$

where Δx_a is the absolute error of the two gridded wind speed values and the mean error (m/s) associated with the remapped wind field is defined by the expression:

$$ME = k_n - k \qquad (4)$$

In addition, given an estimated distance-weighted averaging wind speed k_n and the actual wind speed k at $0.125° \times 0.125°$ grid resolution, the relative difference of the wind field is given by:

$$\Delta x_r = \frac{k_n - k}{k} \qquad (5)$$

where Δx_r is the relative difference of the two gridded wind speed values and the relative error in percentage (%) associated with the remapped wind field is estimated from the given expression:

$$RE = \frac{k_n - k}{k} * 100\% \qquad (6)$$

Using Equations (3)-(6) above, the wind speed biases of the remapped field from the actual field are estimated for a period of 10 years (2005-2014). The estimated annual mean wind speed biases at a grid resolution of $0.125° \times 0.125°$ ($n_x = 97$, $n_y = 97$) are presented in **Figures 15-19**. In addition, **Table 3** summarized the overall minimum (m/s), mean (m/s) and maximum (m/s) wind speed biases of the remapped wind field.

From the validation results of the remapped wind field starting from 00:00:00 UTC Jan-01, 2005 to 18:00:00 UTCDec-31, 2014, the remapped wind field approximated the wind close to the actual wind for a homogenous terrain. For complex terrains with surrounding tall buildings, the remapped wind field based on distance-weighted averaging produced a poor remapping and underestimated the magnitude of the wind flow. However, the remapped gridded field from a $0.25° \times 0.25°$ to $0.125° \times 0.125°$ grid resolution is a good reflection of the prevailing wind originated from the Gulf of Guinea, Chad and Niger when compared with the actual gridded field on $0.125° \times 0.125°$ grid resolution. Furthermore, the wind flow comparison for a 10 years period using gridded datasets on $0.25° \times 0.25°$ and $0.125° \times 0.125°$ resolutions is a significant step for identification of potential sites at wind regimes with no historical records/*in situ* measurements. Because of the limitation of ERA-reanalysis wind records in energy application, the development of state-of-the art mesoscale modeling over the geographical coordinates of the wind field should be considered.

3.3. Grid Interpolation of Gridded Wind Field and Comparisons with Synoptic Observations

Bilinear interpolation was carried out on the remapped and actual wind fields on a grid resolution of $0.125° \times 0.125°$ ($n_x = 97$, $n_y = 97$) to obtain the gridded dataset for each synoptic station. To obtain the gridded datasets, the geographical coordinates of 18 stations in Nigeria was obtained from the World Meteorological Organization (WMO) list containing surface and upper-air weather stations' records for synoptic applications. Using each synoptic station coordinates, the gridded wind speed datasets at the surface layer was obtained performing a bilinear interpolation on the remapped and actual wind fields. The wind speed values of each synoptic station in daily temporal resolution of 00:00:00, 06:00:00, 12:00:00 and 18:00:00 UTC for the period of 10 years were obtained. The comparisons of the mean wind speeds of the interpolated synoptic stations from the remapped and actual wind field are presented in **Figure 20**.

Using Equations (7), (8) above, the minimum, mean and maximum wind speed discrepancies of the remapped field from the actual wind field are estimated as summarized in **Table 3**. In addition, the estimated wind speed discrepancies from the actual wind in terms of the mean error (m/s) and percentage error (%) for each synoptic

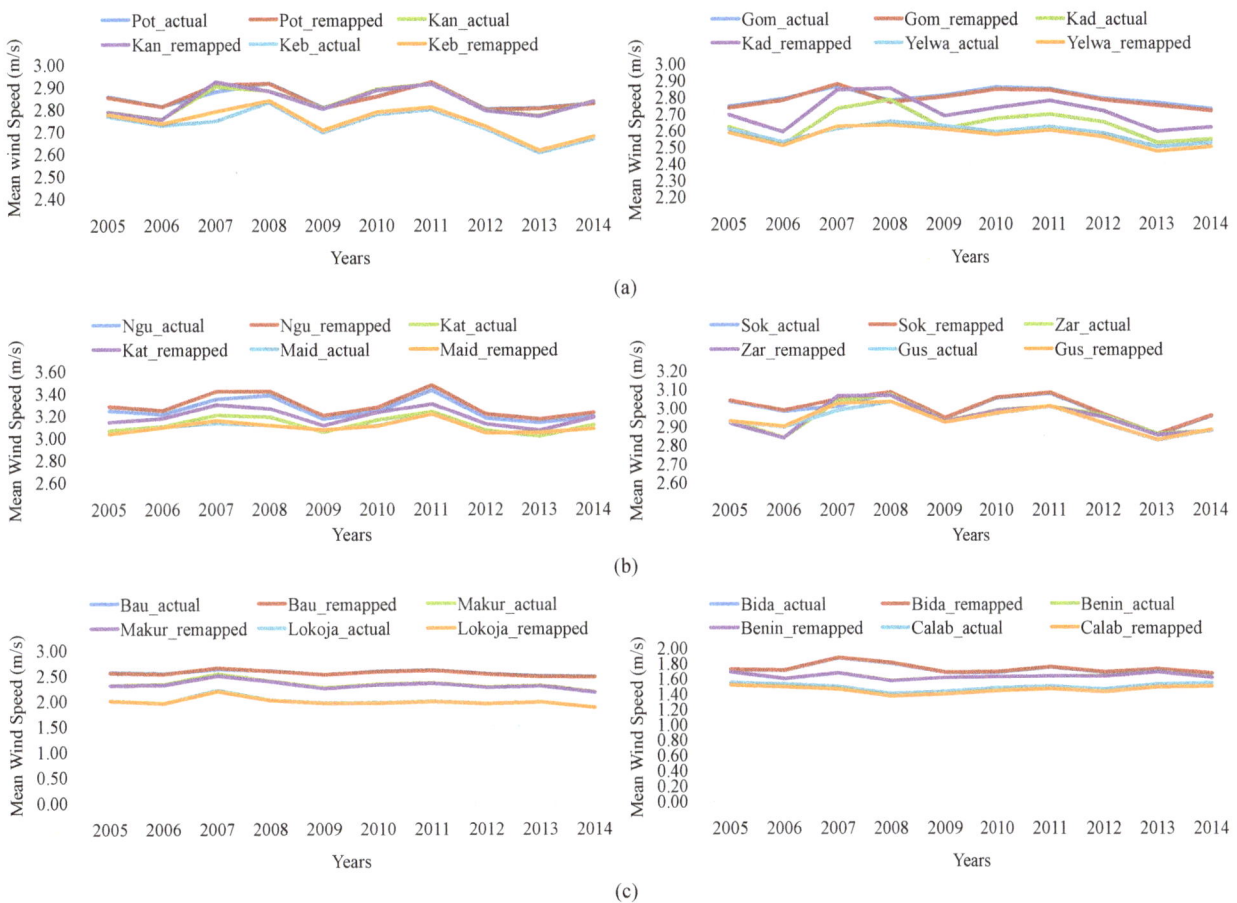

(a)

(b)

(c)

Figure 20. Comparisons of the mean wind speeds of remapped with actual synoptic stations.

Table 3. Summary of the minimum (m/s), mean (m/s) and maximum (m/s) wind speed biases of the remapped field.

Error	2005	2006	2007	2008	2009	2010	2011	2012	2013	2014	Average
Min	−3.048	−1.580	−1.846	−3.903	−2.619	−2.410	−2.236	2.283	−2.602	−4.628	−2.716
Mean	−0.003	−0.002	0.004	−0.001	−0.002	−0.002	−0.001	−0.002	−0.003	0.028	0.002
Max	1.952	2.105	2.137	2.400	2.019	2.386	2.027	2.529	2.427	4.057	2.404

station are summarized in **Table 4**. The results summarized in both tables are useful in evaluating the performance of the distance-weighted remapping of the wind field from one grid resolution to another at different geographical coordinates. From the estimated results, it can be seen that the remapped wind field approximated close to the actual wind field at flat terrain and the discrepancies increased with terrain elevations with contours.

Finally, the hourly surface wind observations from 18 synoptic stations in Nigeria obtained from the National Ocean and Atmospheric Administration (NOAA) are converted to surface monthly mean wind speed because of missing data points at different times and geographical locations. For comparisons, the minimum, mean and maximum wind speeds of each interpolated synoptic stations are plotted in **Figures 21(a)-23(a)**, while the minimum, mean and maximum wind speeds for the surface observations are presented in **Figures 21(b)-23(b)**. **Table 5** summarized the annual minimum (m/s), mean (m/s) and maximum (m/s) surface wind speed observations for the synoptic stations. The comparisons of the ERA wind with the surface wind observations proved that the prevailing wind in Nigeria is higher than the reanalysis wind projection obtained from gridded datasets at resolution of $0.125° × 0.125°$. However, the reanalysis wind simulation is essential in understanding the wind circulation in Nigeria especially at synoptic stations with missing or very poor historical wind records.

4. Discussion

The wind circulation within the remapped field studied for a period of 10 years. The wind circulation in terms of convergence and divergence at different geographical locations in Nigeria can be best understood overlaying **Figures 4-13** on **Figure 1(a)**, **Figure 1(b)**. These figures explain the wind flow divergence and convergence at the wind field of coordinates $λ_x = 3 - 15°E$, $φ_y = 15 - 3°N$ on a grid resolution of $0.125° × 0.125°$. In addition, the magnitude of wind flow covering the longitudinal and latitudinal direction of $3.0 - 15.0°E$ and $15.0 - 3.0°N$ has been graphically analysed using a 10 meter U and V wind speed components at a daily timestamp of 00:00:00, 06:00:00, 12:00:00, and 18:00:00 UTC for period of 2005-2014. A strong wind emerged from the Gulf of Guinea, Chad and Niger into Nigeria, and is deflected by terrain elevations and surrounding tall buildings at different geographical locations within the wind field. Furthermore, the wind converged at some cities in the South-West and South-South regions of Nigeria, and a higher percentage of the wind flow is deflected towards neighbouring countries like Benin, Niger and Cameroun. As a result of the terrain elevation differences at geographical locations within the wind field, the magnitude of the wind flow decreases before reaching the South-West region like Lagos, Ogun, Osun, Oyo, Ondo and Ekiti States. These states have a very poor wind potential for energy conversion in the South-West region of Nigeria. The 18 synoptic stations owned by the Nigerian Meterological Agency (NIMET) and situated at different wind locations have been shown to be subjected to the influenced of the wind flow originated from Gulf of Guinea, Chad and Niger.

Nigeria is located within the northern Hemisphere, and the surface high pressure system in the northern Hemisphere has a clockwise rotation turning the wind outwardly from the highest pressure system towards the lowest pressure system. The pressure gradient force (pgf) causes the air parcel to diverge at the centre of the high pressure near the surface level, leading to sinking air in the centre of a high pressure system. This explains the wind flow in Northern Nigeria through: Maiduguri, Damaturu, Katsina, Sokoto, Nguru, Cashua, Hadejia, Birnin Kebbi and Nguru. In the South-West and South-East of Nigeria where there is wind convergence, the low pressure system has anticlockwise rotation and the wind turns inwardly towards the low pressure. These wind convergence at these cities cause air rising at the centre of the low pressure system near the ground, leading to cloud formation and precipitation occurrence.

The comparison of the mean wind speed chart of Nigeria for a period of 10 years is shown in **Figure 14(a)**, **Figure 14(b)** while **Figures 15-19** summarized graphically the annual mean wind speed biases of the remapped wind field on a grid resolution of $0.125° × 0.125°$. In addition, **Table 3** summarized the annual wind speed biases (m/s) of the remapped field from the actual field for the same period of 10 years. From the comparisons of the figures and table, the wind speed biases were very small at geographical locations with a smooth or flat terrain and increased significantly at locations with high terrain elevations.

The remapped wind field based on the distance-weighted averaging has been interpolated bilinearly for the identification of potential wind sites in Nigeria. The magnitude of wind flow differs from one geographical location to another as seen in **Figures 4-13**. The geographical locations such as Sokoto, Katsina, Nguru, Gusau, Birnin Kebbi, Maiduguri, Zaria, Potiskum, Cashua, Hadejia and Kaduna have a large-scale wind potential for

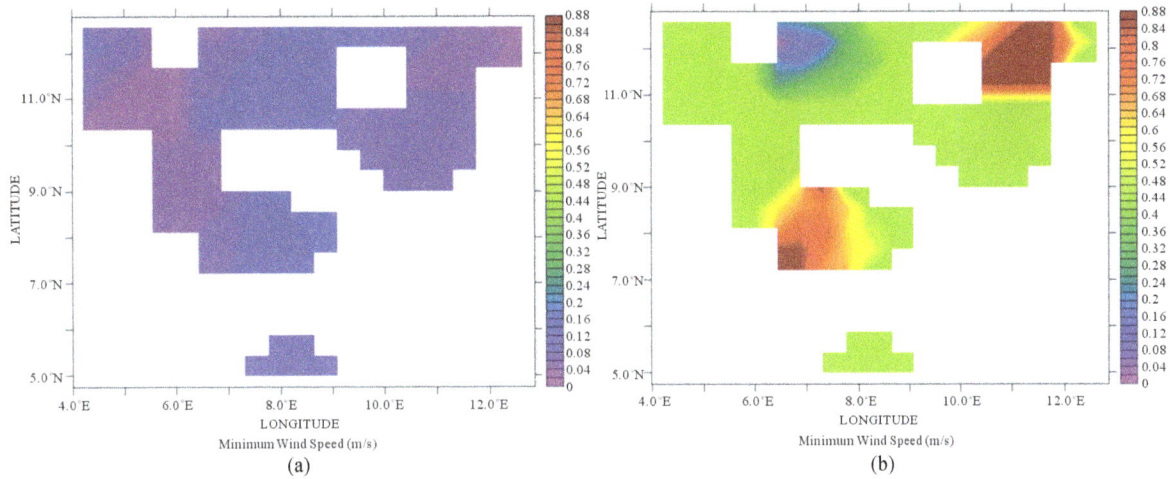

Figure 21. Comparisons of the minimum wind speeds for 18 synoptic stations using: (a) ERA wind simulations; (b) Surface wind observations.

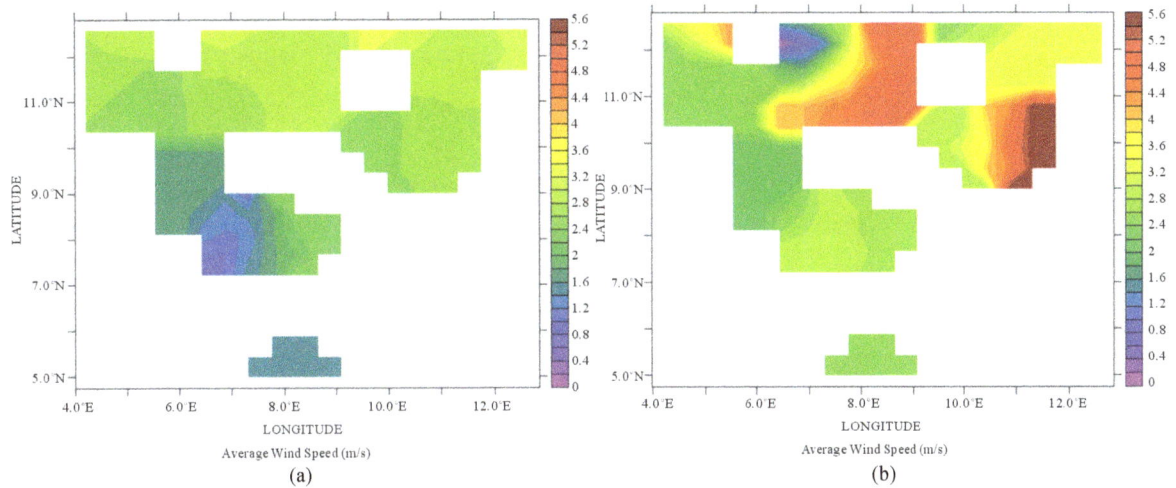

Figure 22. Comparisons of the mean wind speeds for 18 synoptic stations using: (a) ERA wind simulations; (b) Surface wind observations.

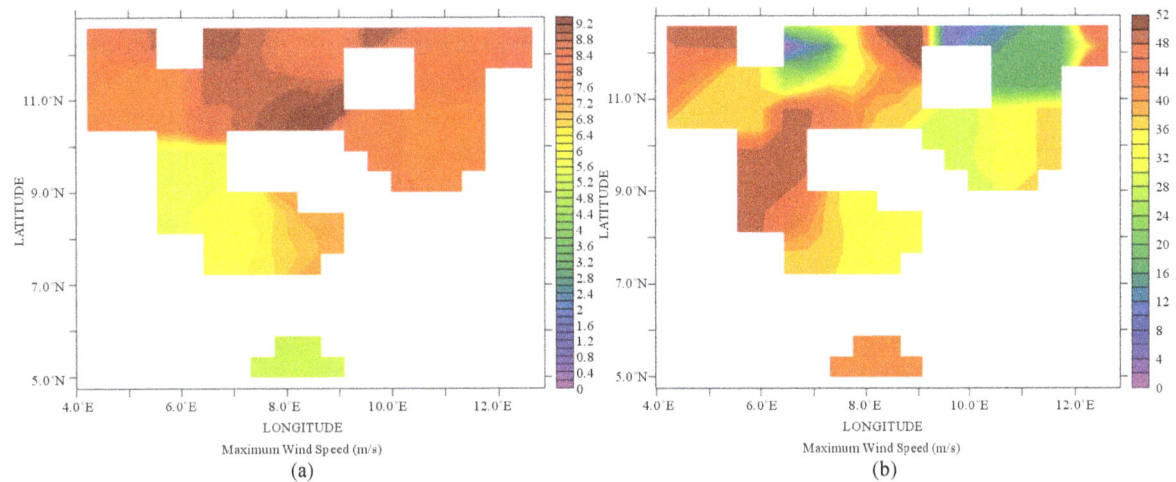

Figure 23. Comparisons of the maximum wind speeds for 18 synoptic stations using: (a) ERA wind simulations; (b) Surface wind observations.

Table 4. Annual mean wind errors (m/s) and percentage errors (%) for 18 synoptic stations on a grid resolution of 0.125° × 0.125°.

State		2005	2006	2007	2008	2009	2010	2011	2012	2013	2014
Ngu.	ME (m/s)	0.038	0.037	0.030	0.092	0.012	0.001	0.054	0.068	0.011	−0.005
	PE (%)	1.167	1.147	0.885	2.698	0.374	0.043	1.555	2.122	0.355	−0.162
Kat.	ME (m/s)	0.073	0.072	0.066	0.107	0.033	0.042	0.090	0.094	0.018	0.023
	PE (%)	2.330	2.273	1.991	3.292	1.064	1.303	2.718	3.012	0.590	0.716
Maid.	ME (m/s)	−0.001	−0.001	0.011	0.012	−0.004	−0.014	0.006	0.013	−0.018	−0.013
	PE (%)	−0.020	−0.035	0.344	0.398	−0.117	−0.458	0.199	0.414	−0.603	−0.410
Sok.	ME (m/s)	0.003	0.004	−0.019	0.059	−0.032	−0.037	0.034	0.035	−0.022	−0.038
	PE (%)	0.112	0.129	−0.628	1.912	−1.099	−1.211	1.113	1.193	−0.778	−1.301
Zar.	ME (m/s)	−0.010	−0.006	−0.047	0.064	−0.034	−0.056	0.016	0.037	−0.041	−0.056
	PE (%)	−0.344	−0.211	−1.523	2.097	−1.178	−1.886	0.546	1.270	−1.443	−1.965
Gus.	ME (m/s)	0.002	0.003	−0.029	0.067	−0.039	−0.043	0.030	0.046	−0.031	−0.047
	PE (%)	0.074	0.119	−0.963	2.202	−1.321	−1.444	1.010	1.583	−1.081	−1.615
Pot.	ME (m/s)	−0.004	−0.001	−0.004	0.037	−0.019	−0.025	0.006	0.016	−0.019	−0.032
	PE (%)	−0.137	−0.048	−0.153	1.256	−0.662	−0.886	0.207	0.576	−0.678	−1.135
Kan.	ME (m/s)	−0.003	−0.004	0.000	0.011	−0.019	−0.017	0.003	0.021	−0.034	−0.030
	PE (%)	−0.108	−0.138	0.011	0.383	−0.685	−0.602	0.106	0.760	−1.225	−1.076
Keb.	ME (m/s)	0.008	0.010	−0.014	0.058	−0.017	−0.022	0.038	0.030	−0.009	−0.018
	PE (%)	0.294	0.353	−0.490	2.052	−0.646	−0.808	1.349	1.086	−0.339	−0.671
Gom.	ME (m/s)	−0.011	−0.010	−0.019	0.039	−0.051	−0.015	−0.006	0.024	−0.042	−0.029
	PE (%)	−0.390	−0.354	−0.669	1.397	−1.805	−0.538	−0.203	0.846	−1.542	−1.076
Kad.	ME (m/s)	0.016	0.018	−0.029	0.110	−0.027	−0.044	0.047	0.062	−0.009	−0.039
	PE (%)	0.580	0.705	−1.009	3.843	−1.008	−1.601	1.676	2.271	−0.348	−1.507
Yelwa	ME (m/s)	−0.023	−0.024	−0.052	0.039	−0.054	−0.032	−0.003	−0.012	−0.037	−0.044
	PE (%)	−0.895	−0.970	−1.994	1.493	−2.089	−1.236	−0.122	−0.463	−1.486	−1.777
Bau.	ME (m/s)	−0.016	−0.012	0.001	0.010	−0.026	−0.009	0.000	0.024	−0.037	−0.016
	PE (%)	−0.615	−0.476	0.043	0.381	−1.028	−0.345	−0.014	0.937	−1.480	−0.659
Makur.	ME (m/s)	−0.011	−0.012	−0.045	0.021	−0.001	−0.082	0.021	−0.001	0.014	−0.079
	PE (%)	−0.469	−0.516	−1.811	0.894	−0.063	−3.567	0.897	−0.028	0.602	−3.693
Lokoja	ME (m/s)	−0.005	−0.003	0.010	−0.005	0.011	−0.021	−0.008	0.004	0.009	−0.053
	PE (%)	−0.261	−0.155	0.466	−0.243	0.545	−1.089	−0.382	0.192	0.478	−2.875
Bida	ME (m/s)	−0.040	−0.038	−0.013	−0.032	−0.051	−0.035	−0.039	−0.025	−0.043	−0.041
	PE (%)	−2.643	−2.534	−0.880	−2.347	−3.639	−2.444	−2.675	−1.7357	−2.897	−2.757
Benin	ME (m/s)	−0.003	−0.003	0.015	−0.017	0.000	−0.003	0.003	−0.003	−0.003	−0.020
	PE (%)	−0.200	−0.209	0.902	−1.062	0.000	−0.160	0.159	−0.188	−0.161	−1.229
Calab.	ME (m/s)	−0.040	−0.038	−0.013	−0.032	−0.051	−0.035	−0.039	−0.025	−0.043	−0.041
	PE (%)	−2.643	−2.534	−0.880	−2.347	−3.640	−2.444	−2.675	−1.736	−2.897	−2.757

Table 5. Summary of the minimum (m/s), mean (m/s) and maximum (m/s) wind speed observations for 18 synoptic stations.

State		2005	2006	2007	2008	2009	2010	2011	2012	2013	2014
Ngu.	Min	-	0.00	0.00	0.00	0.00	0.00	0.00	-	0.00	-
	Mean	-	1.12	0.45	2.17	2.46	1.92	1.04	-	2.01	-
	Max	-	2.68	0.89	5.81	6.26	4.92	2.68	-	4.03	8.94
Kat.	Min	0.89	0.00	-	0.00	0.00	0.00	0.00	0.00	0.00	0.00
	Mean	2.06	4.60	-	3.81	3.42	2.51	2.55	2.75	3.16	2.60
	Max	4.92	9.39	-	9.84	48.75	11.18	11.18	6.71	48.75	29.96
Maid.	Min	0.00	0.00	-	0.00	0.00	0.00	0.00	4.47	0.00	-
	Mean	3.44	4.96	-	4.19	2.62	2.40	2.81	5.22	3.94	-
	Max	30.86	36.23	3.58	7.16	10.29	30.86	7.60	6.71	14.76	-
Sok.	Min	0.00	0.00	2.68	0.00	0.00	0.00	0.00	0.00	0.00	0.00
	Mean	4.19	4.48	8.15	5.09	3.51	3.57	4.05	3.18	3.77	2.89
	Max	28.16	10.73	48.75	11.63	46.52	16.10	10.29	9.39	49.20	11.63
Zar.	Min	-	0.00	0.00	0.00	0.00	0.00	0.00	0.00	0.00	0.00
	Mean	-	3.28	3.89	3.47	2.82	3.57	4.30	6.06	4.27	4.18
	Max	9.84	34.44	16.55	20.13	12.97	15.21	12.08	12.97	36.23	17.89
Gus.	Min	-	-	-	-	-	-	-	-	-	-
	Mean	-	-	-	-	-	-	-	-	-	-
	Max	-	-	-	-	-	-	-	-	-	-
Pot.	Min	-	0.00	3.13	0.00	0.00	0.00	0.00	0.00	0.00	0.00
	Mean	-	3.37	5.93	3.60	3.53	2.99	2.75	2.71	3.15	3.40
	Max	-	17.89	7.16	7.60	7.60	9.84	6.26	6.26	4.92	14.31
Kan.	Min	0.00	0.00	0.00	0.00	0.00	0.00	0.00	0.00	0.00	0.00
	Mean	4.53	4.93	5.74	4.77	4.75	5.05	4.74	4.61	3.71	4.86
	Max	21.02	20.13	15.21	30.86	30.86	30.86	50.54	30.86	46.51	30.86
Keb.	Min	-	-	-	-	-	-	-	-	-	-
	Mean	2.96	2.77	2.65	2.21	2.13	2.23	2.34	-	-	-
	Max	15.21	30.86	47.85	41.14	16.99	16.99	15.21	-	-	-
Gom.	Min	-	-	-	-	-	-	-	2.24	0.00	0.00
	Mean	-	-	-	-	-	-	-	4.92	4.92	4.51
	Max	-	-	-	-	-	-	-	9.39	36.23	35.33
Kad.	Min	0.00	0.00	0.00	0.00	0.00	0.00	0.00	0.00	0.00	0.00
	Mean	3.46	3.64	3.91	3.18	2.49	3.12	3.07	3.29	2.89	3.22
	Max	22.81	48.75	11.18	15.21	10.29	30.86	15.21	46.51	34.88	30.86

Continued

	Min	-	-	-	0.00	0.00	0.00	0.00	0.00	0.00	0.00
Yelwa	Mean	-	-	-	2.43	2.90	2.31	2.63	2.00	1.80	1.46
	Max	-	-	-	36.23	8.05	6.71	12.08	28.62	12.97	6.26
	Min	0.00	0.00	0.00	0.00	0.00	0.00	0.00	0.00	0.00	-
Bau.	Mean	1.73	1.46	0.87	1.20	0.83	1.05	1.56	1.17	1.16	-
	Max	20.57	7.16	3.13	28.62	7.60	8.05	12.97	7.60	4.47	-
	Min	0.00	0.00	0.00	0.00	0.00	0.00	0.00	0.00	0.00	0.00
Makur.	Mean	3.12	2.64	2.80	2.46	2.25	2.33	1.98	1.75	2.20	1.65
	Max	9.39	11.63	9.39	8.50	13.42	33.10	8.05	6.71	33.99	31.75
	Min	0.00	-	0.00	0.00	0.00	0.00	0.00	0.00	0.00	0.00
Lokoja	Mean	1.93	-	1.36	1.18	1.59	1.57	1.25	1.68	1.53	1.59
	Max	5.81	-	4,03	8.05	8.05	9.84	7.60	37.57	7.60	4.92
	Min	-	-	0.00	0.00	0.00	0.00	0.00	0.00	0.00	0.00
Bida	Mean	-	-	1.76	1.71	1.95	2.00	1.94	1.61	1.82	1.89
	Max	-	-	4.03	12.97	48.83	48.30	4.92	8.05	42.93	9.39
	Min	-	0.00	0.00	0.00	0.00	0.00	0.00	0.00	0.00	0.00
Benin	Mean	-	1.85	2.39	1.82	2.10	2.22	1.75	2.28	2.14	2.93
	Max	-	28.18	45.51	7.16	9.39	9.84	15.21	7.16	36.23	30.86
	Min	0.00	0.00	0.00	0.00	0.00	0.00	0.00	0.00	0.00	0.00
Calab.	Mean	2.57	2.73	0.96	2.21	2.11	1.99	2.16	2.08	1.58	1.53
	Max	12.52	8.94	3.13	41.14	25.49	30.86	30.86	9.39	33.54	30.86

grid-tie energy application at a much higher altitude, while Gombe, Bauchi and Minna have fair wind prevalence while wind convergence locations such as Abuja and Bida and South-West Nigeria (Lagos, Ogun, Oyo, Osun, Ekiti and Ondo states) have a very poor wind potential and is classified as a very low wind region. The wind flow study at this poor region using ERA-Interim wind dataset agrees with Fadare [24] classification of Ibadan city as a low wind energy region.

Table 5 summarized the minimum, mean and maximum wind speeds for 18 synoptic stations derived from surface wind observations. The following wind potential locations were identified based on the 10 years wind flow study: Bauchi (Bau.), Gombe (Gom.), Gusau (Gus.), Kaduna (Kad.), Kano (Kan.), Katsina (Kat.), Kebbi (Keb.), Maiduguri (Mad.), Nguru (Ngu.), Potiskum (Pot.), Sokoto (Sok.) and Zaria (Zar.). The wind locations such as Kaduna, Zaria, Bauchi, Katsina, Kano, Nguru, Potiskum and Gombe in northern Nigeria have terrain elevations at a 642.0, 664.0, 609.0, 506.0, 472.0, 344.0, 414.0 and 505.0 m above sea level, respectively. Other northern locations with synoptic stations benefitting from the same wind circulation are: Jos in Plateau state (1285.0 m), Minna (260.0 m) and Bida (143.0 m) in Niger state, Nnamdi Azikiwe in Abuja (342.0 m), Sokoto (302.0 m), Gusau (469.0 m), Birnin Kebbi (220.0 m) and Maiduguri (354.0 m). The local topography (such as surface roughness and terrain elevation) was expected to influence the wind direction at different locations causing resistance and deflection to the wind flow. Hence high wind speed discrepancies were observed at geographical locations with a complex terrain. For example, the direction of wind flow in Kastina and Kano are similar (**Figure 24(b)** and **Figure 24(h)**), while other synoptic stations experienced little or high wind deflection depending on the terrain elevation of the wind location. This applies to the wind flow originated at Gulf of Guinea into the South-West region of Nigeria as deflected toward the Benin Republic, Northern Nigeria as well as some cities in Cameroon. Lagos in the South-West region is a congested and commerical state in Nigeria with

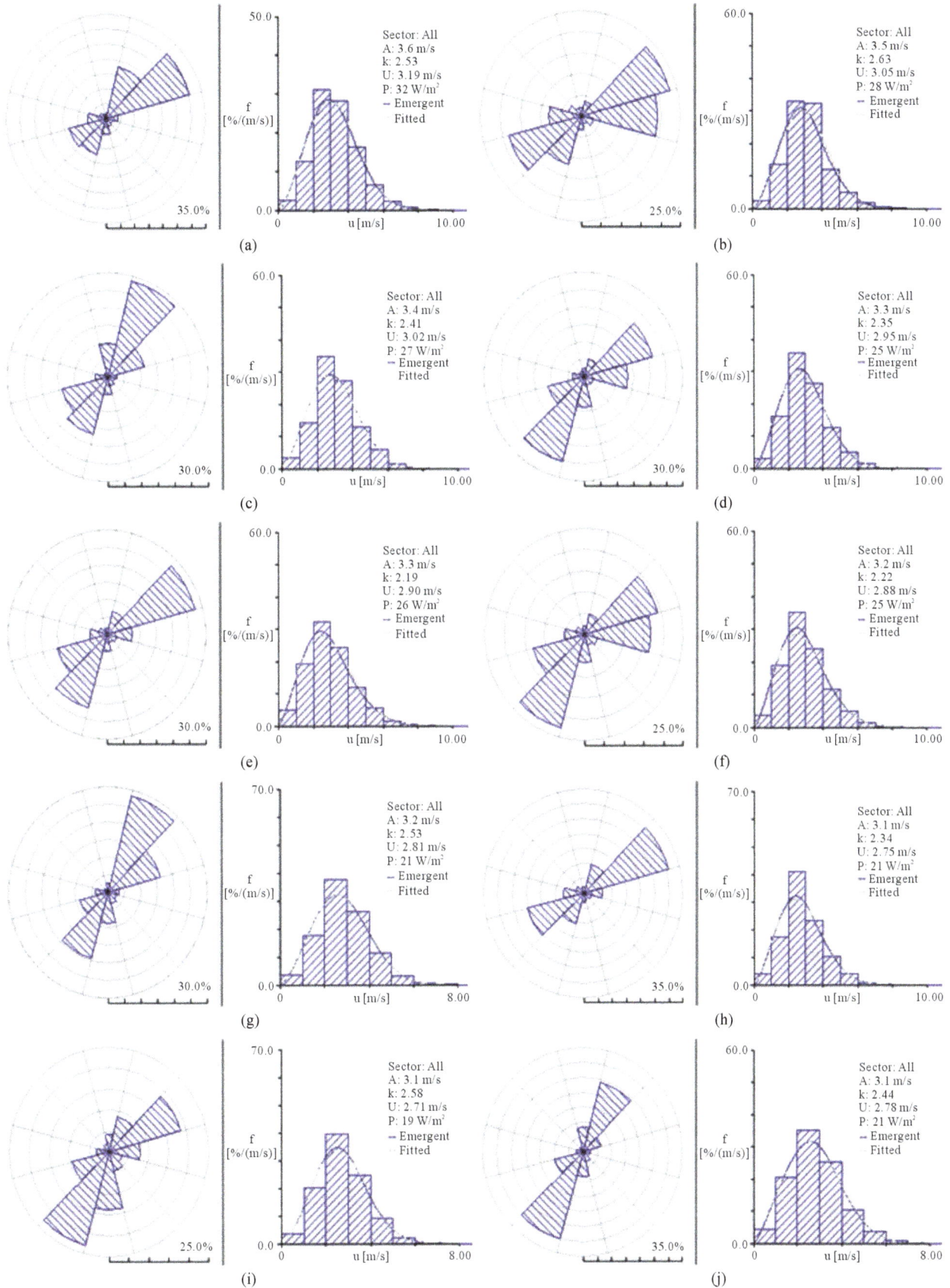

(a)

(b)

(c)

(d)

(e)

(f)

(g)

(h)

(i)

(j)

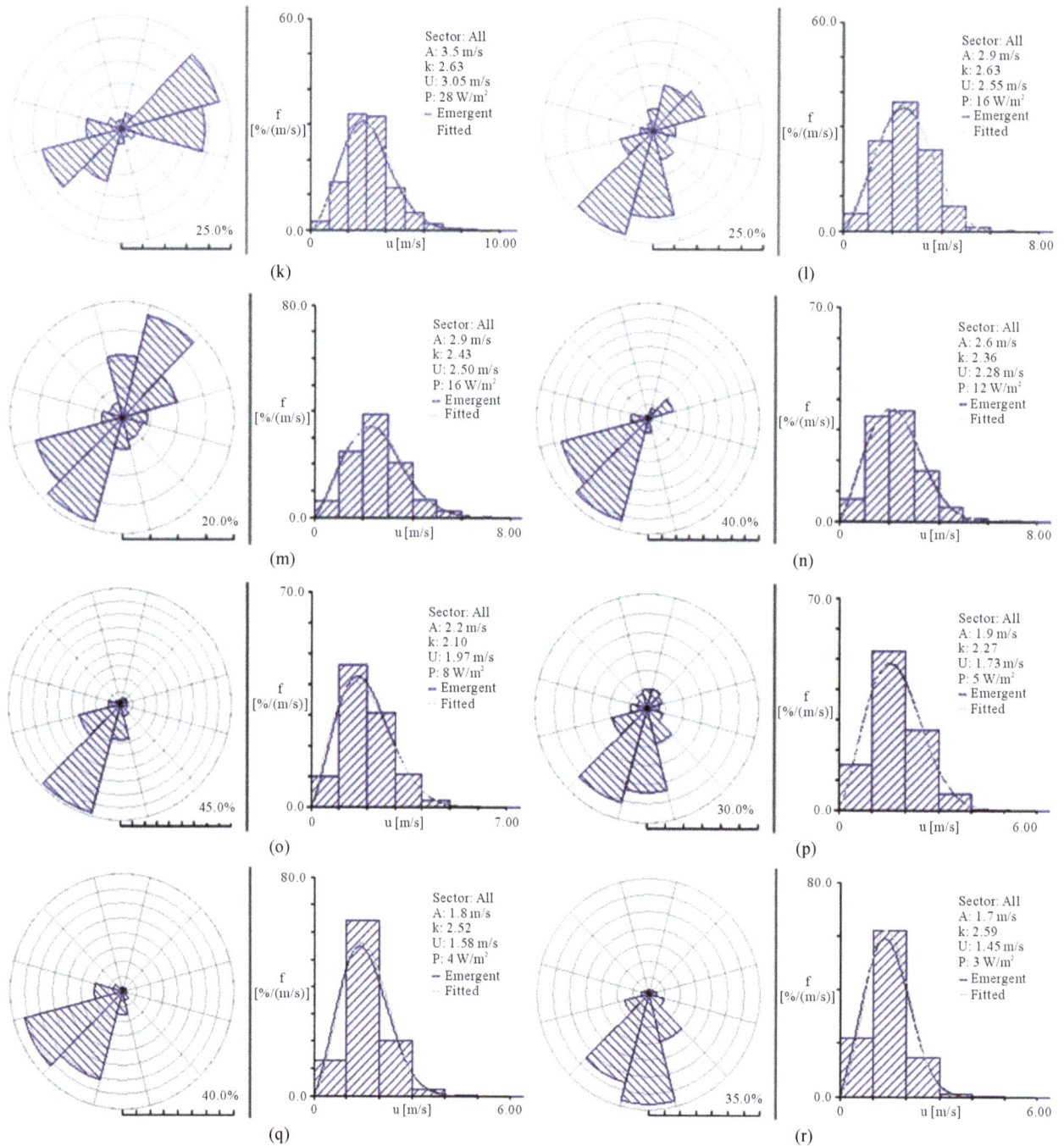

(k)

Sector: All
A: 3.5 m/s
k: 2.63
U: 3.05 m/s
P: 28 W/m²
— Emergent
    Fitted

(l)

Sector: All
A: 2.9 m/s
k: 2.63
U: 2.55 m/s
P: 16 W/m²
— Emergent
    Fitted

(m)

Sector: All
A: 2.9 m/s
k: 2.43
U: 2.50 m/s
P: 16 W/m²
— Emergent
    Fitted

(n)

Sector: All
A: 2.6 m/s
k: 2.36
U: 2.28 m/s
P: 12 W/m²
— Emergent
    Fitted

(o)

Sector: All
A: 2.2 m/s
k: 2.10
U: 1.97 m/s
P: 8 W/m²
— Emergent
    Fitted

(p)

Sector: All
A: 1.9 m/s
k: 2.27
U: 1.73 m/s
P: 5 W/m²
— Emergent
    Fitted

(q)

Sector: All
A: 1.8 m/s
k: 2.52
U: 1.58 m/s
P: 4 W/m²
— Emergent
    Fitted

(r)

Sector: All
A: 1.7 m/s
k: 2.59
U: 1.45 m/s
P: 3 W/m²
— Emergent
    Fitted

Figure 24. (a)-(d) Sectorwise wind directions at 10 m AGL for Nguru, Katsina, Maiduguri and Sokoto, respectively; (e)-(h) Sectorwise wind directions at 10 m AGL for Zaria, Gusua, Potiskum and Kano, respectively; (i)-(l) Sectorwise wind directions at 10 m AGL for Kebbi, Gombe, Kaduna and Yelwa, respectively; (m)-(p) Sectorwise wind directions at 10 m AGL for Bauchi, Makurdi, Lokoja and Bida, respectively; (q), (r) Sectorwise wind directions at 10 m AGL for Benin and Calabar, respectively.

surrounding tall buildings and skyscrapers, causing wind flow obstruction and deflection. Furthermore, the South-West of Nigeria is a very poor region for wind energy investment and should not be considered for grid-tie application except if offshore wind farms would be developed close to Gulf of Guinea and integrated into the transmission network in Lagos.

The comparisons of the mean wind speeds of 18 synoptic stations derived from the bilinear interpolation of the remapped and actual wind fields are presented in **Figure 20**. In addition, the wind and percentage errors in temporal resolution of 00:00:00, 06:00:00, 12:00:00 and 18:00:00 UTC for a 10 years period have been estimated from the remapped and actual wind speed values (see **Table 4**). The annual mean wind errors and percentage errors for each synoptic station differ from one wind location to another. The estimated wind speed discrepancies show that the wind speed dataset on a field of grid resolution of $0.25° \times 0.25°$ is suitable and could be substituted with a gridded dataset on $0.125° \times 0.125°$ grid resolution for smooth and flat terrain. However, the wind speed biases increase at mountains and hill tops with complex topography because of limitation of the distance-weighted averag technique in capturing this terrain complexity during remapping.

The low quality of surface wind observations available for this study makes detailed accuracy comparisons difficult with ERA-Interim reanalysis wind for the synoptic stations. High-quality wind observations at synoptic stations in Nigeria are often not available in long-term period as deployment is often made for routine weather forecasts and not for energy application. However, the hourly surface wind available for 18 synoptic stations were obtained and compared with the interpolated synoptic stations as presented in **Figures 21-23**. The wind speed discrepancies could not be estimated because of the missing surface wind observations at the different time of the year. It is expected that the surface wind observations at the considered synoptic stations would be higher than the ERA-Interim reanalysis wind dataset obtained from the atmospheric model simulation.

5. Conclusion & Recommendation

The remapping of ERA wind field at $0.25° \times 0.25°$ grid resolution based on a distance-weighted averaging for studying the wind flow in Nigeria has been presented. The estimated wind speed biases show that a gridded wind speed dataset at $0.25° \times 0.25°$ resolution is suitable and can be substituted with a gridded dataset at $0.125° \times 0.125°$ grid resolution. The wind potential sites for grid-tie electrification in northern region such as Sokoto, Katsina, Nguru, Gusua, Birnin Kebbi, Maiduguri, Zaria, Potiskum, Cashua, Hadejia and Kaduna have been identified and should be explored for medium-scale energy conversion at a higher altitude AGL. The geographical locations such as Lagos, Ogun, Oyo, Osun, Ekiti and Ondo states are poor wind regime and unsuitable for enery application based on the wind flow study conducted for 10 years period. A high-quality surface wind speed observations at these synoptic stations in Nigeria for long-term period can be acquired *if available* for ascertaining the listed wind sites in North and South-West regions of Nigeria. The integration of other forms of renewable energy can be possible if overhauling of existing transmission lines is done and new networks are built to accommodate different energy sources (*i.e.* energy mix). The procedures for transformation of gridded wind datasets from one resolution to another could be tested and compared with different atmospheric model simulations. Finally, cost analysis model should be developed before wind energy investment in Nigeria.

Acknowledgements

The financial support received from the Ekiti State Government (NG) for this project is acknowledged. The data exchange from the National Ocean and Atmospheric Administration (NOAA) is acknowledged.

References

[1]   Nilsson, K. and Ivanell, S. (2010) Wind Energy. Gotland University, Visby.

[2]   Aidan, J. and Ododo, J.C. (2010) Wind Speed Distributions and Power Densities of Some Cities in Northern Nigeria. *Journal of Engineering and Applied Sciences*, **5**, 420-426. http://dx.doi.org/10.3923/jeasci.2010.420.426

[3]   Ohunakin, O.S. (2011) Wind Characteristics and Wind Energy Potential Assessment in Uyo, Nigeria. *Journal of Engineering and Applied Sciences*, **6**, 141-146. http://dx.doi.org/10.3923/jeasci.2011.141.146

[4]   Ojosu, J.O. and Salawu R.I. (1990) Wind Energy Development in Nigeria. *Nigerian Journal of Solar Energy*, **9**, 29-32.

[5]   Ojosu, J.O. and Salawu R.I. (1990) Survey of Wind Energy Potential in Nigeria. *Solar & Wind Technology*, **7**, 155-167. http://dx.doi.org/10.1016/0741-983X(90)90083-E

[6]   Adekoya, L.O. and Adewale, A.A. (1992) Wind Energy Potential of Nigeria. *Renewable Energy*, **2**, 35-39. http://dx.doi.org/10.1016/0960-1481(92)90057-A

[7]   Anyanwu, E.E. and Iwuagwu, C.J. (1995) Wind Characteristics and Energy Potentials for Owerri, Nigeria. *Renewable Energy*, **6**, 125-128. http://dx.doi.org/10.1016/0960-1481(94)00028-5

[8]   Agbaka, A.C. (1987) Experimental Investigation of the Possible Correction of Wind Speed on Insolation. *Energy Conversion & Management*, **27**, 45-48. http://dx.doi.org/10.1016/0196-8904(87)90051-3

[9]   Igbokwe, M.U. and Omekara C.O. (2002) Stochastic Simulation of Hourly Average Wind Speed in Umudike, South-East Nigeria. *Global Journal of Mathematical Sciences*, **1**, 59-66. http://dx.doi.org/10.4314/gjmas.v1i1.21323

[10] Medugu, D.W. and Malgwi, D.I. (2005) A Study of Wind Energy Potential: Remedy for Fluctuation of Electric Power in Mubi, Adamawa State, Nigeria. *Nigerian Journal of Physics*, **17**, 40-45.

[11] Ngala, G.M., Alkali, B. and Aji, M.A. (2007) Viability of Wind Energy as a Power Generation Source in Maiduguri, Borno State, Nigeria. *Renewable Energy*, **32**, 2242-2246. http://dx.doi.org/10.1016/j.renene.2006.12.016

[12] Oriaku, C.I., Osuwa, J.C., Asiegbu, A.D., Chukwu, G.U. and Kanu, C.O. (2007) Frequency Distribution Analysis of Available Wind Resources in Umudike, Abia State, Nigeria, for Wind Energy Conversion System Design. *Pacific Journal of Science & Technology*, **8**, 203-206.

[13] Garba, A.D. and Al-Amin, M. (2014) Assessment of Wind Energy Alternative in Nigeria from the Lessons of the Katsina Wind Farm. *Civil and Environmental Research*, **6**, 91-94.

[14] Pedersen, B.M., Hansen, K.S., Øye, S., Brinch, M. and Fabian, O. (1992) Some Experimental Investigations on the Influence of the Mounting Arrangements on the Accuracy of Cup—Anemometer Measurements. *Journal of Wind Engineering and Industrial Aerodynamics*, **39**, 373-383. http://dx.doi.org/10.1016/0167-6105(92)90561-N

[15] (2010) Best Practice Guidelines for Mesoscale Wind Mapping Projects for the World Bank. 1-40.

[16] Petersen, E.L., Mortensen, N.G., Landberg, L., Højstrup, J. and Frank, H.P. (1998) Wind Power Meteorology. Part II: Siting and Models. *Wind Energy*, **1**, 55-72.
http://dx.doi.org/10.1002/(SICI)1099-1824(199812)1:2<55::AID-WE5>3.0.CO;2-R

[17] Dee, D.P., Uppalaa, S.M., Simmonsa, A.J., Berrisford, P., Poli, P., Kobayashib, S., Andraec, U., Balmaseda, M.A., Balsamo, G., Bauer, P., Bechtold, P., Beljaars, A.C.M., van de Bergd, L., Bidlot, J., Bormann, N., Delsol, C., Dragani, R., Fuentes, M., Geer, A.J., Haimbergere, L., Healy, S.B., Hersbach, H., Hólm, E.V., Isaksen, L., Kållberg, P., Köhler, M., Matricardia, M., McNally, A.P., Monge-Sanz, B.M., Morcrettea, J.-J., Park, B.-K., Peubeya, C., de Rosnay, P., Tavolato, C., Thépaut, J.-N. and Vitarta, F. (2011) The ERA-Interim Reanalysis: Configuration and Performance of the Data Assimilation System. *Quarterly Journal of the Royal Meteorological Society*, **137**, 553-597.
http://dx.doi.org/10.1002/qj.828

[18] Lavagnini, A., Sempreviva, A.M., Transerici, C., Accadia, C., Casaioli, M., Mariani, S. and Speranza, A. (2006) Offshore Wind Climatology over the Mediterranean Basin. *Wind Energy*, **9**, 251-266. http://dx.doi.org/10.1002/we.169

[19] Jimenez, B., Durante, F., Lange, B., Kreutzer, T. and Tambke, J. (2007) Offshore Wind Resource Assessment with WAsP and MM5: Comparative Study for the German Bight. *Wind Energy*, **10**, 121-134.
http://dx.doi.org/10.1002/we.212

[20] Buzzi, A., Fantini, M., Malguzzi, P. and Nerozzi, F. (1994) Validation of a Limited Area Model in Cases of Mediterranean Cyclogenesis: Surface Fields and Precipitation Scores. *Meteorology and Atmospheric Physics*, **53**, 53-67.
http://dx.doi.org/10.1007/BF01029609

[21] http://apps.ecmwf.int/datasets/data/interim_full_daily/

[22] http://www.ncdc.noaa.gov/

[23] Lucchesi, R. (2008) File Specification for MERRA Products. Global Modeling and Assimilation Office Report, Version 2.1, 8.

[24] Fadare, D.A. (2008) A Statistical Analysis of Wind Energy Potential in Ibadan, Nigeria, Based on Weibull Distribution Function. *Pacific Journal of Science and Technology*, **9**, 110-119.

3

Use of Augmented Reality Methods to Support Legal Conflicts in the Planning Process for Wind Turbines Using the Example of the Landscape Conservation Area "Eulenkopf and Surroundings"

Timo Wundsam, Sascha M. Henninger

Department of Physical Geography, University of Kaiserslautern, Kaiserslautern, Germany
Email: Timo.Wundsam@ru.uni-kl.de, Sascha.Henninger@ru.uni-kl.de

Abstract

The world's growing energy demand poses a serious problem. At the same time fossil fuels are finite, which we must work against. Therefore, the Federal Government of Germany has set itself the goal to push forward the use of renewable energy in order to completely do without the generation of nuclear energy by 2023. There are, however, no specific guidelines from the European Directive on the promotion of electricity from renewable energy sources for the internal electricity market regarding how high each share of the different production method should be and, above all, which specific aim should be achieved by the share of wind energy. Nevertheless, it presents a crucial step toward a nuclear phaseout and a concomitant change of course of the Federal Government of Germany in the spring of 2011 regarding the expansion of renewable energy, taking the nuclear catastrophe in Fukushima into account. Using new legal planning approaches, also including the area of Rhineland-Palatinate, opportunities should be provided to make previously protected land available for setting up facilities for the generation of renewable energy. However, it is important to examine the legal situation regarding the installation of these kinds of constructions more detailed, as no general statements can be made. This will be illustrated using the example of the landscape conservation area "Eulenkopf and surrounding area" in the district of Kaiserslautern. The stated goal of the Social Democrat/Green coalition of the federal state government of Rhineland-Palatinate is to considerably expand the generation of electricity from renewable energy sources so that by 2030 at least the entire electricity demand can be covered by those. Due to the enormous potential of wind power, it is therefore necessary to quintuple its share of electricity generation by 2020, compared to 2011 numbers. In order to achieve the desired political objectives, by 2030 the number of turbines has to be increased to around 2650, representing a capacity of 7500 MW. This increase gives reason for boundary conditions to manage the generation of wind energy to be adjusted. This is intended to facilitate management and simultaneously minimise negative effects, such as the "sprawling" of wind turbines.

Keywords

Renewable Energy, Wind Energy, Turbines, Urban Planning, Landscape Conservation Area, Regulation, Laws, Augmented Reality, Public Interest

1. Introduction

The world's growing energy demand, which is mainly due to increased consumption—which again is due to higher standards of living, increasing mechanization in everyday life as well as due to recreational activities and energy-intensive production processes, poses a serious problem. At the same time fossil fuels are finite, which we must work against. Therefore, the Federal Government of Germany has set itself the goal to push forward the use of renewable energy in order to completely do without the generation of nuclear energy by 2023. Using new legal planning approaches, also including the area of Rhineland-Palatinate, opportunities should be provided to make previously protected land available for setting up facilities for the generation of renewable energy. However, it is important to examine the legal situation regarding the installation of such constructions in more detailed, as no general statements can be made. This will be illustrated using the example of the landscape conservation area "Eulenkopf and surrounding area" in the district of Kaiserslautern.

2. Problem Statement

The latest revision of 2012, the German Act on Granting Priority to Renewable Energy Sources (short: EEG (Erneuerbare-Energien-Gesetz)) regulates the preferred supply of electricity from renewable sources. According to §3 EEG, the following renewable energy sources are used for the generation of electricity:
- Hydropower,
- Geothermal energy,
- Energy from biomass, including biogas, biomethane, landfill gas and sewage gas as well as the biodegradable proportion of waste from households and industry, as well as
- Wind energy.

There are, however, no specific guidelines from the European Directive on the promotion of electricity from renewable energy sources for the internal electricity market (Directive 2001/77/EC), regarding how high each share of the different production method should be and, above all, which specific aim should be achieved by the share of wind energy [1]. Nevertheless, it presents a crucial step toward a nuclear phaseout and a concomitant change of course of the Federal Government of Germany in the spring of 2011 regarding the expansion of renewable energy, taking the nuclear catastrophe in Fukushima into account. The catastrophe of Fukushima illustrated the enormous environmental hazard radiation poses both in the generation of nuclear power as well as in the storage of nuclear waste [2]. As a result, the previously agreed on life-span extension of nuclear reactors to an average of twelve year was discarded. By the end of 2022, all nuclear power plants shall be shut down for good [3]. Particular importance is given to wind energy, because, unlike other renewable energy sources, it is able to cover a large energy demand. Thus it shows great potential for expansion.

2.1. Partial Adjustment of the State Development Programme IV (LEP IV) Regarding Wind Energy Use

The stated goal of the Social Democrat/Green coalition of the federal state government of Rhineland-Palatinate is to considerably expand the generation of electricity from renewable energy sources so that by 2030 at least the entire electricity demand can be covered by those. Due to the enormous potential of wind power, it is therefore necessary to quintuple its share of electricity generation by 2020, compared to 2011 numbers. In order to achieve this, each region of the federal state is to contribute to the target achievement proportionately, taking into account local conditions [4].

In mid-2013, there were about 1300 wind turbines in Rhineland-Palatinate with a total capacity of currently around 2100 megawatts (MW). In order to achieve the desired political objectives, by 2030 the number of turbines has to be increased to around 2650, representing a capacity of 7500 MW [5]. This increase gives reason for boundary conditions to manage the generation of wind energy to be adjusted. This is intended to facilitate management and simultaneously minimise negative effects, such as the "sprawling" of wind turbines. On 11 May 2013, the Partial Adjustment of the State Development Programme IV (Landesen twick lungs programm, or short: LEP IV) came into effect, which states that at least 2% of the land, and of which at least 2% must be woodland areas, should be made available for the use of wind energy. Furthermore, it contains revised guidelines regarding the use of wind energy. Herewith, the development of renewable energy is promoted, and at the same time it is aimed to help achieve climate change and energy policy objectives. It must be pointed out that the Partial Adjustment has transferred a large proportion of the responsibility regarding the management of wind energy use to the municipal level. Site selection for wind turbines, with the exception of a few exclusion areas which have been predefined by land use planning, is carried out at regional planning level. This results in a change in planning hierarchy, which limits regional planning in their authority. The implementation of the expansion of wind energy in Rhineland-Palatinate is therefore transferred to the individual municipalities, as the regulation in regions that are outside of certain areas and priority areas is reserved to urban land-use planning in the form of concentration areas [5]. The aim is to give local authorities additional room to manoeuvre with regards to how wind turbines are disclosed on land development plans. This creates an entirely new scope of duties for local authorities, which must be managed both from an expertise as well as a staffing perspective.

The Partial Adjustment has not only led to changes in responsibilities, but it also features fundamental alterations with regards to prerequisites for the sites of wind turbines. For example, landscape conservation areas, certain parts of nature reserves and biosphere reserves as well as Natura 2000 sites are no longer areas that are out of bounds, and can therefore be considered as potential areas for the use of wind turbines. Areas are therefore modified which must be noted both in local land-use plans as well as in regional development plans. This results not only in changes in planning policies but also changes from a nature conservation and natural law point-of-view.

In order to be able to efficiently expand wind energy, when planning suitable sites must be identified and simultaneously unsuitable sites must be excluded. Protection from neighbours, as well as wind conditions are important factors when selecting suitable locations. Nature conservation issues and landscape conservation are also to be taken into account. These include the protection of species, the protection of the landscape, as well as the recreational value of a landscape [6]. Opening up former prohibited areas does not mean that protected areas are overall conferred right to use, however wind energy generation will be described as admissible if it "(...) is compatible with the conservation aim" [5]. The consequence of this is that a large number of individual decisions will be necessary to clarify whether the conservation aim of certain areas will be impaired by the installation of wind turbines in some locations. In addition, as a result of a complete or partial deregulation of certain areas, from a legal point of view it will be possible to implement wind turbines in certain parts of a conservation area.

2.2. Criticism of the Partial Adjustment of the LEP IV

The problem of the Partial Adjustment is that in future municipalities are to directly conduct the further expansion of renewable energy rather than it being part of local planning authorities, where it really belongs [7]. According to §35 BauGB, it will be necessary to come up with a coherent and comprehensible concept, based on the land development plan, in order to achieve special protection status for wind turbine planning outside of the special protection areas and therewith to be able to obtain a regulation. For this purpose a study of the entire municipal area is carried out and it is to provide evidence that the selected sites for wind turbines are suitable with regards to wind conditions, indicating that therefore the so-called "substantial contribution" to the development of renewable energy is met. The criteria used should be transparent and must not be subjectively controlled [7].

Some local authorities feel capable of coping with this task, as they, among other things, have the qualified personnel and the budget necessary to develop such a concept. The reason that other local authorities are showing some resistance is not so much reluctance or indifference. It is rather because the resulting costs for the development of such concepts pose a crucial problem, since these would amount to about 30.000 - 80.000 Euros at local authority association level, plus similar costs if the adjustment of the land development plan was to be outsourced. The resulting income of the generated wind energy primarily goes towards operators and farmers, and one must be allowed to ask why a local authority association, which often struggles financially itself, should not be the ones benefitting from this [7]. For this reason, the willingness to create such concepts is particularly apparent in regions where more areas are found in which regional planning has given local authorities scope for planning, *i.e.* scope for control. This is, for example, the case in Western Palatinate [7].

3. Objective

The Partial Adjustment of the LEP IV does not contain clear information regarding the questions that have arisen directly from it. In current literature, the deregulation of protected areas for the generation of wind energy is subject of great controversy, and there is no unanimous opinion on it. For that reason, the case example of the landscape conservation area "Eulenkopf and surrounding area" in the district of Kaiserslautern aims to illustrate how best to deal with the "new" planning control of wind turbines, and what risks as well as opportunities for local authorities are associated with the Partial Adjustment of the LEP IV.

4. Boundary Conditions

Below, a general outline of the underlying boundary conditions is presented in order to illustrate the problems and the current legal stipulations.

4.1. Categories of Protected Areas

Due to the planning management of wind energy generation in areas significant from a nature conservation point-of-view, there are various conflicts between nature conservation and climate protection. Depending on the category of protected area, these two are either contrary to the production of wind energy or can be overcome. **Table 1** shows prohibitions and exceptions for the different categories of protected areas.

As this project is in the context of a landscape conservation area within the district of Kaiserslautern, it will be discussed in more detail below.

4.2. Landscape Conservation Areas

Landscape Conservation Areas are, according to national law, legally binding appointed areas that hold particular significance for the conservation of nature and landscape [8]. Compared to nature reserves, these larger areas hold fewer restrictions to use, making it a versatile and flexible category of protected area [9].

Legal Background
The regulation of Landscape Conservation Areas can be found in §36 of the Federal Law of the Protection of Nature (BNatSchG). Hereby three reasons arise for the need of protection:

Table 1. Overview of prohibitions and exceptions of the considered categories of protected areas [10].

Category	Prohibitions	Exceptions
Nature reserve	No changes allowed whatsoever	None, only with special justification
Landscape conservation area	Relative prohibition of changes, adverse effect on the character of the area	If consistent with conservation aim and character of the area
Natura 2000-area	Potentially inconsistent with preservation and conservation aim	Consistent with preservation and conservation aim, depending on the stipulation of the selected area
Biosphere reserve	Depending on the zone, absolute or relative ban on changes	Depending on zone stipulation of nature reserve or landscape protection area
Nature park	Depending on the zone, absolute or relative ban on changes	Depending on zone stipulation of nature reserve or landscape protection area

1. "To maintain or restore the effectiveness and functionality of the ecosystem or the ability for regeneration and sustainable use of natural resources, including the protection of biotopes and habitats of certain species of wildlife, both fauna and flora,
2. because of the diversity, uniqueness and beauty or because of the special cultural and historic significance of the area or,
3. because of their importance for recreation" [11].

Therefore, landscape conservation areas are usually large sections of the countryside which have satisfied at least one of the three protective features mentioned. Thus landscape areas can be identified that are characterized by human use and also have an importance for nature conservation and landscape management, even though they do not have the prerequisites of a nature reserve [8].

4.3. Setting up Wind Turbines in Landscape Conservation Areas

In order to meet the ambitious goals of Rhineland-Palatinate of promoting wind energy in the coming years, pressure is rising in the selection of sites for wind turbines, as more areas will be required. This raises the question to what extent it is possible to set up wind turbines in landscape conservation areas, or whether these areas should be kept clear from such use. Numerous dissentients are arguing that wind turbines have a negative impact on the landscape, which suggests that wind turbines and landscape conservation cannot coincide.

However, areas subject to the conservation status of a landscape conservation area do not enjoy complete protection, but there is a change clause which prohibits actions which, according to § 26 Section 2 BNatSchG, alter the character of the area or run contrary to its special conservation aim. In accordance with § 26 Section 1 No. 1-3 BNatSchG, it is also prohibited to introduce foreign bodies, such as wind turbines, into the area. It is therefore the prevailing opinion that is not possible to derive a general exclusion of wind turbines in landscape conservation areas [12]. In particular cases, it is necessary to review if the installation of wind turbines is consistent with the conservation aim, as the conservation aim and therefore the designation of a landscape conservation area is based on its individual characteristics and qualities [13].

If the sole conservation aim of a landscape conservation area is that of conservation, development or restoration of the ecosystem's productivity and efficiency in accordance with § 26 Section 1 No. 1 Federal Law of the Protection of Nature (§26 Section 1 No. 1 BNatSchG), it is possible to set up wind turbines but only if it is consistent with the conservation aim [2]. It is of particular importance to ensure that the conservation aim does not go back to wind-sensitive animal species in the area [2].

Changes in the area's character are not permitted either, because the unique identity of the landscape must be preserved. If elements are added which are not in line with the features of the landscape in terms of their dimensions, height, material etc., then they are incompatible with the area [2]. According to Nohl [14], wind turbines are perceived to be "totally inappropriate and foreign bodies to the landscape", which means, that the area would lose its unique character. Scheidler [12] considers the setting up of wind turbines in landscape conservation areas to be an unlawful act. According to §35 Section 3 (1) No. 5 BNatSchG, privileged utilisations are only opposed to the interests of the overall appearance of the landscape, if deformation is to be expected. Impairment is therefore not generally sufficient to counteract the establishment of wind turbines. There is, however, still a ban on impairments in landscape conservation areas, as the area is under particular protection according to the Federal Law of the Protection of Nature §36 Section 2 in conjunction with § 36 Section 1 No. 2 Federal Law of the Protection of Nature (§36 Section 2 BNatSchG combined with §36 Section 1 No. 2 BNatSchG) [12]. Although no general exemption may be granted, there may however be the possibility of individual cases to be issued with an exception or exemption for the construction of wind turbines, according to §67 Federal Law of the Protection of Nature [15] [16]. This can only arise from overriding grounds of public interest, and if arguments for the construction of wind turbines in an individual case outweigh the reasons against it. During consideration, the interest in providing continuous supplies of energy is relevant, although priority of renewable energy over the protection of the environment cannot be derived from this [17]. In such an assessment of individual cases, however, the case-specific grounds must play a crucial role and the result must remain open [17]. In addition, there is the possibility of obtaining the approval for wind turbines in a specific area by changing the regulations of a landscape conservation area. To justify this amendment, authorities must provide objective reasons that make such an amendment necessary, in order to weigh this up against the interests of nature conservation, where the principle of proportionality has to be respected [6]. In accordance with §22 Section 1 (3) of the Federal Law of the

Protection of Nature, a zoning of protected areas may be carried out, in order to define individual areas within a landscape conservation area in which the construction of wind turbines is permitted. This ensures that the conservation aim, which is not affected by the construction of wind turbines, remains [6] and a partial revocation of landscape conservation areas can be avoided [18].

5. Landscape Conservation Area "Eulenkopf and Surrounding Area"

5.1. Classification

The landscape conservation area Eulenkopf is located in the north-western district of the independent city of Kaiserslautern and extends over an area of 3550 ha, reaching across the districts of the association of municipalities Weilerbach and Otterbach. The settlements of the municipalities Sulzbach, Frankelbach, Erzenhausen and Eulenbis all lie entirely within the landscape conservation area as **Figure 1** shows [19].

The landscape conservation area "Eulenkopf and surroundings area" was provisionally placed under protection in 1963 by the Rhineland-Palatinate rights of nature. On August 30, 1977, the regulation on the landscape conservation area "Eulenkopf and surrounding area" became final. The protected status was due to its characteristic landscape worthy of conservation, in which the typical character of this region is formed by cultivated landscape for agricultural and forestry use. These again are characterised by agriculturally used tablelands, numerous valley cuttings and slopes [21]. Above all, observation points, which allow a panoramic view across large parts of the protected area, are to be emphasised. The municipality Eulenbis can be classed as a well-developed tourist area due to its location on a plateau, resulting in attractive views across the protected area. In addition, it is a designated and recognised tourist destination [22] [23]. It offers numerous cycling and hiking trails, as well as the Eulenkopf tower as a lookout point, all of which are popular tourist attractions.

5.2. Wind Energy Use and Wind Conditions

In the district of Kaiserslautern, wind energy is—with about 47%—the most subsidized form of renewable energy [24]. These figures put the district of Kaiserslautern into eighth place in comparison with other districts, bearing in mind that not all districts exhibit the same regional requirements for the generation of wind energy. According to the Wind Atlas Rhineland-Palatinate, the landscape conservation area "Eulenkopf and surrounding area" exhibits the following wind speeds:

In 100 m above ground, **Figure 2** shows mainly average wind speeds under 5 m/s prevail, with slight small scale increases of speeds from 5.4 to 6.0 m/s. Sporadically measurements are taken of wind speeds up to 6.0 to 6.4 m/s [25].

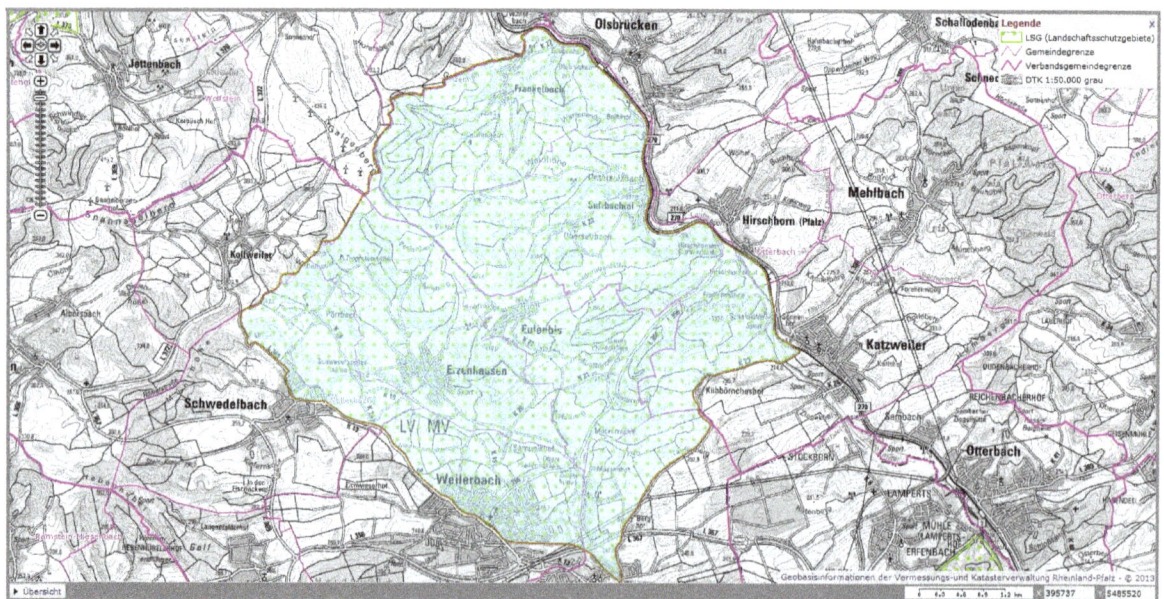

Figure 1. General map of the landscape conservation area "Eulenkopf and surrounding area" [20].

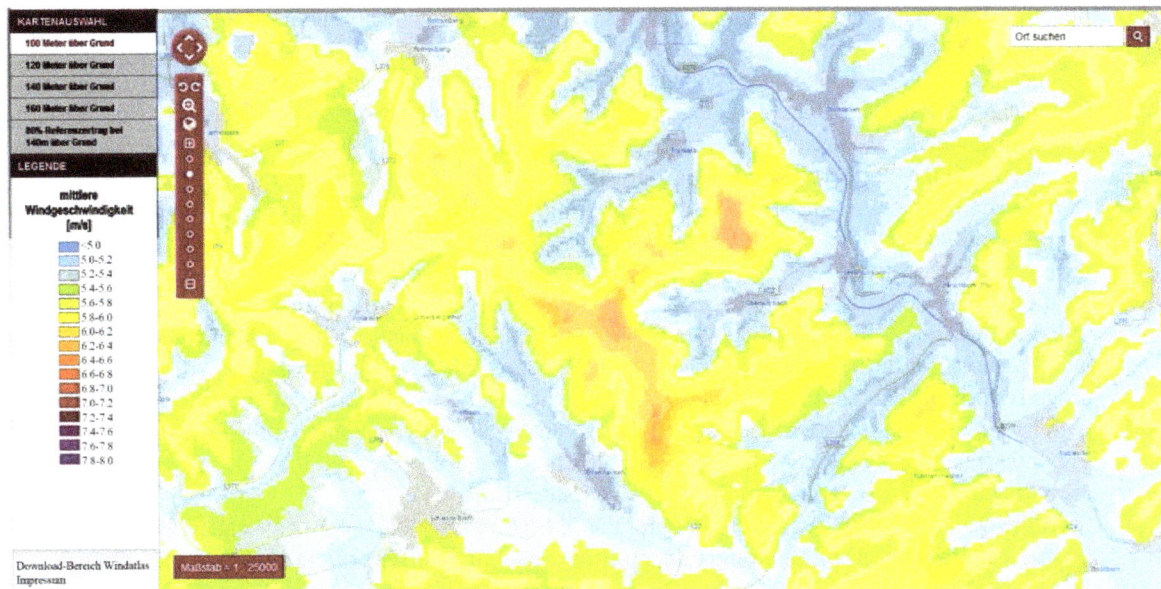

Figure 2. Wind Speed in the landscape conservation area "Eulenkopf and surrounding area" in 100 m above ground [26].

In 120 m above ground there are also mainly average wind speeds of 5.4 to 5.6 m/s. Wind speeds of 6.4 to 6.6 m/s occur very rarely [25].

Medium wind speeds from 5.2 to 6.0 m/s are expected in 140 m above ground. Only rarely do wind speeds reach 6.4 - 6.8 m/s [25].

In 160 m above the ground, wind speeds range from 5.8 to 6.0 m/s, although in individual areas wind speeds of 6.4 to 6.8 m/s can be achieved [25].

5.3. Statements from the Landscape Regulation

Legal regulation of the protected area defines the conservation aims of the characteristic landscape as quality of landscape, recreation as well as the protection of natural resources in accordance with § 3a-c of the landscape conservation regulation (LSchVO) Eulenkopf. All conservation aims have been fulfilled in accordance with § 36 Section 1, 1 - 3 Federal Law of the Protection of Nature. According to § 4 No. 1 LSchVO, it is not permitted to "erect or extend structural works of any kind". However, an exemption may be granted if this would not compromise the conservation aim of the area or if it cannot be attenuated or adjusted by conditions or requirements (§ 4 Section 2 LSchVO Eulenkopf). If it is not possible to prove that the necessary measures of attenuation or adjustment can be provided within the proposed development, an exemption cannot be granted.

According to the current regional planning programme Westpfalz (Western Palatinate), landscape conservation areas such as "Eulenkopf and surrounding area" are among the excluded area categories. Thus they are not available for the erection of wind turbines [27]. This, however, is questioned by the Partial Adjustment of the LEP IV.

Already in 2001, consideration was given to use the conservation area as a location for wind turbines. At that time, a land owner intended to build a wind energy plant with 65 m hub height in the district of Sulzbach. This plan was rejected by the country care authority of the district administration Kaiserslautern. In response, the owner submitted a lawsuit to the responsible administrative court in Neustadt a.d.W. This, however, confirmed the decision of the country care authority and considered it to be inadmissible to build wind turbines in that area, because it would contradict the conservation aim and would bring about a massive intervention into the cultivated landscape of this protected area [21]. Thus, it would lead to a strong technical overprinting of the landscape, which would be mainly due to the topography of the hilltop location.

Furthermore, the verdict also points out that the erection of wind turbines within the landscape conservation area must not be automatically approved, as wind turbines which have already been erected in the boundary area of the adjacent district Rothselberg in the district of Kusel, already provide a certain pollution level [21]. It is clear that the adjacent wind turbines do not negatively affect the visual connections of the landscape conservation area "Eulenkopf and surrounding area" due to their topographic arrangement and height. Moreover, it is not at a location which offers a lookout point from where a panoramic view is possible.

Since the proposed development contradicts the conservation aim, and no reduction, prevention or compensation of the adverse effects could be determined by constraints, the decision of the country care authority via the administrative court Neustadt a.d.W. was regarded as legitimate [21]. Furthermore, the decision points out that the projects in the outdoors may not be arbitrarily undertaken at any location, but that a continuously considerate exploitation of resources must be ensured [21]. Furthermore, there it is no atypical case of unintended hardship [21]. The only point in favour of erecting wind turbines is the public interest in using renewable energies. This, however, is not sufficient for an exemption because wind turbines are not site-specific and can be built at other locations in the outskirts, and not necessarily within the landscape conservation area [21].

In contrast, the "Rundschreiben Windenergie" (Newsletter Wind Energy), which was published in Rhineland-Palatinate in 2013, states that on one hand it is possible to grant permissions and exemptions, if it is on grounds of overriding public interest [28]. On the other hand it is determined that the necessary permission shall be reissued on a regular basis because the use of renewable energy always overrides other considerations [28]. This presents an essential contradiction, in which the "Rundschreiben Windnergie" (Newsletter Wind Energy) according to Kusche [29] does not follow the current jurisdiction.

At this point, it is not possible to interpret the judicature obviously. As a consequence, it is quite difficult, to inform the citizens and residents about the potential upcoming impacts of the development proposal.

6. Details of the Proposed Project

Based on the Partial Adjustment of the LEP IV, the municipalities Sulzbachtal and Frankelbach and the association of municipalities Otterbach are planning a cooperation with the JUWI AG, which already operates a number of wind turbines in Germany, in order to make new advances towards the planning of new wind turbines in the landscape conservation area "Eulenkopf and surrounding area" [29]. At the meeting of the Board of Association of Municipalities on 16 August 2013, the amendment of the land utilisation plan was initiated in order to establish legal planning requirements for the construction of wind turbines [29]. The intention was to build seven wind turbines within the municipal area of Sulzbachtal and Frankelbach with a total height of 200 meters each [22].

The three municipalities Sulzbachtal, Frankelbach and Eulenbis, all of which are located in the landscape conservation area "Eulenkopf and surrounding area", are in dispute over these proposed plans. The local authority Eulenbis is using the conservation area as a tourist destination and expresses great interest in the preservation of the landscape. The local authority of Sulzbachtal, on the other hand, can hardly benefit from its use as a tourist attraction because of its topographical location in the valley. Visitors tend to stay on the plateau of Eulenbis and therefore also mainly frequent the local gastronomy there. Thus, the municipality Sulzbachtal does not benefit but merely has to battle against the restrictions arising from the regulations of the conservation area [10]. The driving force in connection with the planning of wind turbines is represented by the mayor of Sulzbachtal, Mr. Zinsmeister, whose concern it is to diminish the indebted financial situation of his local authority. Sulzbachtal would benefit the most from the installation of wind turbines, but the local authority Erzenhausen, which also adjoins to the plateau, would also endeavour setting up wind energy plants in their district if a permit was secured.

However, the local population has mixed views towards the project. On the one hand there is the group of people that benefits from the landscape conservation area and therefore are against the setting up of wind turbines. On the other hand, supporters of the project can see no positive effect coming from the protected area or are even limited by its restrictions. Although no citizen initiatives have yet been formed, the mayor of Eulenbis, Mr. Bürgner, together with the local council, is very dedicated in his campaign against wind turbines.

In the course of fighting the construction of wind turbines, image manipulations were used to raise people's awareness of the proposed plans and to visually bring forward consequences and impacts on a previously untouched protected area. For this, maps and graphical material was used to illustrate the locations and size of the different projected wind turbines.

To expand this procedure further still, and to inform as many people as possible about the serious consequences, the University of Kaiserslautern, or more specifically the teaching and research department Physical Geography, developed a way to make this illustration available for everyone via smartphone or tablet. By means of "Augmented Reality", images with 3D models are overlaid by wind turbines. Previously, coordinates had been placed on those 3D models, in order to comprehend the planned future location as accurately as possible.

In Augmented Reality "(...) real situations are equipped with additional digital information so that the relevant objects and items can enter into communication with computer systems. Thus, a conceptual design, generated as a computer graphic similar to a traditional photomontage, may be overlaid by inserting a real life situation. This can be done automatically by using a computer equipped with positioning technology (GPS-receiver), which executes the image overlay geometrically exact with regards to position and orientation" [30].

A clear advantage of this is the possibility that, firstly, the 3D models can be accurately assigned to the proposed locations using coordinates. Secondly, it is therefore also possible to simulate the exact height of the wind turbine and illustrate it on the screen of the mobile device.

For the simulation of this project, the application (abr.: app) "LayAR" is used, which is freely available on the platforms Android, iOS and BlackBerry. In addition, an adapted database is accessed, which was developed by the department Computer-aided Planning and Design Methods of the University of Kaiserslautern. Here coordinates are placed in a map and connected to the previously developed 3D model, which has a total height of about 200 m and is offered in **Figure 3**. Additionally, more information of the model or the project can be stored, however, this is currently irrelevant for the specified project.

This can help interested citizens, as well as political decision makers, gain an impression of the extent of the impacts such a development would have, without having to actually carry out a project.

In practice, however, this type of illustration presented some problems difficult to exert some direct influence on. For example, it became apparent that the GPS sensors and gyroscopes (tilt sensors) of different smartphone and tablet manufacturers show differing accuracies. The consequence of this was that the models were in parts not positioned at the correct, predetermined coordinates, and therefore deviated from the expected positions. It is important to ensure that the devices are calibrated before use and the view of the sky is as unobstructed as possible, so that a high number of GPS satellites can be used for position determination. The more satellites are reached, the higher is the accuracy of the location.

Furthermore, it should be mentioned that it is not yet currently possible to display the shadow of the blades within the app, which would certainly be helpful in the illustrated case. The so-called disco effect of turbines of this size would for sure play a crucial role for local residents when looking at the impact on the public interest.

Another possible measure may be a marker-based illustration of the 3D models. These so-called markers, indicated in **Figure 4**, must be affixed to the particular position at which a wind turbine is to be installed. The markers can be either simple photos or, for example, QR codes. By linking this marker with the 3D model, this will be illustrated as an overlay directly on-site. One problem of this method is that the 3D models of the wind turbines are about 200 m high and can therefore not be fully represented on a smartphone or tablet display. This is because an illustration of the model is only possible as long as the camera of the smartphone or tablet is

Figure 3. Illustration of the 3-dimensionally modelled wind energy plant at Eulenbis.

Figure 4. Illustration of a Marker Grid with a 3D-model [31].

directed at the marker. However, since one is situated directly at the location of the 200 m high wind turbine, a complete illustration is not possible.

It would, however, be conceivable to use this technique to develop a complete spatial 3D model of the municipality, in which house structures and roads, as well as the digital terrain model (DTM) are stored in addition to the wind turbines. If such a model was recorded on a marker, it would be possible to re-enact the overall effect on the screen of your smartphone or tablet from many different perspectives, such as the bird's eye view.

Nonetheless, this illustrated method is not resorted to, as the implementation and the development of the 3D-models is too expensive and time-consuming. In addition, no clear statement can be made as to whether any current smartphones meet the requirements and whether a correct representation of the model can be guaranteed.

According to the shown up technology, it is possible for every citizen and resident, to illustrate the planned wind turbines on the display of mobile devices that are equipped with the required application and a camera. In this way, the user gets the opportunity, to demonstrate visually the upcoming planning intentions.

7. Project Evaluation

The proceedings regarding the examination of the locations of wind turbines within the landscape conservation area "Eulenkopf and surrounding area", which were resumed due to the Partial Adjustment of the LEP IV, show that, despite the newly gained competencies of the communities, setting up wind turbines in protected areas remains difficult. The municipalities of Frankelbach and Sulzbachtal have so far merely initiated the adjustments in the land utilisation plan. What is still missing, however, is the request at the nature conservation authority of the district administration Kaiserslautern whether a concentration zone for the use of wind energy can be successful in the area of the landscape conservation area "Eulenkopf and surrounding area". It is important to clarify the question whether the land utilisation plan stands much chance of being approved, taking into consideration legal aspects of immissions, and whether planning in this kind of area would be possible [10]. The conservation aim of the landscape conservation area is to preserve the typical characteristics of the landscape of the "North Palatine hill country", as well as the protection of the ecosystem. Previous impacts exist through the existing facilities at the border area, at which repowering is to take place in future, which means that the level of impact on the protected area will increase further [10]. While this may be justification to release border areas of the protected area, however, there is no legal obligation to do so.

The aim of the landscape conservation area "Eulenkopf and surrounding area" is to preserve the typical character of the area [10]. The responsibility of granting permission for the erection of wind turbines within the protected area lies with the nature conservation authority of the district of Kaiserslautern. As a reference for the authority's decision in the current case, the previously mentioned verdict of the Administrative Court of Neustadt a.d.W. is cited, in which neither permission nor a waiver were granted for the project of erecting wind turbines with a hub height of 65 m. These circumstances must now, however, be re-examined.

With regard to the exemption, however, it is to be referred to the conservation aim of the regulation. If this is in conflict with the project, no exemption shall be granted. In the event that seven 200 meter high wind turbines will be built, it can be assumed that this would lead to a technical overprinting of the landscape, which again would lead to a transformation of the cultivated landscape into a technologically shaped landscape [32]. Since the construction of wind turbines is to be implemented in the heart of the landscape conservation area, it is clear that the plans contradict the conservation aim, which in turn has the consequence that an exemption cannot be granted [32]. It now needs to be examined whether a waiver can be considered. According to §67 Section 1 BNatSchG, two reasons can be cited here:

1. Predominant public interest and,
2. an unreasonable burden (on an individual basis).

The existence of an unreasonable burden must be consistent with the interests of nature conservation in order to be worth a waiver. However, Dein certified that this is no atypical case of unintended hardship [32]. In the present case, the harshness that leads to the exclusion of the proposed plans is intended. This ensures that the regulations of the protected area continue to fulfil its purpose. Public interest in the construction of wind turbines is indeed given, but it must be balanced with other interests and thus cannot as a rule lead to a waiver. Otherwise, the procedure must be declared as illegitimate, as the decision of the legislative body would be administratively overridden [32]. Furthermore, granting a waiver must not lead to the landscape protection regulation (LSchVO) to become inoperable.

As it can currently not be assumed that an exemption or a waiver will be granted, a partial or full cancellation of the regulations of the protected area is seen as a last chance to revise planning of the protected area [32]. After the construction of wind turbines, the regulations would no longer be applicable, also in view of restriction of the past for other uses. A partial abolition is discarded in the present case, as the major part of the protected area surrounding the Eulenkopf is affected. In the case of a full revocation of the landscape conservation area status, the designation needs to be examined by applying the same criteria from the opposite viewpoint. A conservation aim does not need to exist. The discretion lies with the regulatory authority [10].

Thus it needs to be determined that the legal situation regarding the application of the regulation of the landscape conservation area "Eulenkopf and surrounding area" does not change because of the Partial Adjustment of the LEP IV. Wind turbines are not permitted in this area, as the construction is in opposition to the conservation aim. This would have a potentially negative impact on the landscape. The wind turbines are contrary to the conservation aim to preserve the distinctive and diverse landscape, as well as to the protection of general nature-orientated recreation (§ 3 Section 1a and c LSchVO Eulenkopf). A panoramic view is greatly disturbed by the construction of wind turbines, and lookout points with a good view into the far distance will lose their attraction for recreation seekers. Therefore, the compatibility of wind turbines within the core area of the protected area is to be classed as very low, as a deterioration of the circumstances is to be expected.

To date, the area does not exhibit any zoning, thus the conservation aims apply area-wide. In principle, by changing the protected area regulations, zoning, which allows implementing wind turbines in the predisposed border areas of the protected area, would indeed be conceivable. However it would be appropriate to aspire to attain an exemption for the entire protected area, as the conservation area is relatively small in size. The construction of wind turbines in the core of the area would lead to a significant deterioration of the landscape and could potentially result in the loss of its status being worthy of protection as well as its unique features. Furthermore, the wind speeds within the area are even below the economic limit, meaning that the location cannot be classed as particularly suitable. The Partial Adjustment of the LEP IV stipulates that especially areas with good wind conditions are to be given priority, which is not met in the landscape conservation area "Eulenkopf and surrounding area". Moreover, wind turbines are not tied to a location, thus excess planning of the protected area is not imperative, as there are plenty of areas in the district of Kaiserslautern, which are outside the protected area and exhibit good enough wind conditions. These areas should therefore be favourably considered for the use of wind energy.

Should the JUWI AG, as well as the surrounding municipalities of Sulzbachtal and Frankelbach, still wish to proceed with their plans, it will be in their duty to provide citizens with enough information about the consequences of the construction of wind turbines, and to raise their awareness so that they gain an impression of what an implementation of wind turbines would mean for the landscape conservation area "Eulenkopf and surrounding area". Although it is on the one hand understandable that, due to their financial situation, local authorities need to balance their budgets and improve their economic outlook, but on the other hand neither nature itself nor affected citizens should be excluded from this decision.

Using the illustration of the proposed wind turbines in the form of 3D models, it is possible to provide laymen with a tool which allows them to gain a better idea of the proposed plans, which, above all, aims to expand the conventional type of comprehensive plans with an additional illustration method. For this, no explicit expert knowledge is necessary in order to get one's bearings on maps or plans, and to understand proposed plans. Rather, it can even be simulated right on-site, where future plans of wind turbines are to be implemented.

Due to the demonstrated legal position, it is actually fairly difficult, to make a prediction concerning about the planning purposes. It indicates that the Partial Adjustment of the State Development Programme IV (LEP IV) will still be opposed to a great number of legal obstacles, whereby a contemporary execution of the project is hard to imagine.

References

[1] Fest, P. (2010) Die Errichtung von Windenergieanlagen in Deutschland in seiner ausschließlichen Wirtschaftszone—Genehmigungsverfahren, planerische Steuerung und Rechtsschutz an Land und auf See, Duncker & Humblot, Berlin.

[2] Scheidler, A. (2011) Errichtung von Windkraftanlagen in naturschutzrechtlich festgesetzten Schutzgebieten. *Natur und Recht*, **12**, 848-856. http://dx.doi.org/10.1007/s10357-011-2182-z

[3] Braune, T. and Ismar, G. (2011) Schwarz-Gelb beschließt Atomausstieg bis 2022. http://www.welt.de/politik/article13401638/Schwarz-Gelb-beschliesst-Atomausstieg-bis-2022.html

[4] Ministerium für Wirtschaft, Klimaschutz, Energie und Landesplanung, Ministerium für Finanzen, Ministerium für Umwelt, Landwirtschaft, Ernährung, Weinbau und Forsten, Ministerium des Innern, für Sport und Infrastruktur Rheinland-Pfalz (2013) Hinweise für die Beurteilung der Zulässigkeit der Errichtung von Windenergieanlagen in Rheinland-Pfalz (Rundschreiben Windenergie), Mainz.

[5] Ministerium für Wirtschaft, Klimaschutz, Energie und Landesplanung Rheinland-Pfalz (2013) Windatlas Rheinland-Pfalz, Mainz.

[6] Fischer-Hüftle, P. (2012) Windenergieanlagen und Landschaftsschutz. *Bayerische Verwaltungsblätter*, **23**, 709-715.

[7] Spannowsky, W. and Hofmeister, A. (2012) Naturschutzgerechte Steuerung der Windenergienutzung durch die gesamträumliche Planung. Lexxion Verlag, Berlin.

[8] Fischer-Hüftle, P. (2003) §26 Landschaftsschutzgebiete. In: Schumacher, J. and Fischer-Hüftle, P., Eds., *Bundesnaturschutzgesetz Kommentar*, Kohlhammer-Verlag, Stuttgart, 412-425.

[9] Heugel, M. (2010) §26 Landschaftsschutzgebiete. In: Lütkes, S. and Ewer, W., Eds., *Bundesnaturschutzgesetz Kommentar*, C. H. Beck, München, 293-297.

[10] Langenbahn, E. and von der Au, L. (2014) Analyse der Teilfortschreibung Erneuerbare Energien des LEP IV in Rheinland-Pfalz hinsichtlich naturschutzfachlicher sowie, rechtlicher Konflikte bei der planerischen Steuerung von Windenergieanlagen anhand ausgewählter Beispiele. Master Thesis, University of Kaiserslautern, Kaiserslautern.

[11] Federal Agency for Nature Conservation (2010) Bundesnaturschutzgesetz. http://www.gesetze-im-internet.de/bnatschg_2009/index.html

[12] Scheidler, A. (2012) Windräder in Natura 2000-Gebieten? *Deutsches Verwaltungsblatt*, **4**, 217-221.

[13] Thyssen, B. (2011) Rückenwind? Bewältigung der Herausforderungen in Genehmigungsverfahren zum Ausbau der Windenergie. *Zeitschrift für Immissionsschutzrecht und Emissionshandel I + E*, **2011**, 134-145.

[14] Nohl, W. (2010) Landschaftsästhetische Auswirkungen von Windkraftanlagen. *Schönere Heimat*, **2010**, 3-12.

[15] Barth, S., Baumeister, H. and Schreiber, M. (1997) Leitfaden für die kommunale Planung unter besonderer Berücksichtigung von Naturschutzbelangen. Rhombos-Verlag, Berlin.

[16] Greiving, S. and Schröder, M. (2003) Neue Herausforderungen bei der planerischen Steuerung von WEA. *Umwelt und Planungsrecht*, **2003**, 13-17.

[17] Attendorn, T. (2013) Berücksichtigung der Belange der Energiewende bei der Anwendung des Naturschutzrechts. *Natur und Recht*, **35**, 153-262. http://dx.doi.org/10.1007/s10357-013-2415-4

[18] Kühnau, C., Reinke M., Blum, P. and Brunnhuber, M. (2013) Standortfindung für Windkraftanlagen im Naturpark Altmühltal. Erstellung eines Zonierungskonzepts. *Naturschutz und Landschaftsplanung*, **45**, 271-278.

[19] Landschaftsinformationssystem der Naturschutzverwaltung Rheinland-Pfalz (2014) Verordnung über das Landschaftsschutzgebiet "Eulenkopf und Umgebung" vom 30. August 1977. http://www.naturschutz.rlp.de/dokumente/rvo/lsg/07-LSG-7335-010.pdf

[20] LANIS Rheinland-Pfalz (2014) Landschaftsschutzinformationssystem der Naturschutzverwaltung, Rheinland-Pfalz. http://map1.naturschutz.rlp.de/mapserver_lanis

[21] Administrative Court Neustadt an der Weinstrasse (2001) Verdict of 24. September 2001, Reference No. 4K2104/00. N.W.

[22] Die Rheinpfalz (2012) Wieso mit 10,000 Euro zufriedengeben. 320, 14.

[23] Association of Municipalities Weilerbach (2014) Ortsgemeinde Eulenbis. http://www.weilerbach.de/ortsgemeinden/eulenbis/index.html

[24] Die Gemeindeverwaltung in Rheinland-Pfalz (2013) Nordwesten von Rheinland-Pfalz bei erneuerbaren Energien vorn. Windkraft vor Fotovoltaik-Überragende Bedeutung des ländlichen Raums. *Die Gemeindeverwaltung in Rheinland-Pfalz*, **9**, 252-256.

[25] Ministry of Economic Affairs, Climate Protection, Energy and Land Use Planning Rhineland-Palatinate (2013) Windatlas Rheinland-Pfalz, Mainz.

[26] Ministry of Economic Affairs, Climate Protection, Energy and Land Use Planning Rhineland-Palatinate (2014) Windatlas Rheinland-Pfalz. http://www.windatlas.rlp.de/windatlas/

[27] Planning Association Westpfalz (2013) Westpfalz-Informationen—Potentialrechner Erneuerbare Energien fpü die Region Westpfalz, Wie viel Strom aus Erneuerbaren Energien kann in der Region Westpfalz erzeugt werder? Ein Leitfaden für Landkreise, kreisfreie Städte und Kommunen.

[28] Ministry of Economic Affairs, Climate Protection, Energyand Land Use Planning; Ministry of Finance, Ministry of the Environment, Rural Affairs, Food, Viniculureand Forestry, Ministry of theInterior, Sport and Infrastructure Rhineland-Palatinate (2013) Hinweise für die Beurteilung der Zulässigkeit der Errichtung von Windenergieanlagen in Rheinland-Pfalz (Rundschreiben Windenergie), Mainz.

[29] Kusche, K.L. (2013) Errichtung von Windenergieanlagen im Landschaftsschutzgebiet Eulenkopf, Beschlussvorlage. Kreisverwaltung Kaiserslautern, 11-14.

[30] Streich, B. (2011) Stadtplanung in der Wissensgesellschaft-Ein Handbuch. VS-Verlag, Wiesbaden. http://dx.doi.org/10.1007/978-3-531-93164-7

[31] Noll, R. (2012) Der Einsatz von Augmented Reality Methoden zur Kommunikation bei Konversionsprojekten. Bachelor Thesis, University of Kaiserslautern, Kaiserslautern.

[32] Dein, A. (2012) Windkraft-Planung im Landschaftsschutzgebiet, Eulenkopf und Umgebung. Rechtliche und fachliche Aspekte.Niederschrift der 26. Sitzung des Kreistages vom 12.11.2012, 56-71.

4

Theoretical Study of Wind Turbine Model with a New Concept on Swept Area

Sagarkumar M. Agravat[1*], N. V. S. Manyam[2], Sanket Mankar[3], T. Harinarayana[1]

[1]Gujarat Energy Research and Management Institute, Gandhinagar, India
[2]Department of Electrical Engineering, University of Petroleum and Energy Studies, Dehradun, India
[3]School of Mechanical and Building Sciences, Vellore Institute of Technology, Vellore, India
Email: [*]sagar.a@germi.res.in

Abstract

Commercially available wind-turbines are optimized to operate at certain wind velocity, known as rated wind velocity. For other values of wind velocity, it has different output which is lower than the rated output of the wind plant. Wind mill can be designed to provide maximum power output at different wind velocities through modification of swept area to match with the wind speed available at the moment. This can result in higher power output at all the velocities except that at rated wind speed because of limitation of generator. This results in increased utilization of generation capacity of wind mill compared to its commercially designed counterpart. A theoretical simulation has been done to prove a new concept about swept area of wind turbine blade which results in a significant increase in the power output through the year. Simulation results of power extracted through normal wind blade design and new concept are studied and compared. The findings of the study are presented in graphical and tabular form. Study establishes that there can be a significant gain in the power output with the new concept.

Keywords

Cut-In Wind-Speed, Cut-Out Wind-Speed, CUF, Swept Area, Radius, Chord, Aerofoil, Axial Flow Induction Factor, Inflow Factor, Actuator Disc Concept, Momentum Theory

1. Introduction

Wind turbine generators are producing power by exchanging momentum with air particles in motion. While harnessing wind energy, wind turbine does not damage environment and is a clean source of energy. As per

---

[*]Corresponding author.

CEA 2013 report, 8% of the installed power plant capacity is to be attributed to the wind energy installation. This only contributes 1.6% of total electricity generation [1]. Percentage share of wind energy as part of renewable energy installations in India is close to 70%. Moreover, the PLF observed for different sites are reported to be in the range of 15% - 30% [2].

Wind turbines are selected based on operating conditions of site. In other words, wind turbines are designed to provide maximum power for maximum occurring wind speed for which rotor and turbine are optimized and tested for. Like any other renewable energy sources, wind energy also has different intensity during different parts of the day and also for different months. With decline in wind energy, output decreases, below cut-off wind speed, output reduces proportionately till cut-in wind speed and thereafter no generation can be found. This is because at this instance momentum of wind particle is insufficient to rotate wind turbine and hence generator. The power output is proportional to the swept area of the blade and also to the pitch angle. During field conditions, momentum and RPM of turbine are adjusted to provide optimum torque and rpm. Traditionally, this is done by varying pitch angle and variable speed wind-turbine [3] [4]. As an alternate to this practice, wind farm need to be planned optimally, so as to take into account different types of wind characteristics [5].

There are number of methods reported in literature on design optimization of wind mill [6]-[13]. As per Muljadi, E. *et al.* [14] limit output power by suitable control limit in high wind speed regime and similarly try to extract more power by modifying pitch and speed of turbine to provide more torque from aerodynamic power. However, in all the cases, momentum of the blade causes loss of captured power. Talavera Juan, A. *et al.* [15] reported dynamic pair of overlapping blade segments used to control swept band area. M/s. Powersail® is in the development of adjustable swept area wind turbine model [16]. The company claims to increase the output by two-fold for an increase in swept area by 25%. No technical study of their proprietary technology is reported. Mark Dawson *et al.* [17] of *Energy Unlimited Inc.* made first practical effort to overcome this limitation by changing swept area of rotor by adjusting the length of the blade. In his design Mark increased the length of the blade during low wind area and blade length is reduced during high wind area. US granted patent on June 7, 2005 vide US patent No. 6902370 B2 for variable speed wind turbine blade. In his design, Mark Dawson built 11 m long blade and a controller to change the blade length. Mark reported that the new design is beneficial for about 6000 hours in a year when the wind speed is lower than required. Except few efforts as reported below, not much R&D work has been reported in literature or patent search based on the variable swept area wind mill as an effective tool to enhance power output. Moreover, the proprietary/patented technologies referred above are also only increasing the length of the blade. This way, aerodynamic shape is distorted. Our new design proposes to increase or decrease swept area so as to retain aerodynamic shape of the blade at all the time and thus extract more output.

2. Materials and Methods

To prove the new concept, new simulation is carried out using Q-Blade, V 8.0 software. This is open software developed by Hermann Fottinger Institute of Technical University (TU), Berlin. The software helps to evaluate the wind turbine blade profile on the basis of Blade Element Momentum theory. The main advantage with Blade Element Momentum theory is that it takes less time compared to that of Computational Fluid Dynamic analysis. Because of this advantage, a rapid evaluation of various blade designs is possible. The software helps user instantly to design the custom aerofoil and compute the performance polar and also has the capability to directly include the new design into rotor and simulate the power output of the wind generator. The software has necessary functions to simulate blades for HAWT and VAWT as well. NREL wind turbine model that is available in software library has been taken as reference blade design for further modifications. The blade profile that exist in Q-Blade V 8.0 trial version for NREL-5MW are provided in **Table 1**.

The first column indicates normal profile of original blade with 63 m radius, used as reference profile. Subsequent columns indicate the blade profile modified for its radius with radius of 70 m, modified blade profile of 63 m blade with increased chord, 71.5 m blade profile with increased chord from tip to root in five sectors respectively. Increase in radius of reference blade was restricted to 71.5 m, in order not to exceed the critical radius that defines the fatigue strength of the blade. The Blade profile diagrams are shown in **Figure 1**.

As per Actuator Disc concept, analysis of the aerodynamic behavior of wind turbines can be done without any specific turbine design but through Energy Extraction Process. In **Figure 2**, the energy extraction process through Actual Disc concept is shown. Upstream of the disc the stream-tube has a cross-sectional area smaller than

Table 1. NREL-5MW reference blade profile with modified blade profiles for simulation[*].

Normal		Increased radius		Increased chord		Increased radius & chord	
Length	Chord	Length	Chord	Length	Chord	Length	Chord
0.00	3.20	0.00	3.20	0.00	3.20	0.00	3.20
1.36	3.54	1.36	3.54	1.36	3.54	1.36	3.54
4.10	3.85	4.10	3.85	4.10	3.85	4.10	3.85
6.83	4.17	6.83	4.17	6.83	4.17	6.83	4.17
10.25	4.55	10.25	4.55	10.25	4.55	10.25	4.55
14.35	4.65	14.35	4.65	14.35	4.65	14.35	4.65
18.45	4.46	18.45	4.46	18.45	4.46	18.45	4.46
22.55	4.25	22.55	4.25	22.55	4.25	22.55	4.25
26.65	4.01	26.65	4.01	26.65	4.01	26.65	4.01
30.75	3.75	30.75	3.75	30.75	3.75	30.75	3.75
34.85	3.50	34.85	3.50	34.85	3.50	34.85	3.50
38.95	3.26	38.95	3.26	38.95	3.26	38.95	3.26
43.05	3.01	43.05	3.01	43.05	3.01	43.05	3.01
47.15	2.76	47.15	2.76	47.15	2.76	47.15	2.76
51.25	2.52	51.25	2.52	51.25	4.55	51.25	4.55
54.67	2.31	54.67	2.31	54.67	4.17	54.67	4.17
57.40	2.09	57.40	2.09	57.40	3.85	57.40	3.85
60.13	1.40	60.13	1.40	60.13	3.54	60.13	3.54
61.50	0.70	70.00	0.70	61.50	3.20	70.00	3.20

[*]All dimensions are in meter.

that of the disc and an area larger than the disc downstream. The expansion of the stream-tube is because the mass flow rate must be the same everywhere. The mass of air which passes through a given cross section of the stream-tube in a unit length of time is $\rho A U$, where ρ is the air density, A is the cross-sectional area and U is the flow velocity. The mass flow rate must be the same everywhere along the stream tube and so

$$\rho A_\infty U_\infty = \rho A_d U_d = \rho A_w U_w \tag{1}$$

The symbol ∞ refers to conditions far upstream, d refers to conditions at the disc and w refers to conditions in the far wake.

The wind energy after actuator disc is given by,

$$U_w = (1 - 2a)U_\infty \tag{2}$$

It is usual to consider that the actuator disc induces a velocity variation which must be superimposed on the free-stream velocity. The stream-wise component of this induced flow at the disc is given by $-aU_\infty$, where "a" is called the axial flow induction factor, or the inflow factor. At the disc, therefore, the net stream-wise velocity is

$$U_d = U_\infty (1 - a) \tag{3}$$

Following the momentum theory using the Bernoulli's principle applied to the upstream and downstream sections of the stream tube, separate equations are calculated for energy upstream and downstream as the total

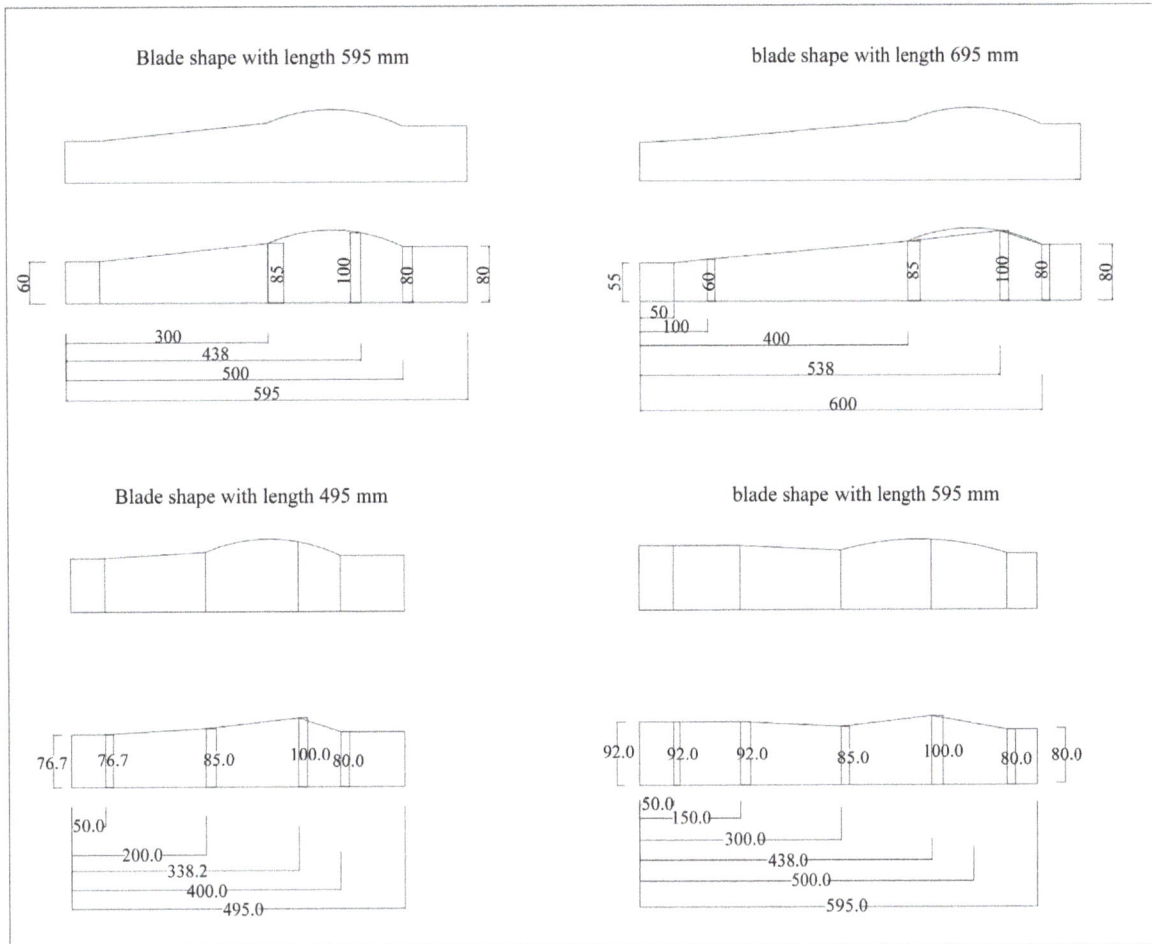

Figure 1. Blade profile diagrams.

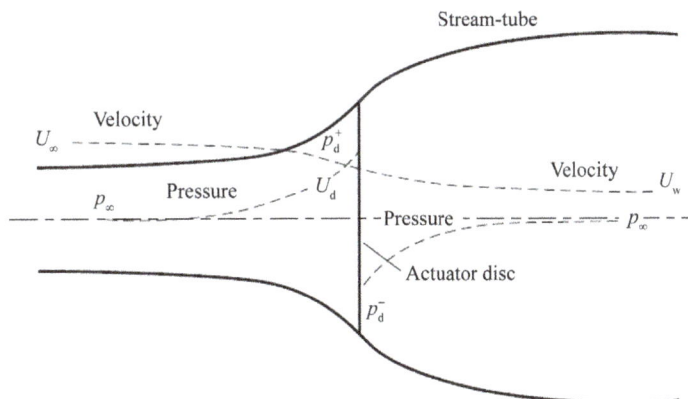

Figure 2. An energy extracting actuator disc and stream tube.

energy is different for both the streams. From further calculations we derive Equation (1).

Using Equations (1) & (3), we obtain

$$\left(P_d^+ - P_d^-\right) = \left(U_\infty - U_w\right)\rho A_d U_\infty \left(1-a\right) \tag{4}$$

As this force is concentrated at the actuator disc the rate of work done by the force is FU_d and hence the power extraction from the air is given by

$$\text{Power} = FU_d = 2\rho A_d U_\infty^3 a(1-a)^2 \tag{5}$$

The power coefficient is defined as

$$C_p = \frac{\text{power}}{\dfrac{1}{2}\rho U_\infty^3 A_d} \tag{6}$$

C_p will be max when

$$\frac{dC_p}{da} = 0$$

$$\therefore 4(1-a)(1-3a) = 0$$

$$a = \frac{1}{3}$$

$$\therefore C_{p\max} = \frac{16}{27} = 0.593$$

The coefficient of power $\left(C_p\right)$ is then used to calculate power output.

3. Results & Discussion

The percentage increase in power output for different radius (s) in steps of 2 m from 63 m radius as reference blade are simulated against power output and summarized in **Table 2**. **Figure 3** graphically summarizes the simulation result given in **Table 2**. The right extreme column in **Table 2** indicates percentage gain for 71.5 m radius blade over reference radius.

Similarly, percentage gain in output power delivered according to the chord variations is summarized in **Table 3**.

Figure 4 summarizes change in power output for variation in chord of the blade proportion to the radius and also for varying the chord alone.

A graph demonstrating measurements, taken by Mark Dawson et al., for extended length (only), wind blade design is reproduced here in **Figure 5** for comparison.

Table 2. Simulation result for power output for variation in radius[*].

Wind speed	Existing radius—63 m	Radius—64.5 m	Radius—66.5 m	Radius—68.5 m	Radius—71.5 m	Benefit over existing
3	1.0	1.1	1.2	1.2	1.3	25.42892
4	3.8	4.2	4.4	4.4	4.8	26.60006
5	8.5	9.5	10.0	10.0	10.8	26.94602
6	15.5	17.5	18.4	18.4	19.8	27.21183
7	24.9	27.8	29.1	29.1	31.2	25.19004
8	37.2	41.3	44.1	44.1	47.2	27.05546
9	52.9	59.3	62.7	62.7	66.6	25.87414
10	72.4	80.3	84.1	84.1	90.1	24.45874
11	95.9	100.0	100.0	100.0	100.0	4.30885
12	100.0	100.0	100.0	100.0	100.0	0
13	100.0	100.0	100.0	100.0	100.0	0
14	100.0	100.0	100.0	100.0	100.0	0

[*]All values are in percentage (%) of simulated output against output for reference turbine at rated capacity.

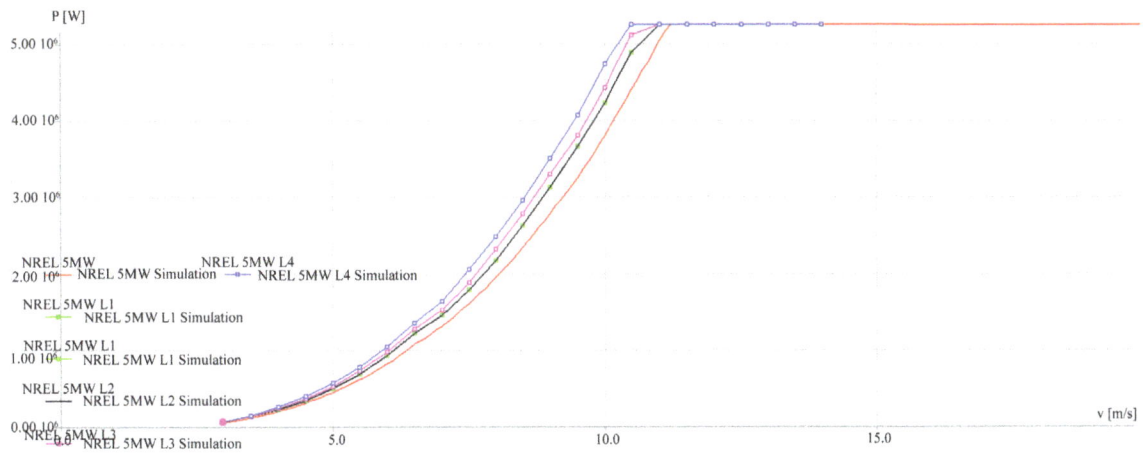

Figure 3. Graphical summary of simulation for variation in radius.

Table 3. Simulation result for power output for variation in radius & chord simultaneously and variation in chord alone[*].

Wind speed	Radius & Chord—63 m & from tip to root by 21% proportionately till five steps	Chord—from tip to root by 21% proportionately till five steps
3	0.0	0.1
4	3.3	2.9
5	9.3	0.8
6	18.4	14.9
7	30.6	24.4
8	45.7	36.6
9	46.2	52.2
10	89.8	71.7
11	100.0	95.6
12	100.0	100.0
13	100.0	100.0
14	100.0	100.0

[*]All values are in percentage (%) of simulated output against output for reference turbine at rated capacity.

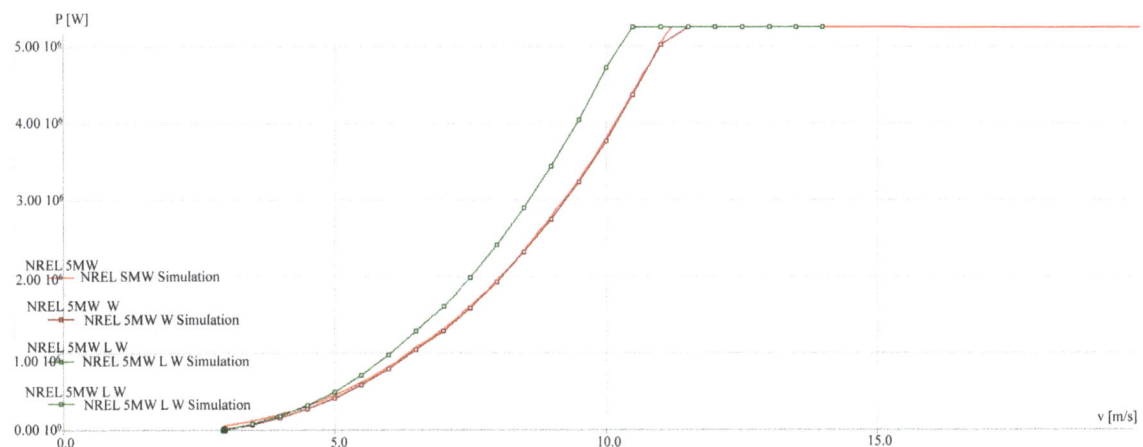

Figure 4. Summarizes change in power output for variation in chord of the blade proportion to the radius and also for varying the chord alone.

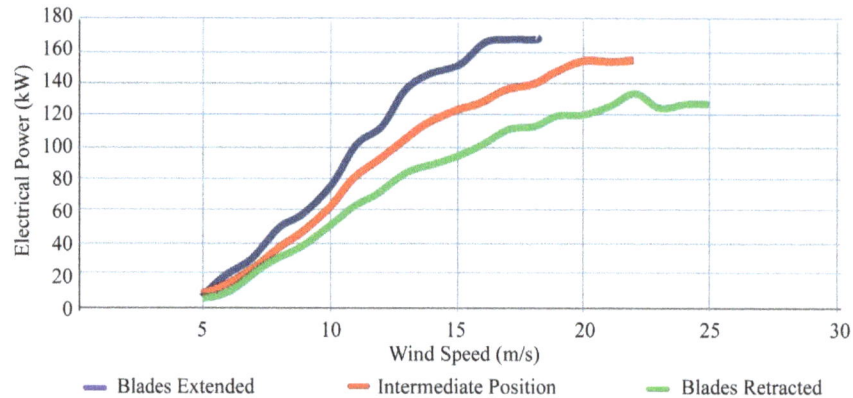

Figure 5. Reproduced experimental, proof-of-concept power curve for variable length wind turbine by Mark H. Dawson *et al.* [17].

In **Figure 3**, the right extreme line (red color) represents the power curve related to actual rotor. The curves to the left represent power curves with increased radius of rotor in various orders. In all the cases, output delivered by modified blade dimensions is found to be greater than the output received w.r.t. reference blade. The left extreme power curve (violet color) in **Figure 3** indicates the power output for 71.5 m. It is evident from the simulation result that increase in radius is favored at all the wind velocity including cut-in wind speed.

The right extreme line (red color) in **Figure 4** shows two curves exactly coinciding with each other. The only difference observed is for the wind speed corresponding to rated wind speed. It can be seen that no significant increase in power output observed except close to rated wind-speed that elevates the rated wind velocity value. The left extreme line (green color) in **Figure 4** demonstrates the power curve correspond to simultaneous & proportionate increase in tip chord and radius of the blade. The curve rises at all the wind speed over power curve corresponding to reference blade. It yields more output at higher wind speed, reaches the maximum power earlier than the existing and hence, increases the utilization of the wind generator CUF.

The simulation results confirm significant gain in power output corresponding to the increase in radius of the rotor alone. Maximum gain of 27% is observed for radius equivalent to the critical radius *i.e.* 71.5 m compared to the radius of reference rotor *i.e.* 63 m. There is no gain observed for increase in the chord of the blade alone.

Best results are obtained when both chord and radius both are increased simultaneously. It can be inferred that the wind turbine shown promising results for the output power generated at both the lower and higher wind speed. Major impact observed in the generator reaches the synchronous speed at an earlier (*i.e.*, at lower than the rated wind speed) stage than the existing.

4. Conclusions

Output of the wind turbine increases with increase in chord radius (length of blade) at all wind velocities; the power output becomes maximum at rated wind velocity. Thereafter output is stable for any wind velocity up to cut-off wind velocity. Higher gains can be obtained with more increase in blade length from lower to higher wind speed. It is possible to achieve state of rated power output at comparatively lower wind velocity than the rated wind velocity by increasing swept area. That means, more power can be harnessed on yearly basis. We can logically conclude that reduction in swept area can also generate power from wind mill above cut-off wind velocities wherein otherwise wind mill will stop for safety. Trend in both the cases matches well with each other.

Additional gain can be obtained by increasing width also proportionately which is not experimented before. It can be concluded that in order to gain more output from wind turbine, complete aerodynamic shape should be changed. We recommend that experiments need to be conducted on the basis of these simulation studies to identify the percentage gain and also to identify practical constraints that can lower the gain.

Acknowledgements

Authors like to acknowledge management team of Gujarat Energy Research & Management Institute (GERMI), Gandhinagar for providing all necessary support to conduct present study. Special acknowledgements to Vice

Chairman & Managing Trusty (VGMT)—GERMI and Managing Director of M/s. Gujarat State Petroleum Corporation (GSPC), Gandhinagar for providing necessary funding under Summer Internship Project (SIP) scheme.

References

[1] CEA Report (2013) http://cea.nic.in/reports/yearly/annual_rep/2012-13/ar_12_13.pdf

[2] Madhu, S. *et al.* (2014) A Review of Wind Energy Scenario in India. *International Journal of Environmental Sciences*, **3**, 87-92.

[3] Sing, C. (2012) Variable Speed Wind Turbine. *International Journal of Engineering Science*, **2**, 652-656.

[4] Li, H. and Chen, Z. (2009) Design Optimization and Site Matching of Direct-Drive Permanent Magnet Wind Power Generator Systems. Renew. *Energy*, **34**, 1174-1185. http://dx.doi.org/10.1016/j.renene.2008.04.041

[5] Eminoglu, U. and Ayasun, S. (2014) Modeling and Design Optimization of Variable-Speed Wind Turbine Systems. *Energies*, **7**, 402-419. http://dx.doi.org/10.3390/en7010402

[6] Fingersh, L., Hand, M. and Laxson, A. (2006) Wind Turbine Design Cost and Scaling Model; Technical Report NREL/TP-500-40566. National Renewable Energy Laboratory (NREL), Golden.

[7] Diveux, T., Sebastian, P., Bernard, D., Puiggali, R.J. and Grandidier, J.Y. (2001) Horizontal Axis Wind Turbine Systems: Optimization Using Genetic Algorithms. *Wind Energy*, **4**, 151-171. http://dx.doi.org/10.1002/we.51

[8] Fuglsang, P., Bak, C., Schepers, J.G., Bulder, B., Olesen, A., van Rossen, R. and Cockerill, T. (2010) Site Specific Design Optimization of Wind Turbines; Technical Report JOR3-CT98-0273. National Laboratory Risø, Roskilde.

[9] Collecutt, G.R. and Flay, R.G. (1996) The Economic Optimization of Horizontal Axis Wind Turbine Design Parameters. *Journal of Wind Engineering and Industrial Aerodynamics*, **61**, 87-97.
http://dx.doi.org/10.1016/0167-6105(96)00048-7

[10] Fuglsang, P. and Bak, C. (2002) Site-Specific Design Optimization of Wind Turbines. *Wind Energy*, **5**, 261-279.
http://dx.doi.org/10.1002/we.61

[11] Kongam, C. and Nuchprayoon, S. (2010) A Particle Swarm Optimization for Wind Energy Control Problem. *Renewable Energy*, **35**, 2431-2438. http://dx.doi.org/10.1016/j.renene.2010.02.020

[12] Fuglsang, P. and Madsen, H.A. (1999) Optimization Method for Wind Turbine Rotors. *Journal of Wind Engineering and Industrial Aerodynamics*, **80**, 191-206. http://dx.doi.org/10.1016/S0167-6105(98)00191-3

[13] Maki, K., Sbragio, R. and Vlahopoulos, N. (2012) System Design of a Wind Turbine Using a Multi-Level Optimization Approach. *Renewable Energy*, **43**, 101-110. http://dx.doi.org/10.1016/j.renene.2011.11.027

[14] Muljadi, E. and Butterfield, C.P. (1999) Pitch-Controlled Variable-Speed Wind Turbine Generation. *IEEE Transactions on Industry Applications*, **37**, 240-246.

[15] Talavera Juan, A. and Cassarrubios Francisco, J. (2005) Swept Area Regulation for Increased Energy Output in Off-Shore Wind Turbine.
http://wind.nrel.gov/public/SeaCon/Proceedings/Copenhagen.Offshore.Wind.2005/documents/papers/Poster/J.A.Talavera_SweptAreaRegulationforIncreasedEnergy.pdf

[16] http://www.powersails.com/powersails/adjustable-swept-area-asa/

[17] Dawson Mark, H. (2006) Variable Length Wind Turbine Blade. Project Report.

5

Impact of the Fermeuse Wind Farm on Newfoundland Grid

Seyedali Meghdadi, Tariq Iqbal

Department of Electrical Engineering, Memorial University of Newfoundland, St. John's, Canada
Email: s.meghdadi@mun.ca, tariq@mun.ca

Abstract

This paper aims to study the impact of the Fermeuse wind farm (46°58'42"N 52°57'18"W) through simulation of wind turbines driven by doubly fed induction generator which feed AC power to the isolated utility grid of Newfoundland. The focus is on the determination of both voltage and system stability constraints. The complete system is modeled and simulated in Matlab Simulink environment.

Keywords

Renewable Energy Systems, Fermeuse Wind Farm, Wind Turbine, Doubly Fed Induction Generator, Wind Energy Conversion Systems

1. Introduction

Every day decreases in the cost of renewable energy generators, increasing demand for renewable energy sources to provide a sustainable future, and worldwide regulations to reduce greenhouse gas emissions have made renewable energy sources become the strongest candidates to substitute for oil/gas power plants. The demand for more power combined with interest in clean technologies has driven researchers to develop distributed power generation units using renewable energy sources [1]-[2]. The employment of these distributed generations in power systems can offer benefits as follows:

- Reliable Power Supply:

  Any unexpected events, such as grid faults occurring in the upstream power lines, result in a disconnection of the distributed generation unit from the grid, causing black-outs. However, in this situation, the autonomous operation of a group of distributed generation units can provide power to local loads.

- Power Loss Compensator:

  Many rural communities in Canada are generally connected to the central power stations through long transmission lines or obtain their power supply through diesel generators. Delivering power to those communities from a central power station causes a significant amount of loss. In Newfoundland, the power loss in the transmission system is about 9% [3]. Therefore, the formation of small renewable energy plants around these rural areas can significantly reduce the amount of power.

• Reduction in Transmission System Expansion:

According to the Newfoundland and Labrador Hydro's (NLH) 2010 long term planning load forecast, power demand is predicted to grow at 1.3 percent per year through 2029. This increase requires more power transmission infrastructure, which may not be economically feasible because of the higher cost involved in new transmission lines installation and maintenance for rural areas. Therefore, the installation of renewable plants near the user load eliminates the requirement of re-designing new long transmission lines, resulting in cost savings.

This paper presents a dynamic model of a 27 MW wind farm located in Fermeuse, Newfoundland, Canada. The wind farm system model is simulated in Matlab/Simulink with the purpose of investigating the effects of wind speed variations on voltage and frequency of the grid.

2. System Components

2.1. Wind Power Generation System

The Fermeuse wind farm model consists of 9 dynamic models of variable-speed, doubly-fed induction generator based wind energy conversion systems (WECS). Each variable speed, doubly-fed induction generator contains the model of a variable pitch wind turbine rotor and a wound rotor asynchronous generator. The rotor winding is connected to the grid using a back-to-back pulse width modulated (PWM) voltage source converter (GEC and SEC), where the stator is directly connected to the grid. The Vestas-90 WECS is utilized in the proposed system, with the topology shown in **Figure 1**. The decouple control technique is incorporated to control the converters in the rotor side [4].

2.2. Wind Turbine

In steady state, the mathematical power developed by the wind turbine P_m, can be expressed by the following set of equations:

$$P_m = C_p\left(\lambda,\beta\right)\frac{\rho A}{2}\upsilon^3$$

$$C_p\left(\lambda,\beta\right) = C_1\left(\frac{C_2}{\lambda_i} - C_3\beta - C_4\right)e^{\frac{-C_5}{\lambda_i}} + C_6\lambda$$

$$\frac{1}{\lambda_i} = \frac{1}{\lambda + C_7\beta} - \frac{C_8}{\beta^3 + 1}$$

$$\lambda = \frac{R\omega}{\upsilon}$$

Mechanical torque T_m, is the ratio of mechanical power to turbine speed.

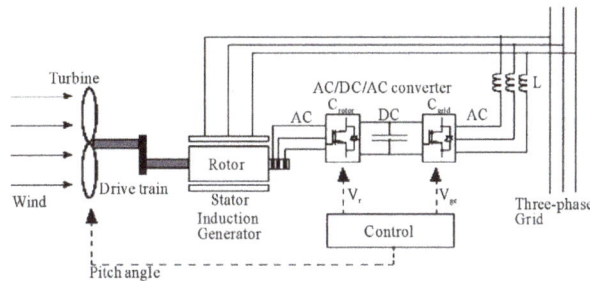

Figure 1. Schematic of a variable speed doubly-fed induction generator based wind energy conversion system.

$$T_m = \frac{P_m}{\omega}$$

where $C_p(\lambda, \beta)$, A, υ, and β are power coefficient, sweep area, wind speed, and pitch angle respectively.

2.3. Two-Mass Model of Drivetrain

In [5] it is indicated that the three-mass model can be reduced to a two-mass model by considering an equivalent system with an equivalent stiffness and damping factor. The moment of inertia for the shafts and the gearbox wheels can be neglected because they are small compared with the moment of inertia of the wind turbine or generator. Therefore the resultant model is essentially a two mass model connected by a flexible shaft. Only the gearbox ratio has influence on the new equivalent system.

The dynamic equations can be written in two points: the wind turbine side with the influence of generator component through the gearbox and on the generator side respectively. The equivalent system on the generator side is shown in **Figure 2**.

The dynamic equations of the drive-train written on the generator side are:

$$T'_{wtr} = J'_{wtr} \frac{d\Omega''_{wtr}}{dt} + D'_e \left(\Omega'_{wtr} - \Omega''_{gen}\right) + K'_{se} \left(\theta'_{wtr} - \theta_{gen}\right)$$

$$\frac{d\theta'_{wtr}}{dt} = \Omega'_{wtr}$$

$$-T'_{gen} = J_{gen} \frac{d\Omega_{gen}}{dt} + D'_e \left(\Omega_{gene} - \Omega'_{wtr}\right) + K'_{se} \left(\theta_{gen} - \theta'_{wtr}\right)$$

$$\frac{d\theta_{gen}}{dt} = \Omega_{gen}$$

where T'_{wtr}, J'_{wtr}, Ω'_{wtr}, D'_e, and K'_{se} respectively are wind turbine torque, wind turbine, moment of inertia, wind turbine mechanical speed, damping constant, and spring constant. The equivalent stiffness K'_{se} is given by:

$$\frac{1}{K'_{se}} = \frac{1}{\dfrac{K_{wtr}}{K^2_{gear}}} + \frac{1}{K_{gen}}$$

And the equivalent moment of inertia for the rotor is:

$$J'_{wtr} = \frac{1}{K^2_{gear}} \cdot J_{wtr}$$

2.4. Generator

The dynamic equivalent (d-q) circuit of a doubly-fed induction machine in a synchronously rotating reference frame is represented in **Figure 2**. Using d-q axis transformation in the synchronously rotating reference frame, the output voltages for the doubly-fed induction generator are implemented based on **Figure 3** [6].

2.5. Converter

The rotor side of the doubly-fed induction generator based wind generator system is connected to the grid through a back-to-back PWM voltage source converter (**Figure 4**). In order to eliminate fluctuations in current, and so output power of wind farm caused by triggering converter, the average model of wind turbine detailed in [7] is used. This model exploits an AC/DC/AC IGBT-based PWM converter modeled by voltage sources. The stator winding is connected directly to the 60 Hz grid while the rotor is fed at variable frequency through the AC/DC/AC converter. The DFIG technology allows extracting maximum energy from the wind for low wind

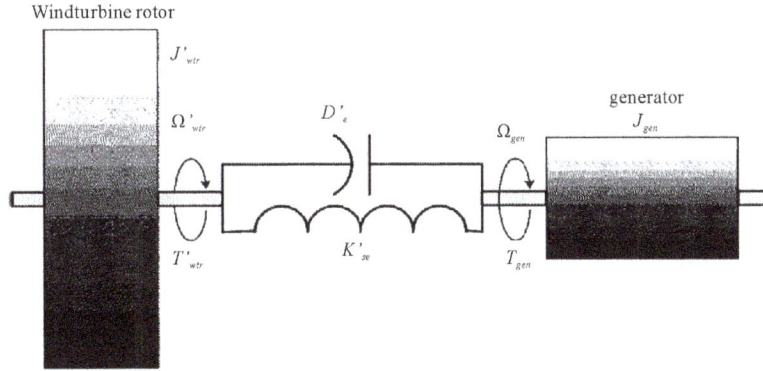

Figure 2. Equivalent diagram of the wind turbine drive train on the generator side.

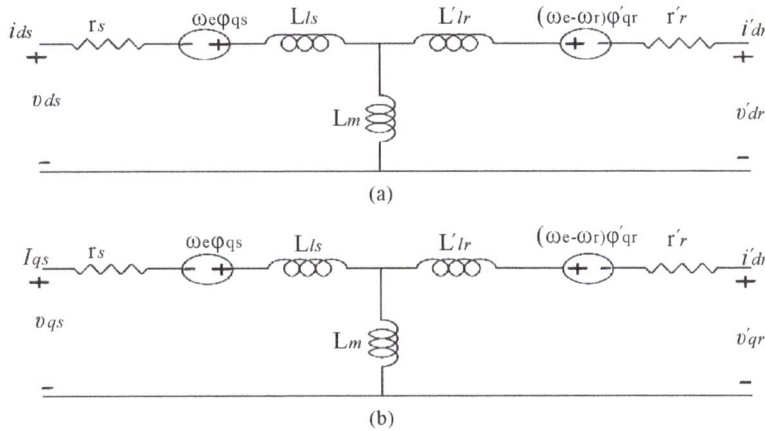

Figure 3. Equivalent circuit representation of an induction machine in synchronously rotating referenced frame: (a) d-axis, (b) q-axis.

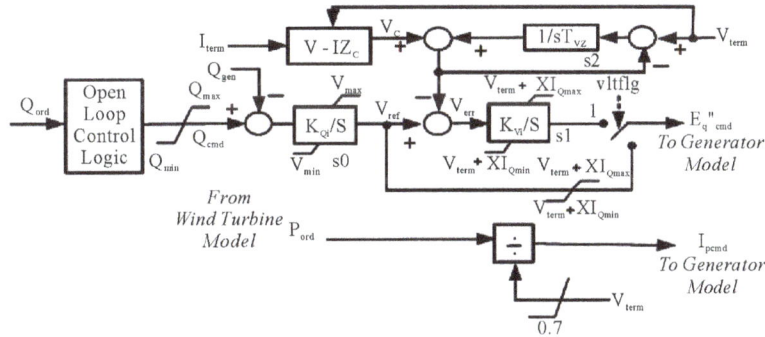

Figure 4. Electrical Control model.

speeds by optimizing the turbine speed, while minimizing mechanical stresses on the turbine during gusts of wind. **Figure 3** depicts the control diagram used in the referenced model [7] [8]. Based on inputs from the turbine model (P_{ord}) and from the supervisory VAR controller (Q_{ord}), the Excitation Control Model dictates the active and reactive power to be delivered to the system.

3. Location of the Wind Farm

Fermeuse is a small rural community located on the southern shore of the Avalon Peninsula on Newfoundland Island in the province of Newfoundland and Labrador, Canada (**Figure 5**).

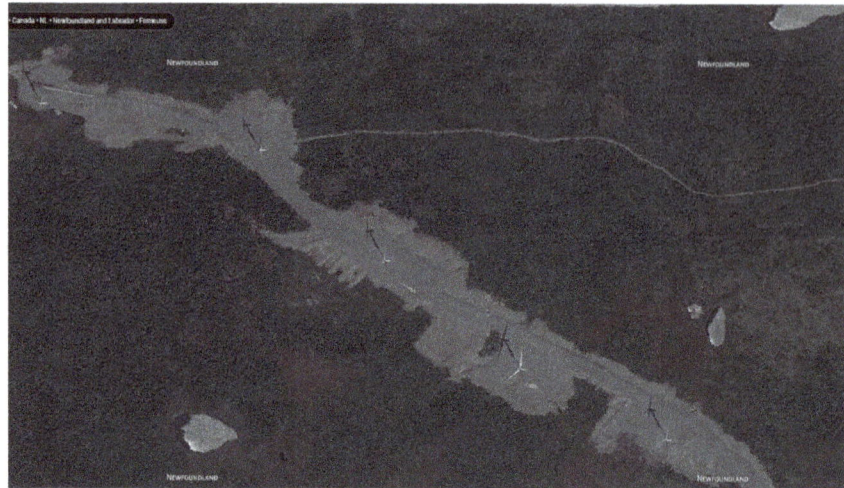

Figure 5. Bird-eye view of a part of the wind farm.

4. Power System Planning and Operation Criteria

Connection of a wind farm to the grid has a large impact on grid stability. The increased penetration of wind energy into the power system over the last decade has therefore led to serious concern about its influence on the dynamic behavior of the power system. An overview of the national grid requirements in several countries is provided in [9] and main attention is drawn to fault ride-through (*i.e.* requirements imposed to avoid significant loss of wind turbine production in the event of faults) and power control capability (*i.e.* regulating active and reactive power) and performing frequency and voltage control on grid [9].

4.1. Voltage Criteria

Under normal conditions the transmission system is operated such that the voltage is maintained between 95% and 105% of nominal. During contingency events the transmission system voltage is permitted to vary between 90% and 110% of nominal prior to operator intervention. Following an event, operators will take steps (*i.e.* re-dispatch generation, switch equipment in or out of service, curtail load or protection) to return the transmission system voltage to the 95% to 105% normal operating range.

4.2. Stability Criteria

The frequency of a power system can be considered as a measure of the balance or imbalance between production and consumption in the power system. With nominal frequency, production and consumption including losses in transmission and distribution are in balance. If the frequency is below nominal, the consumption of electric energy is higher than production and if it is above nominal the consumption is lower than production [10]. Control of frequency on the island system is the responsibility of NLH's generating stations. Adding non-dispatchable generation to the island may result in fewer on NLH's dispatchable generation resources being on line. As fewer generators are left to control system frequency, frequency excursions become magnified for the same change in load. A theoretical point can be reached where the slightest increase in load will cause the system to become unstable.

5. Single Line Diagram of the System

The nine WTs together are placed in the Fermeuse wind farm block in **Figure 6**. Each of the WTs in the wind farm is a 3 MW variable-speed doubly-fed induction generator wind turbine. The distance between each wind turbine generator is small and neglected. Each wind turbine generator is connected to the line through a transformer, TW, and a short transmission line. The transmission line (TL3) of the wind farm is connected to bus 4, which is eventually connected to bus 2 using a power transformer, T1 (**Figure 7**). Data for the buses of the proposed system are provided in **Table 1**.

Moreover, the data for transmission lines and transformers are provided in **Table 2**.

Figure 6. View of the wind farm from its transformation station.

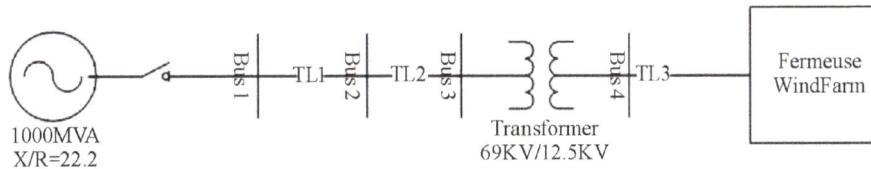

Figure 7. Fermeuse grid connection single line diagram.

Table 1. Fermeuse buses.

BUS	P [MW]	Q [MVAR]	V [KV]
1	ΔP	ΔQ	69
2	ΔP	ΔQ	69
3	0	0	69
4	27	12	12.5

Table 2. Fermeuse transmission Lines and Transformers.

Component	V [KV]	Z [Ω]	S [MVA]
TL1	69	0.002 + j0.0032	40
TL2	69	0.0416 + j0.0663	40
TL3	12.5	0.0374 + j0.3741	40
T1	69/12.5	0.05 + j0.5	40
TW	12.5/1	0.006 + j0.0625	5

6. Simulink Model

The Simulink model is a mathematical interpretation of the explained single line diagram of the system (**Figure 8**). This includes model of wind turbine, transformers, Pi model of transmission lines, utility grid and some measurement blocks. These blocks are connected together to build the model of the system. In the end, injected current and voltage form the wind farm delivered by TL1 to the grid are measured and the output power of the wind farm is calculated through ten seconds of running the simulation.

In order to accurately model wind turbines, data in **Table 3** and **Table 4** were used based on [11].

(a)

(b)

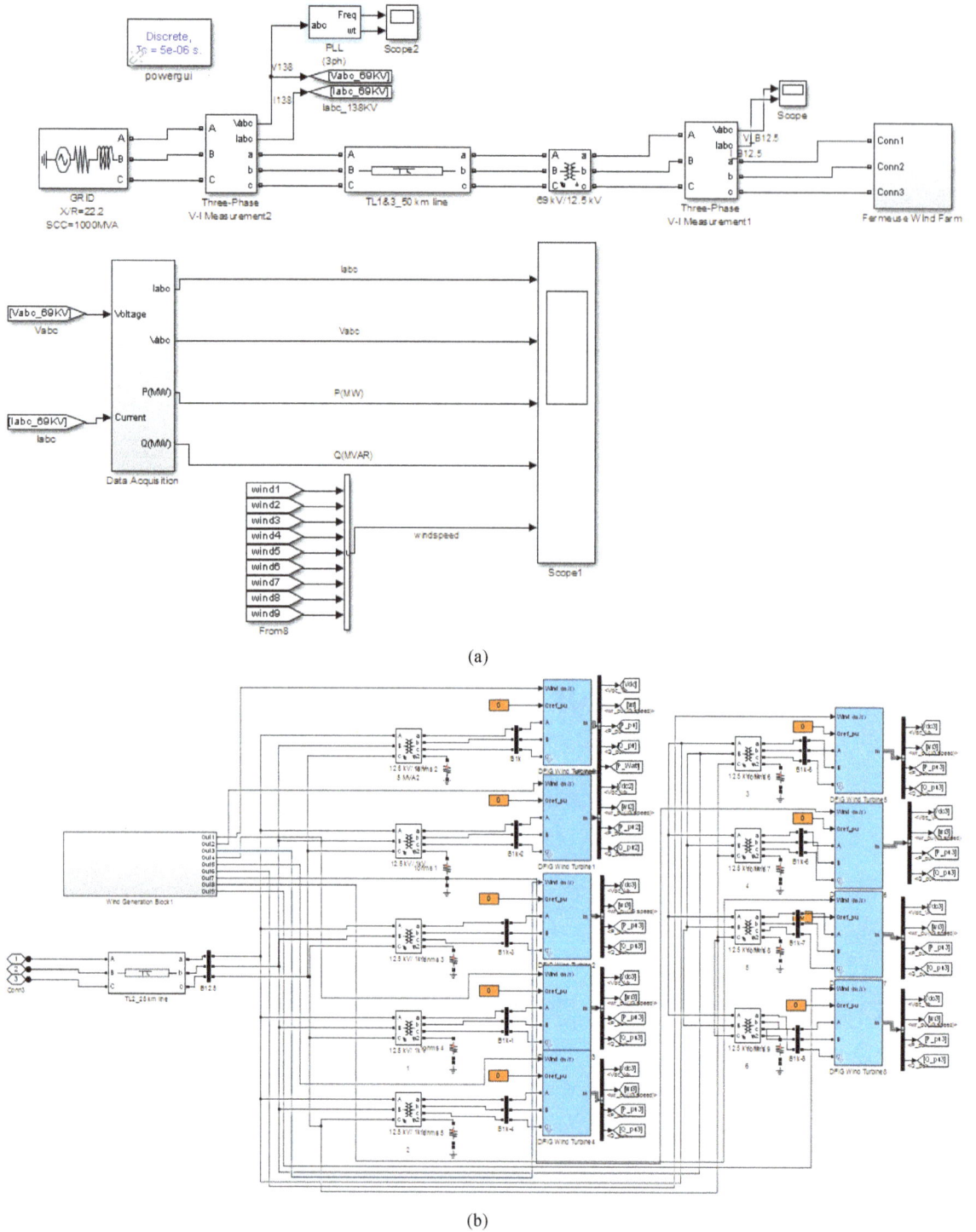

Figure 8. Simulink model of the wind farm: (a) general view of the model. (b) Fermeuse wind farm model.

Where r_s, r_r' are stator and rotor winding resistance of induction machine, L_{ls}, L_{lr}' are leakage inductance of stator and rotor winding, L_m is magnetizing inductance.

7. Results

Three different scenarios are considered: 1) at constant wind speed, 2) at variable wind speeds, and 3) when the wind farm trips due to a fault and then reconnected to the grid.

Firstly, results of the system working under variable wind speeds are shown as below in **Figure 9** and **Figure 10**. Secondly, the results of the system under constant wind speed at 15 m/s are as follows:

Comparing frequency and current, and so power, in these two scenarios, it is clear that variable wind speeds cause small fluctuations in the frequency and the current injected to the grid, meaning the grid is quite stiff.

Thirdly, the results in case of a fault happening at t = 3 s and the wind farm trips then it is reconnected to the grid at t = 5 s. **Figure 11** indicates measured and calculated values at receiving end of the line, at 69 KV bus.

Figure 12 depicts the voltage and current when fault happens measured from the sending end of the line, at 12.5 KV bus.

8. Conclusion

A Simulink model of the 27 MW wind farm, including 9 wind turbines with the power of 3 MW, transformers, transmission lines, utility grid, and measurement blocks, located in Fermeuse, NL, Canada, was presented in this paper and simulation results for three different scenarios were analyzed. This paper concludes that the developed model is able to represent the dynamics of the wind farm and its effect on the power system.

Table 3. Wind turbine specifications.

Rated power	3 MW
Rotor diameter	90 m
Nominal speed	16.1 rpm
Rotor speed range	9.9 - 18.4 rpm
Number of blades	3
Blade length	44 m
Gearbox ratio	1:109
Cut-in wind speed	4 m/s
Cut-out wind speed	25 m/s

Table 4. Induction generator parameters.

Type	DFIG
Nominal power	3 MW
Voltage	1000 V
Rated speed	1758 rpm
Number of poles	4
Total inertia constant	3.02 s
Friction coefficient	0.01
r_s, r_r'	2.35 mΩ, 1.67 mΩ
L_{ls}, L_{lr}'	0.151 H, 0.1379 H
L_m	2.47 H

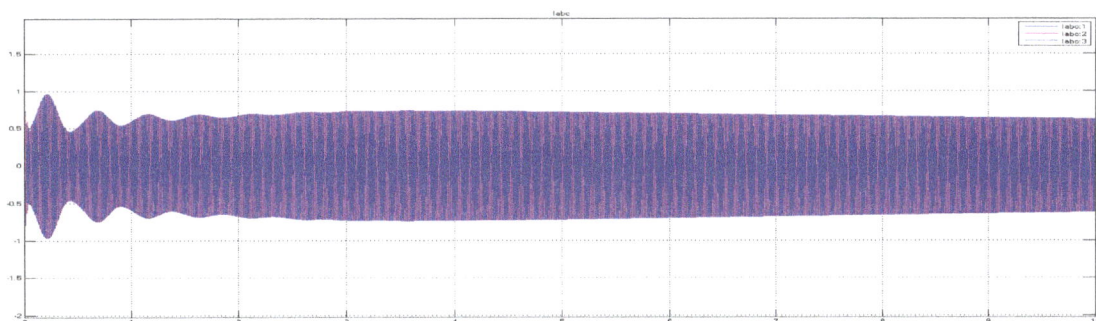

(a)

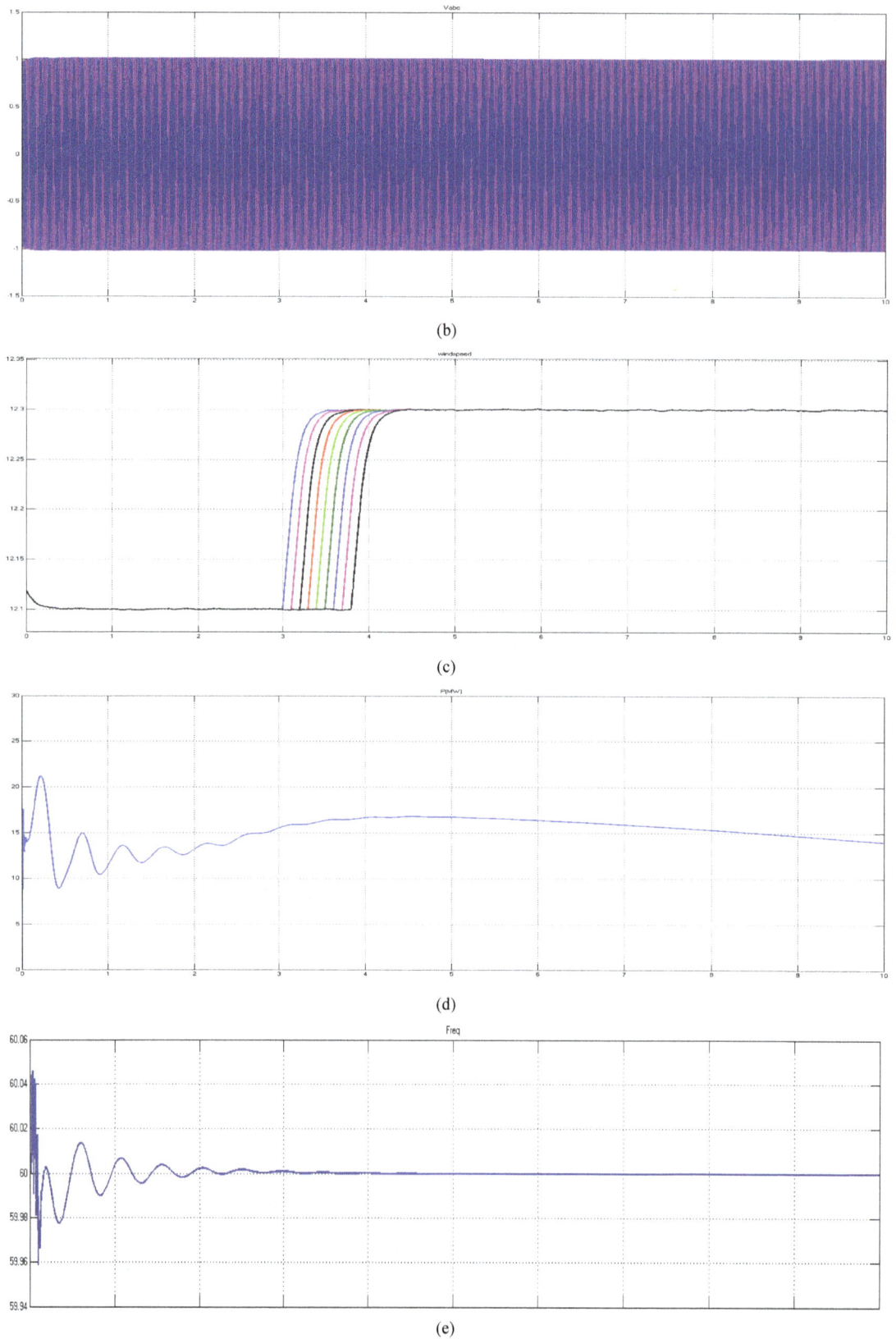

Figure 9. (a) to (d): Simulink model results for variable wind speeds. (a): Current; (b): Voltage; (c): Wind speed variations; (d): Power output; (e): Frequency.

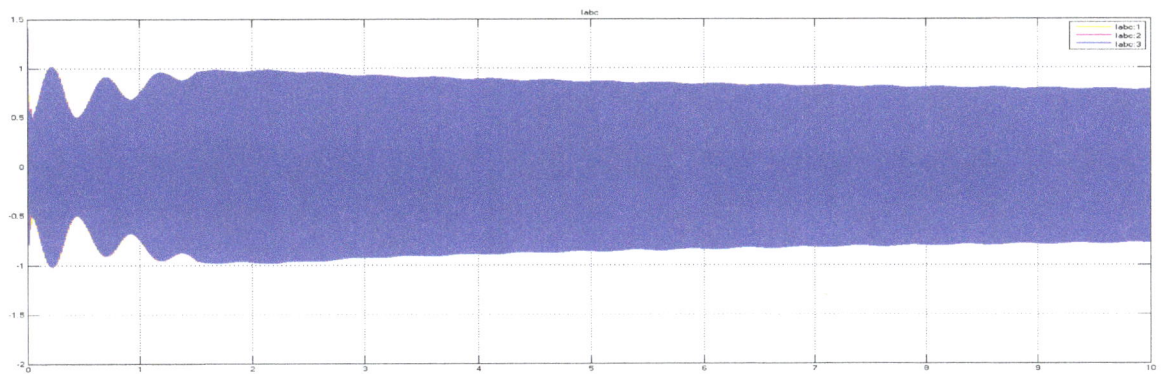

(a)

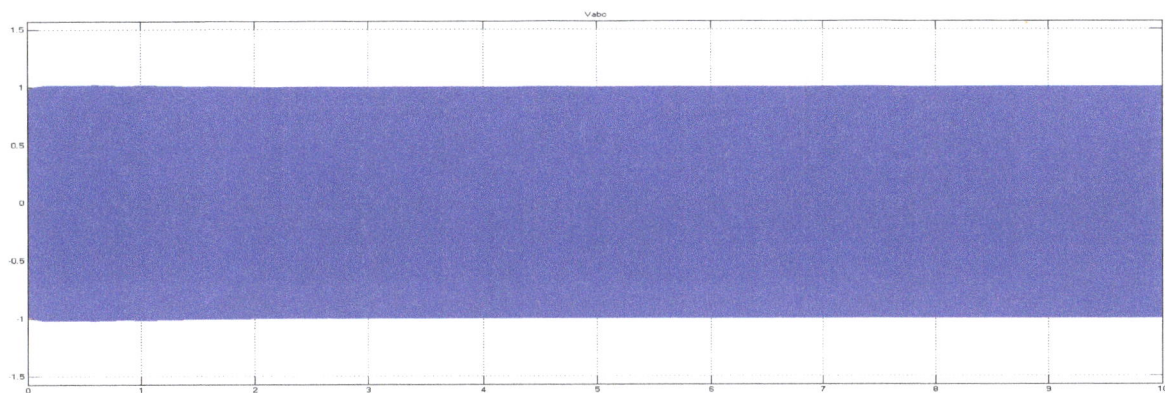

(b)

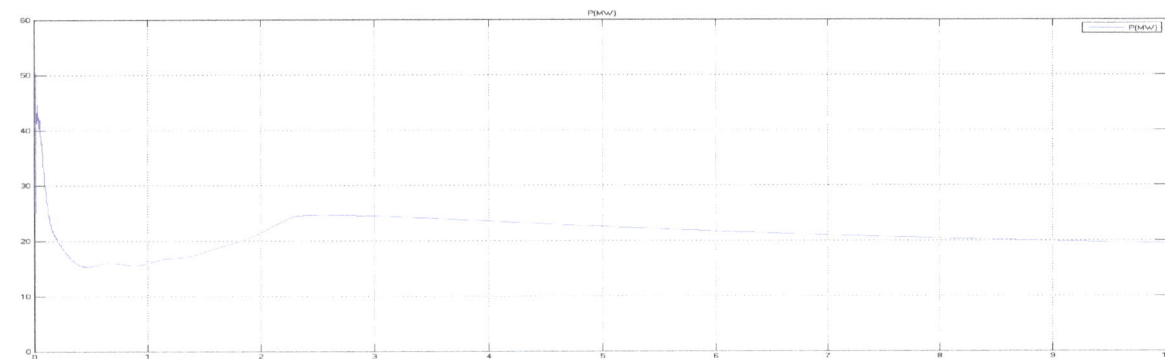

(c)

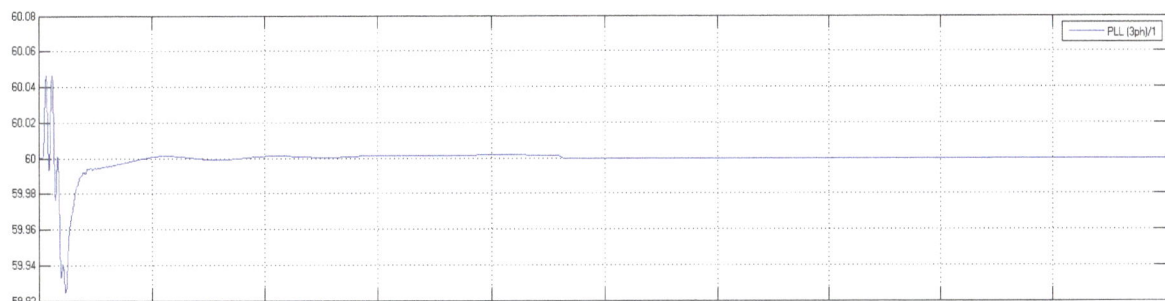

(d)

Figure 10. (a) to (d): Simulink model results for constant wind speed. (a): Current; (b): Voltage; (c): Power output; (d): Frequency.

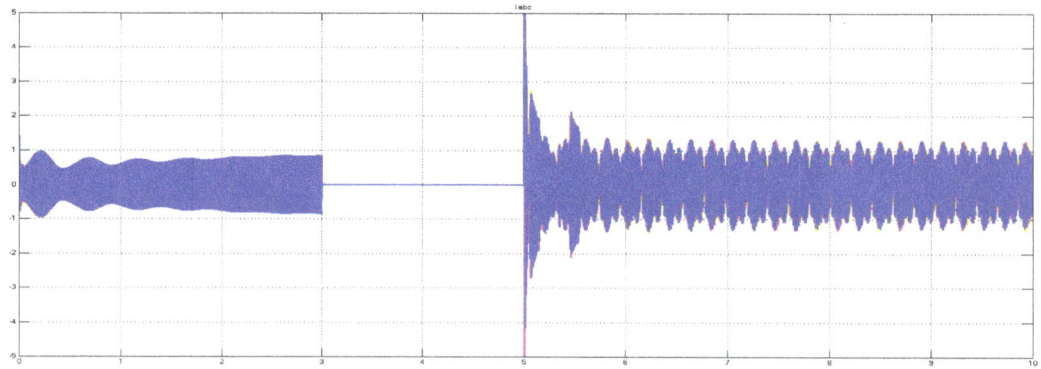

(a)

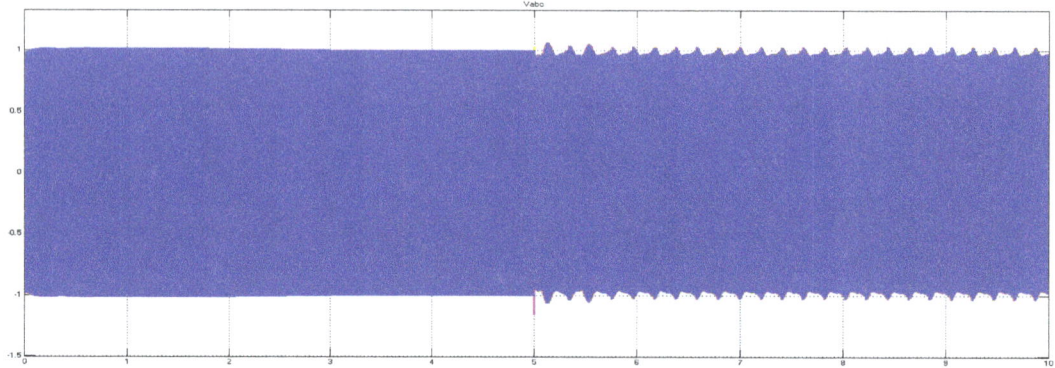

(b)

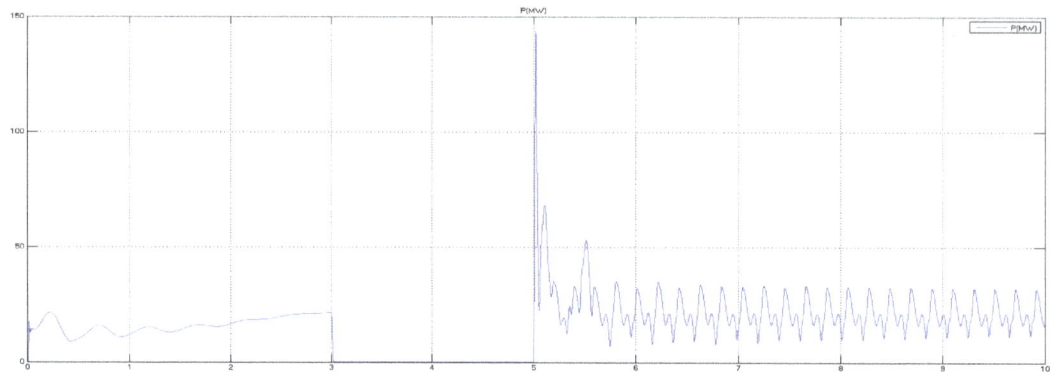

(c)

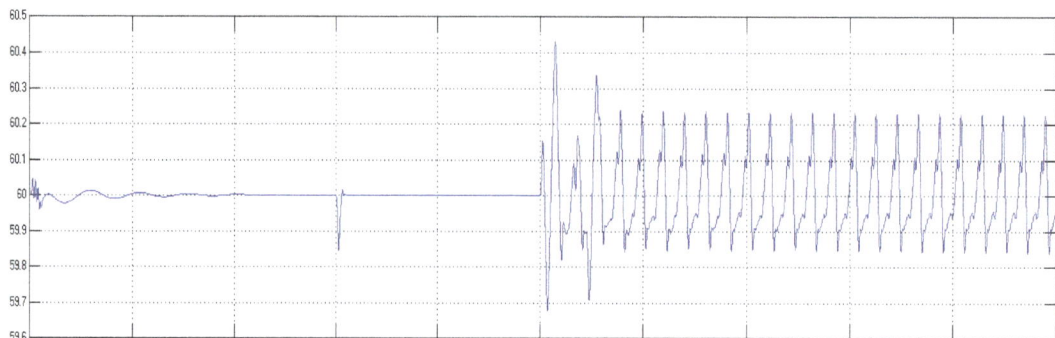

(d)

Figure 11. (a) to (d): Simulink results when fault occurs measured at receiving end. (a): Current; (b): Voltage; (c): Power output; (d): Frequency.

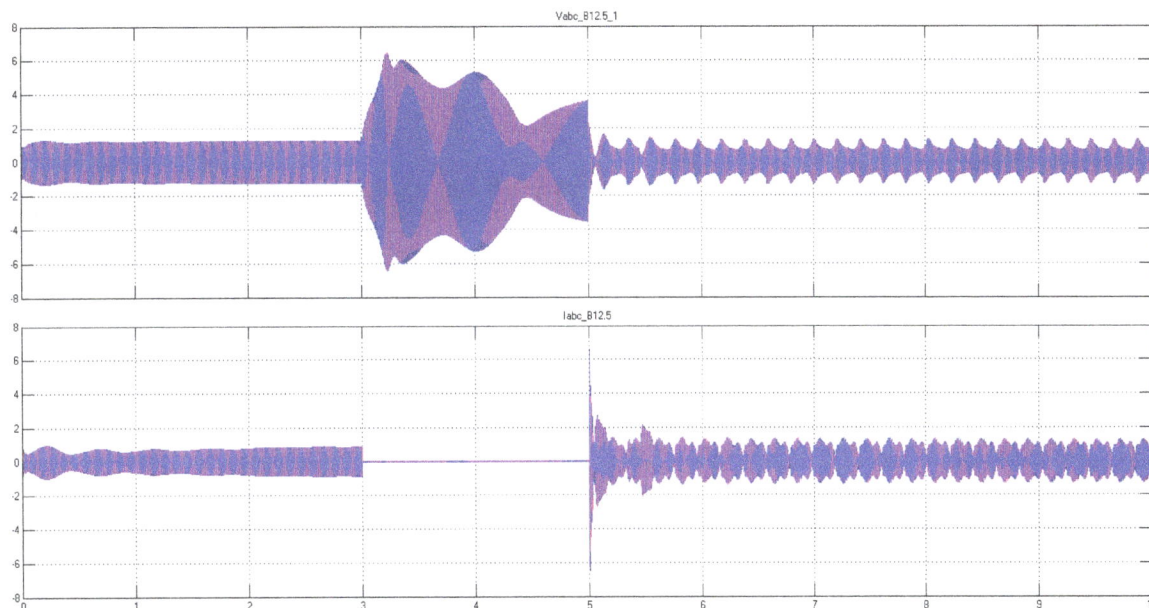

Figure 12. Simulink results when fault occurs measured at sending end.

References

[1] Blaabjerg, F., Teodorescu, R., Liserre, M. and Timbus, A.V. (2006) Overview of Control and Grid Synchronization for Distributed Power Generation Systems. *IEEE Transactions on Industrial Electronics*, **53**, 1398-1409.
http://dx.doi.org/10.1109/TIE.2006.881997

[2] Abbey, C., Katiraei, F., Brothers, C., Dignard-Bailey, L. and Joos, G. (2006) Integration of Distributed Generation and Wind Energy in Canada. *IEEE Power Engineering Society General Meeting PES'06 Proceedings*, Montreal, 1-7.
http://dx.doi.org/10.1109/pes.2006.1709430

[3] Expert Estimates, Based on Transmission Losses Published in Energy Statistics Handbook. A Joint Publication of Statistics Canada and Natural Resources.
http://www.statcan.gc.ca/access_acces/alternative_alternatif.action?teng=57-601-x2012001-eng.pdf&tfra=57-601-x2012001-fra.pdf&l=eng&loc=57-601-x2012001-eng.pdf

[4] Schauder, C. and Mehta, H. (1993) Vector Analysis and Control of the Advanced Static VAR Compensators. *IEE Proceedings C: Generation, Transmission and Distribution*, **140**, 299-306.

[5] Iov, F., Hansen, A.D., Sørensen, P. and Blaabjerg, F. (2004) Wind Turbine Blockset in Matlab/Simulink. Aalborg University, Aalborg.

[6] Krause, P.C., Wasynczuk, O. and Sudhoff, S.D. and Pekarek, S. (2013) Analysis of Electric Machinery and Drive Systems, Vol. 75. John Wiley & Sons, Hoboken.

[7] http://www.mathworks.com/help/physmod/sps/examples/wind-farm-dfig-average-model.html

[8] Miller, N.W., Price, W.W. and Sanchez-Gasca, J.J.. (2003) Dynamic Modeling of GE 1.5 and 3.6 Wind Turbine-Generators. GE-Power Systems Energy Consulting.

[9] Iov, F., Hansen, A.D., Sørensen, P. and Cutulis, N.A. (2007) Mapping of Grid Faults and Grid Codes. Riso Report, Riso-R-1617(EN), Risø National Laboratory, Roskilde.

[10] Ackermann, T. (2005) Wind Power in Power Systems. John Wiley and Sons, Ltd., Hoboken, 58-110.

[11] http://www.vestas.com/Files/Filer/EN/Brochures/ProductbrochureV90_3_0_UK.pdf

6

The Influence of Radial Area Variation on Wind Turbines to the Axial Induction Factor

Kedharnath Sairam, Mark G. Turner

School of Aerospace Systems, University of Cincinnati, Cincinnati, USA
Email: kdbeggar@gmail.com

Abstract

Improvements in the aerodynamic design will lead to more efficiency of wind turbines and higher power production. In the present study, a 3D parametric gas turbine blade geometry building code, 3DBGB, has been modified in order to include wind turbine design capabilities. This approach enables greater flexibility of the design along with the ability to design more complex geometries with relative ease. The NREL NASA Phase VI wind turbine was considered as a test case for validation and as a baseline by which modified designs could be compared. The design parameters were translated into 3DBGB input to create a 3D model of the wind turbine which can also be imported into any CAD program. Design modifications included replacing the airfoil section and modifying the thickness to chord ratio as a function of span. These models were imported into a high-fidelity CFD package, Fine/TURBO by NUMECA. Fine/TURBO is a specialized CFD platform for turbo-machinery analysis. A code-geomturbo was used to convert the 3D model of the wind turbine into the native format used to define geometries in the Fine/TURBO meshing tool, AutoGrid. The CFD results were post processed using a 3D force analysis code. The radial force variations were found to play a measurable role in the performance of wind turbine blades. The radial component of the blade surface area as it varies in span is the dominant contributor of the radial forces. Through the radial momentum equation, this radial force variation is responsible for creating the streamline curvature that leads to the expansion of the streamtube (slipstream) that is responsible for slowing the wind velocity ahead of the wind turbine leading edge, which is quantified as the axial induction factor. These same radial forces also play a role in changing the slipstream for propellers. Through the design modifications, simulated with CFD and post-processed appropriately, this connection with the radial component of area to the radial forces to the axial induction factor, and finally the wind turbine power is demonstrated. The results from the CFD analysis and 3D force analysis are presented. For the case presented, the power increases by 5.6% due to changes in airfoil thickness only.

Keywords

Wind Energy, Axial Induction Factor, Radial Area Variation, Wind Power, 3DBGB, Force Analysis

1. Introduction

Wind energy is one of the fastest growing energy sectors today. Apart from being a very promising source of energy, it also has other advantages. It is a source of green power and hence is a very critical part of the solution to the global climate change crisis. It is affordable and cost effective. The cost of manufacturing and monitoring wind turbines has reduced drastically over the last few decades owing to advancements in manufacturing technology. It improves economic development by providing a range of job opportunities in the communities where the wind farms are located. It is a sustainable source of energy and widely available [1].

The power output of a wind turbine is directly related to the swept area of the rotors [2]

$$P_{\text{wind}} = 0.5\rho A V_z^3 \tag{1}$$

$$P_{\text{mech}} = 0.5\rho A V_z^3 C_p \tag{2}$$

where P_{wind} is the power available in the kinetic energy of the wind.

The power coefficient depends on the overall aerodynamic and mechanical efficiency of the wind turbine design. Therefore, the power output of a wind turbine at a prescribed wind velocity can be improved by increasing the swept area of the rotor. However, not all wind turbine installation sites are suitable for large rotors and a large part of the capital cost is related to the rotor size. One of the key aspects of wind turbine design is the ability to extract as much energy from the wind as possible and reducing aerodynamic and mechanical losses to a minimum.

A general illustration of 2D wind turbine aerodynamics is given in **Figure 2**. Wind turbine design nomenclature is provided in **Figure 3** and conventional turbine nomenclature is provided in **Figure 4**. When the on-coming wind hits the wind turbine blade, a low pressure field is created on the suction side of the airfoil and a high pressure field is created on the pressure side of the airfoil. This creates a resultant force to act on the blade which can be represented as components of the force that are normal to each other i.e the lift and drag on the airfoil. The drag force is small compared to the lift force at the high reynold's number of modern wind turbines. This lift causes the blade to rotate.

Current wind turbine design approaches are based on the Blade-Element-Momentum theory, a 2D method which is explained in detail in [2]. 2D force analysis is a great way to understand the fundamentals of aerodynamics. However, in reality, complex 3D effects occur which need to be analyzed in order to optimize the blade design. A post-processing code was developed for 3D force analysis on the blade. This post-processing capa-

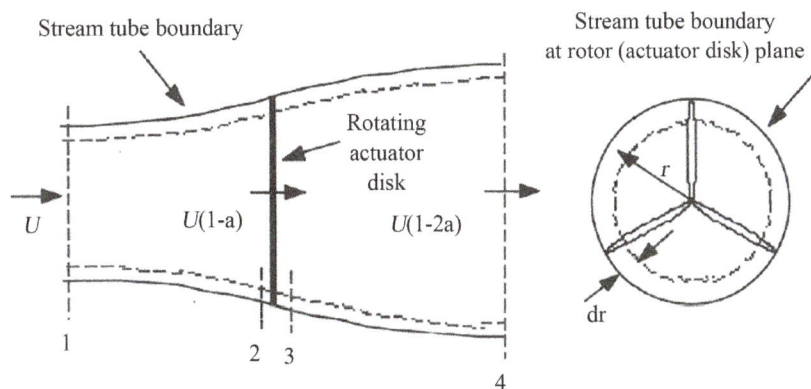

Figure 1. Actuator disc model [3].

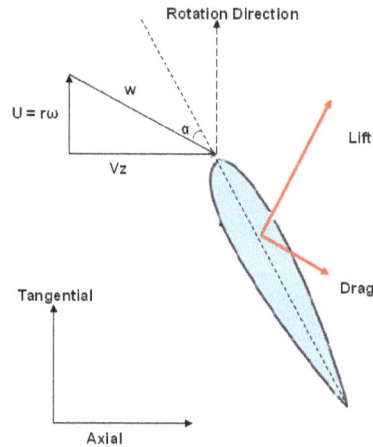

Figure 2. Wind turbine aerodynamics in 2d [4].

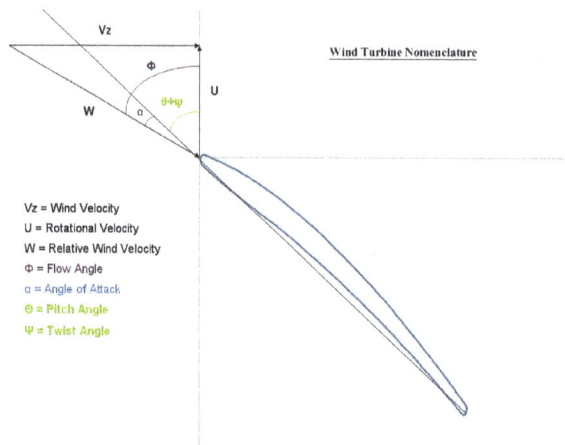

Figure 3. Wind turbine nomenclature [4].

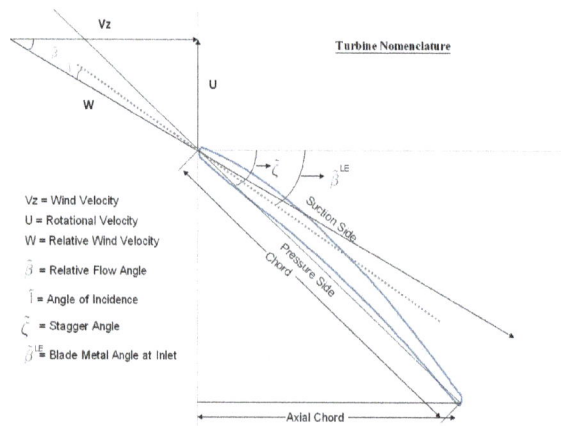

Figure 4. Conventional turbine nomenclature [4].

bility allows the interpretation of forces relevant to a wind turbine. The effect of radial force variation on the performance of the blade was discovered to be significant. The radial forces have been found to act as a main driver for the stream tube area expansion. New blades were designed to further explore the effect of radial forces in detail by manipulating the geometry.

Access to the 3D parametric blade geometry builder-3DBGB [5]-[7], a gas turbine blade geometry generation tool, proved to be a useful way to approach the design of wind turbines with complex geometries. Using the built-in airfoil definition, tip-offset, sweep and rake options, it was possible to manipulate the rotor geometry. CAD models of the blade were created using the output from 3DBGB. The design was then analyzed using a 3D CFD package-Fine/Turbo. Post-processing was carried out by the 3D force analysis code-Calc- PdA and other tools. The NREL Phase VI wind turbine was used as a starting point because of the availability of detailed design and analysis data [8].

Very little investigation has been done on manipulating the geometry to modify the radial forces on the blades. This paper aims to explore improvements in performance due to thickness to chord ratio variation of not only wind turbines, but also propellers and other turbo-machinery applications as well. Wind turbine geometry definition using 3DBGB and 3D CFD analysis using Fine/Turbo are explained in subsequent sections. A detailed explanation of the 3D force analysis code is provided. The results and comparison for the wind turbine test cases are presented and effect of radial area variation on wind turbine rotor blades is explained.

2. Wind Turbine Geometry Definition Using 3DBGB

3DBGB [5]-[7] is a blade geometry generation tool that uses a parametric approach to design blades for turbomachinery applications. Due to its parametric nature, a wide range of blades can be designed with relative ease. 2D airfoil sections are created using geometric and aerodynamic input quantities. The sections are stacked in a 3D cylindrical space and are transformed to cartesian space. This can be imported into any CAD package to create 3D solid models. Conventional wind turbines are stacked radially, but other input can be used to define a stacking axis and location on the blade for stacking. The geometry builder is also capable of creating complex geomtries such as bent tips and split tips with minimum change to the inputs.

2.1. Blade Design Process

3DBGB uses the input file "3dbgbinput.dat" to extract the input parameters and create 3D sections. The number of blades, scaling factor and number of streamlines are specified in the first few lines of the input file. A switch to use non-dimensional actual chord values is available to enable reverse engineering of known blades. At each streamline the blade inlet and outlet angles, relative mach number at inlet, non-dimensional actual chord, thickness to chord ratio, incidence, deviation and secondary flow angle are specified. The blade stagger angle can be input through the inlet and outlet leading edge and trailing edge angles.

3DBGB has the capability to generate a variety of airfoils. Due to its parametric nature it is easy to define different airfoils at different sections which makes the design process highly flexible. Presently, the code has the capability to create circular sections, NACA 4-digit airfoils, NREL S809 airfoils and the default mixed camber airfoil. New airfoils can also be included in the code.

3. 3D CFD Analysis

FINE/Turbo[TM] is a Computational Fluid Dynamics (CFD) analysis tool from NUMECA International [9]. It stands for Flow Integrated Environment and it is a complete CFD package which includes grid generation, flow solver and post-processing capabilities. FINE/Turbo is specialized to simulate internal, rotating and turbo-machinery flows for all types of fluids. The package has a fully hexahedral and highly automated grid generation module *AutoGrid*[TM]. The package uses a 3D Reynolds Averaged Euler and Navier Stokes flow solver EURANUS. *CFView*[TM] is a post-processing module which is also part of the package.

3.1. Grid Setup

The grid generation module, AutoGrid, provides the option of either importing and linking a CAD file or importing a ".geomturbo" file, which is the native blade section file format. The blade generation code 3DBGB has the capability to output the blade geometry in the ".geomturbo" file format, which makes importing the blade geometry simple as it initializes the grid parameters and geometry definition. The built-in wind turbine wizard mode in AutoGrid was used to create good grids with the appropriate blade topologies. **Figure 5** shows the grid created using AutoGrid. A complete description of all the parameters and definitions can be found in the NUMECA User Manuals [10].

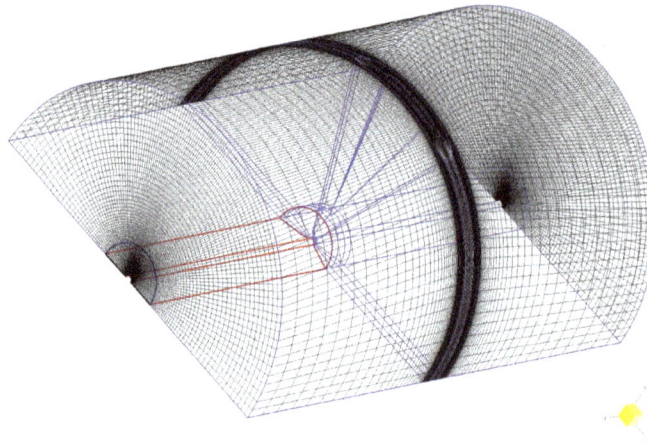

Figure 5. 3D grid.

3.2. Flow Setup and Solution

The Fine/Turbo flow solver module is an intuitive platform to set up the flow properties and to run a simulation. Once the boundary conditions are appropriately assigned using the IGG module, the path to the IGG file is specified in the flow solver module and the flow parameters are specified. The parameters used for the wind turbine case are given below.
• *Configuration
--Fluid Model-Air-(perfect gas)
--Flow Model-Characteristics of the flow:
* -Time Configuration-Steady
* -Mathematical Model
• Turbulent Navier-Stokes
• Spalart-Allmaras Turbulence Model
* -Low Speed Flow- $M < 0.3$
--Rotating Machinery-72 RPM
• *Optional Models—Abu-Ghanam-Shaw Transition Model
• *Boundary Conditions—The boundary condition already generated in Autogrid are available for additional input definition.

Once the flow properties are set, the flow solver EURANUS can be started and the convergence can be monitored. Fine/Turbo also provides the option of running the solution in batch processing mode. The convergence history can be viewed graphically for the global residuals, torque, efficiency and axial thrust. Using the CFView module, the solution can be post-processed to visualize the flow feild properties. **Figure 6** shows the 3D static pressure contour for the NREL blade using CFView. Further post-processing was done using Asgard. Asgard is a collection of visualization/post-processing tools. The asgard library serves as the backbone for file manipulations, calculations, and feature extractions.

4. 3D Force Analysis

The flow field around a wind turbine is complex and no matter how close a 2D approximation is, it will still be innacurate and fail to account for all the details. This was the reason a code was developed using MatLab for 3D force analysis. From the CFD results, the data from each grid point on the blade was extracted using Asgard. This data was processed to arrive at the aerodynamic forces acting on elemental slabs of the wind turbine blade and these forces were integrated over the span to arrive at the global force, torque and power output of the blade.

4.1. Code Formulation

Using the data extracted from Asgard, the blade was divided into the individual sections. Each section and the section above it in the radial direction were treated as a 3D slab. **Figure 7** shows two consecutive 3D slabs. Each

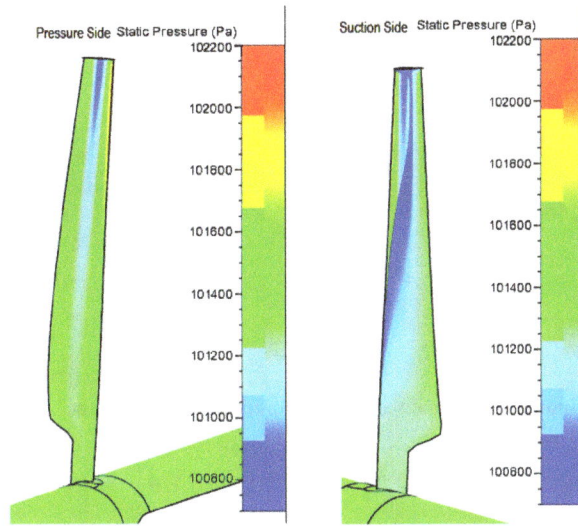

Figure 6. 3d Static pressure contour–pressure, side and suction side-nrel blade.

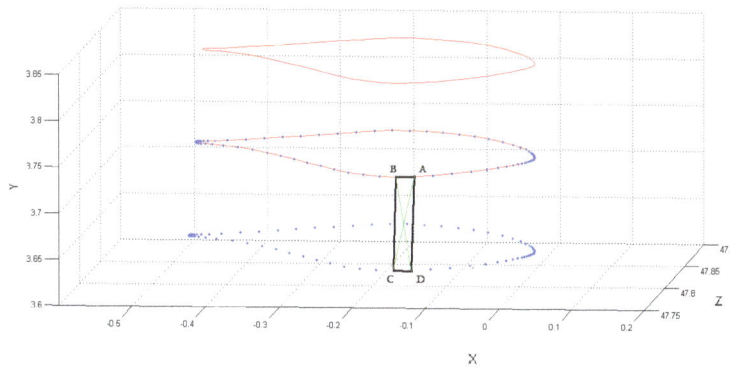

Figure 7. 3D slabs created using CFD data.

slab is divided into elemental areas. Points A,B,C and D enclose one such area. The static pressure at each point is known. The average static pressure acting on the elemental area is found.

$$\bar{p} = \left(p_A + p_B + p_C + p_D \right)/4 \tag{3}$$

The area vector of the elemental area is found by using the diagonal vectors which is then multiplied with the average pressure to obtain the force (f) acting on the elemental area. The resultant force (F) acting on the slab is the summation of the forces acting on all the "n" elemental areas on the slab.

$$\mathrm{d}A = \frac{1}{2}\left(AC \times BD \right) \tag{4}$$

$$f = \bar{p}\mathrm{d}A \tag{5}$$

$$F = \sum_1^n \bar{p}\mathrm{d}A = \sum_1^n f \tag{6}$$

Figure 8 shows two slabs with diagonal vectors and force vectors on each elemental area. The cartesian force vectors are transformed into cylindrical force vectors. The position vector for the transformation is taken as the centroid of the slab. The centroid is found using an open source MatLab function-Polygeom [11].

$$\theta = \tan^{-1}\frac{y_{\text{centroid}}}{x_{\text{centroid}}} \tag{7}$$

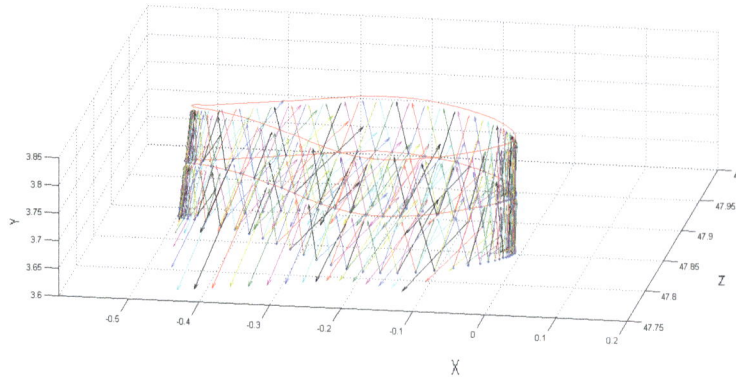

Figure 8. Diagonal vectors and force vectors on, a slab.

$$\begin{bmatrix} F_r \\ F_\theta \\ F_z \end{bmatrix} = \begin{bmatrix} \cos\theta & \sin\theta & 0 \\ -\sin\theta & \cos\theta & 0 \\ 0 & 0 & 1 \end{bmatrix} \begin{bmatrix} F_x \\ F_y \\ F_z \end{bmatrix} \tag{8}$$

The product of the tangential force and the local radius gives the torque of the slab. Summing all the torques yeilds the global torque of the blade. The product of the global torque and the rotational speed of the blade gives the power output output of the blade which can be multiplied with the total number of blades to give the total power output of the wind turbine.

$$Q = \sum_{hub}^{tip} F_\theta r \tag{9}$$

$$P = BQ\Omega \tag{10}$$

It should be noted that shear forces are not taken into account in this analysis as they were found to be very negligible owing to the high Reynold's number.

5. Results and Comparison

The 3D force analysis was carried out on the NREL Phase VI blade and used as a baseline. CFD results for the NREL blade predicted a power output of 4.8 kilowatts for a Vz of 7 m/s. This was validated using the NREL design report [8] as seen in **Figure 9** which produced a power output of 4.8 kilowatts at 7 m/s in a test as interpreted from the graph. In order to analyze the effect of the radial area variation in great detail, the NREL blade's S809 airfoil was substituted with a NACA 2420 airfoil since this was a close approximation to the S809 airfoil. The dependence of the radial forces associated with the blade on the radial component of the cross sectional area can be seen mathematically through the integral form of the radial momentum equation.

$$\frac{\partial}{\partial t}\iiint \rho v_r r\,d\theta dr dz + v_r d\dot{m} + pr d\theta dz = \iiint \frac{p}{r} r\,d\theta dr dz + \iiint \frac{\rho v_\theta^2}{r} r\,d\theta dr dz \tag{11}$$

The third term on the left hand side is the radial force term and "$rd\theta dz$" is the radial component of the surface area. This shows the dependence of the radial force on the radial component of the surface area. A blade was designed to have varying airfoil sections with span using the NACA 24XX family of aifoils. This was done to vary the thickness to chord ratio of the airfoil at various span locations to change the radial component of the surface area.

The NACA 2420 airfoil was found to be a very close approximation to the S809 airfoil in both thickness and lift coefficient characteristics. The difference in the lift coefficient is within a tolerable margin in the angle of attack range from 0 to 7 degrees for the S809 and NACA 2420 airfoils, which is the region where the effective angle of attack lies for the blade. The maximum difference in C_l is 0.04 at 0 degrees and 7 degrees. It is for these reasons that the NACA 2420 airfoil was chosen to be the baseline for further investigation into the effect of radial area variation on the blade.

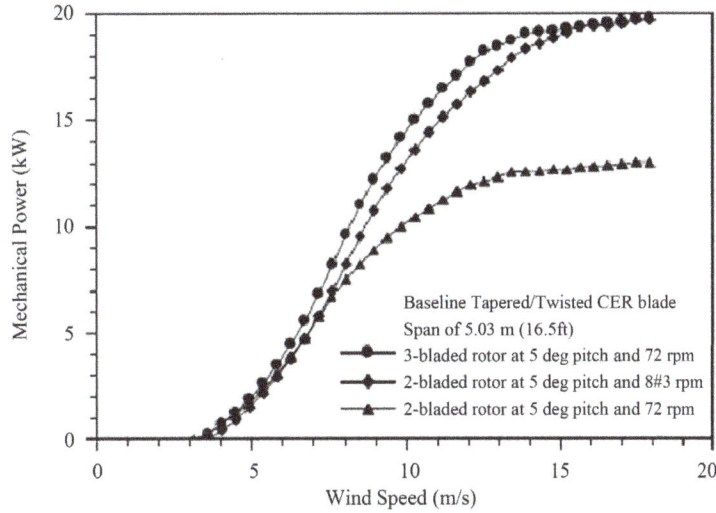

Figure 9. NREL power curve [12].

The addition of the NACA 4-digit airfoil generation capability in the 3DBGB code provided a tremendous range of airfoil sections that could be included in the design. The thickness to chord ratio of the airfoils is the key parameter that is used to manipulate the radial component of the surface area. **Figure 10** shows the lift coefficient characteristics obtained using XFOIL [4] of a few NACA airfoils that have been used in the blade design. It can be seen that they all have very similar C_l vs. α curves. NACA 2404 does not follow the same characteristcs beyond an angle of attack of 6 degrees, but the tip region, where the NACA 2420 airfoil is used, lies in the effective angle of attack range of 0 to 2 degrees.

5.1. NACA 2420 vs. NACA 2420-2404

The NACA 2420-2404 blade consists of the NACA 2420 airfoil till 50% span and then gradually decreases to NACA 2404 at the tip as shown in **Figure 11**. Due to the greater change in the radial component of the area, the NACA 2420-2404 blade has higher radial forces acting on the blade than the NACA 2420 blade. The jump in the radial force at 50% span for the NACA 2420-2404 blade can be clearly seen in **Figure 12**. The tangential forces, axial forces and torque are also greater for the NACA 2420-2404 blade than the NACA 2420 blade as seen in **Figure 13-15**.

Figure 12 shows the radial force comparison between the NACA 2420 and NACA 2420-2404 blades using gauge pressure instead of static pressure. This is done by substituting the pressure term in the force equation with $\left[p - p_{ref} \right]$ where p_{ref} is the reference static pressure in the far-field-101295 pascals. The NACA 2420-2404 blade performs better than the NACA 2420 blade because it bends the streamlines more than the NACA 2420 blade as shown in **Figure 15**. It has a higher torque starting at 50% span, precisely where the thickness to chord ratio variation starts and this trend continues as we move towards the tip. The bending of the streamlines causes the Vz to go down in front of the blade (**Figure 16**), improving the effective angles of attack (**Figure 18**); thereby improving the lift coefficient (**Figure 19**). **Figure 17** shows the axial induction factor. The axial induction factor is a vital performance parameter of a wind turbine. It clearly shows what percentage of the free stream wind's kinetic energy is being converted into useful work. Even though there is a theoretical limit imposed on the axial induction factor [13], current wind turbine designs are still short of achieving axial induction factors close to the limit.The area averaged axial induction factor for the blades are:

$$\frac{\int a_{2420}(r) r \mathrm{d}r}{\int r \mathrm{d}r} = 0.1637 \tag{12}$$

$$\frac{\int a_{2420-2404}(r) r \mathrm{d}r}{\int r \mathrm{d}r} = 0.1798 \tag{13}$$

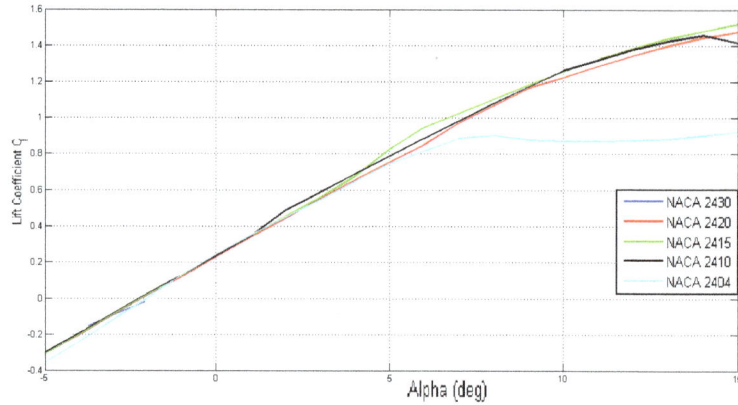

Figure 10. Lift coefficient vs. Angle of attack-NACA series.

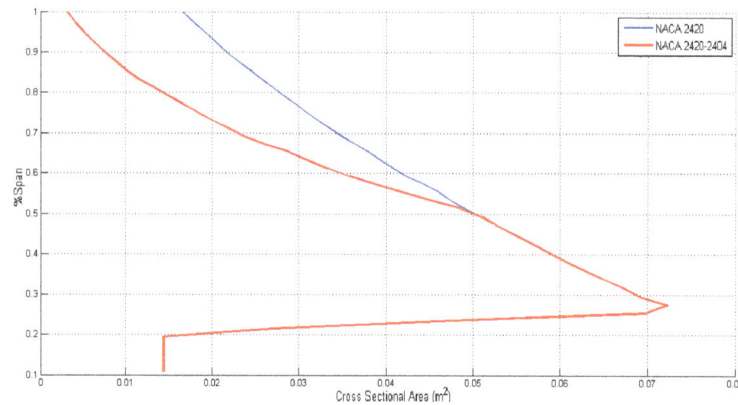

Figure 11. Cross sectional area vs. span.

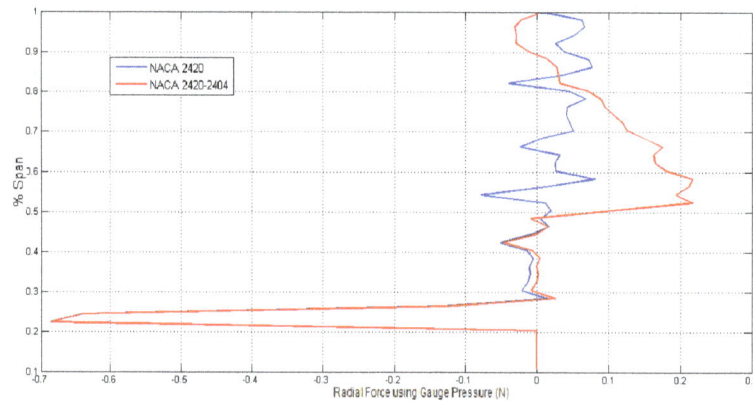

Figure 12. Radial force using gauge pressure vs. span.

So the axial induction factor for the NACA 2420-2404 blade is 9.8% higher than the NACA 2420 blade. The mass flow rate is constant through a streamtube for each design. From the actuator disc theory [2] shown in **Figure 1**, subscript "1" corresponds to the station far upstream and subscript "2" corresponds to the station just in front of the leading edge.

$$\dot{m}_{2420} = \rho A_{1-2420} V_{1-2420} = \rho A_{2-2420} V_{2-2420} \tag{14}$$

$$\dot{m}_{2420-2404} = \rho A_{1-2420-2404} V_{1-2420-2404} = \rho A_{2-2420-2404} V_{2-2420-2404} \tag{15}$$

The density is essentially constant and $A_{2-2420} = A_{2-2420-2404}$ and $V_{1-2420} = V_{1-2420-2404} = U_1$.

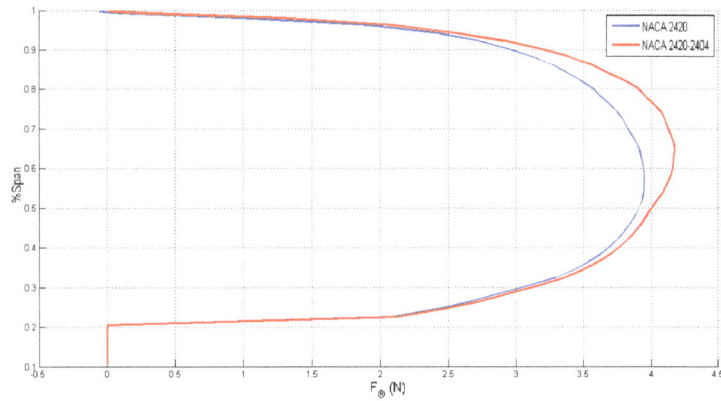

Figure 13. Tangential force vs. span.

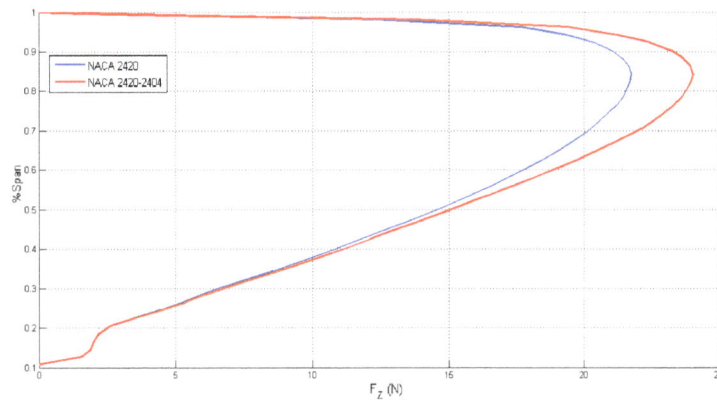

Figure 14. Axial force vs. span.

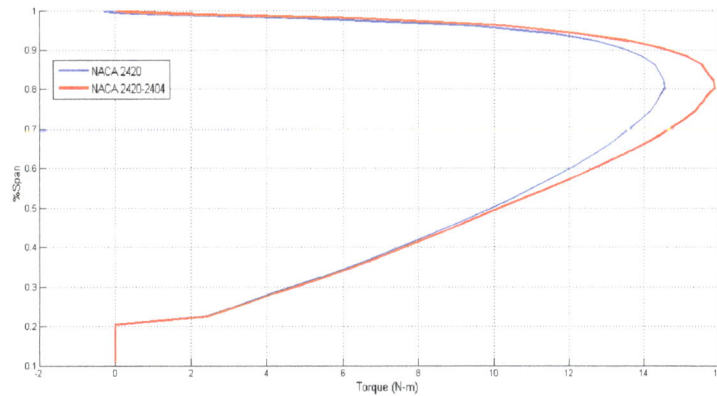

Figure 15. Torque vs. span.

$$\frac{A_{1-2420}}{A_{1-2420}} = \frac{V_{2-2420}}{V_{2-2420-2404}} = \frac{(1-a_{2420})U_1}{(1-a_{2420-2404})U_1} \tag{16}$$

$$\frac{A_{1-2420}}{A_{1-2420-2404}} = \frac{V_{2-2420}}{V_{2-2420-2404}} = \frac{(1-0.1637)}{(1-0.1798)} = 1.0196 \tag{17}$$

The subtle and gradual radial area variation between the NACA 2420 and NACA 2420-2404 blades has resulted in a 1.96% decrease in the streamtube area far upstream of the blade. This area change results in a lower

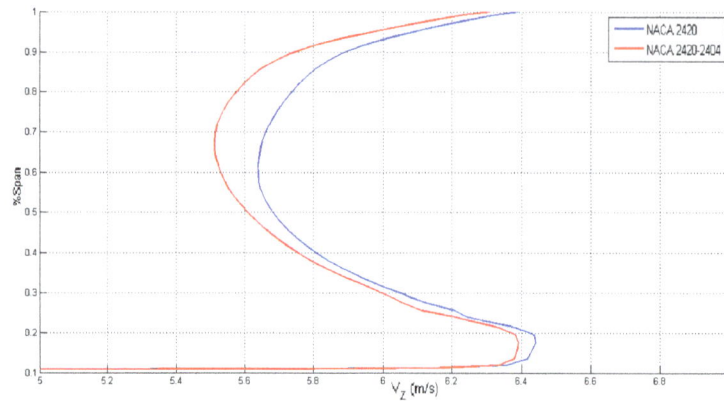

Figure 16. Area averaged vz in front of blade vs. span.

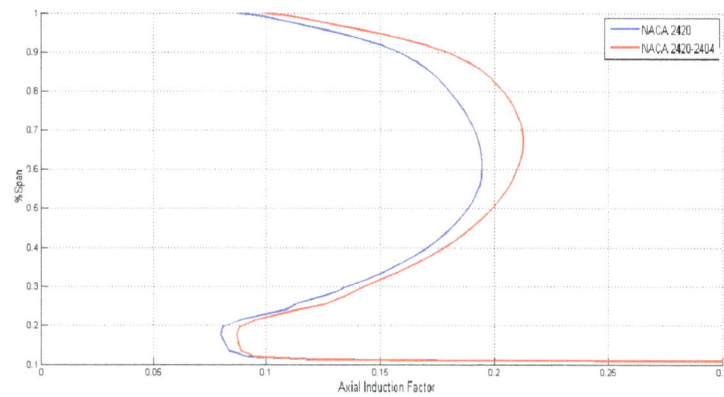

Figure 17. Axial induction factor vs. span.

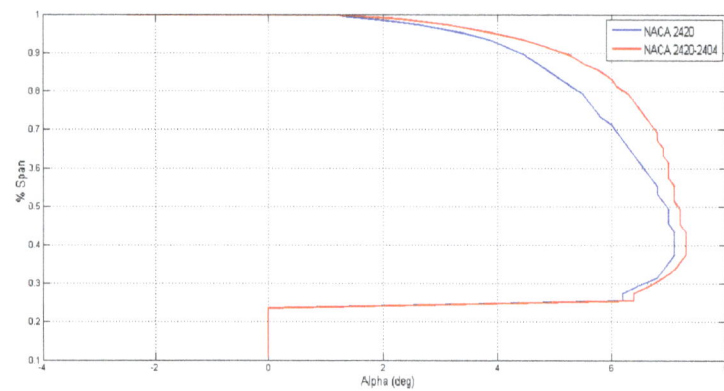

Figure 18. Effective angle of attack vs. span.

Vz in front of the blade, increasing the effective angles of attack over the span of the blade and directly contributes to the increase in the tangential forces acting on the blade. The increase in tangential forces ultimately improves the power output of the blade.

$$C_P = 4a(1-a)^2 \qquad (18)$$

$$C_{P2420} = 4 \times 0.1637(1-0.1637)^2 = 0.4579 \qquad (19)$$

$$C_{P2420-2404} = 4 \times 0.1798(1-0.0.1798)^2 = 0.4838 \qquad (20)$$

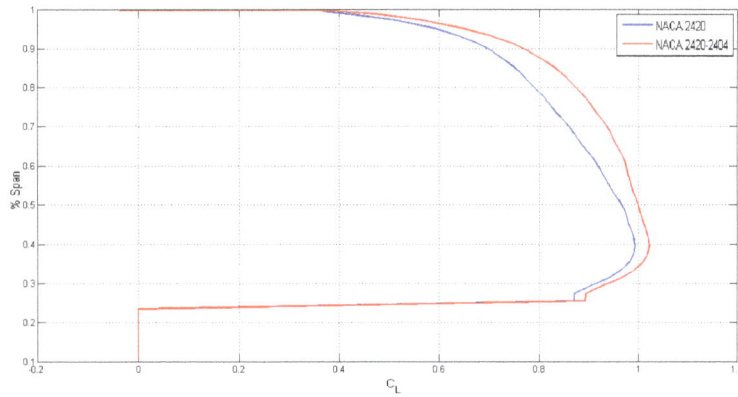

Figure 19. Lift coefficient vs. span.

$$\frac{C_{P2420-2404}}{C_{P2420}} = 1.05661 \tag{21}$$

Theoretically, the power output of the NACA 2420-2404 blade is 5.661% more than the NACA 2420 blade. The power output of the blades was calculated from the torque.

$$Q_{\text{Blade}} = \sum Q_{\text{slab}} \tag{22}$$

$$P = BQ_{\text{Blade}}\Omega \tag{23}$$

The NACA 2420 blade has a power output of 6.0117 Kilo Watts which is 25.8% more than the NREL original blade. This increase in the power output is due to the slightly better C_l characteristics of the NACA 2420 airfoil than the S809 airfoil. The NACA 2420-2404 blade gives the best performance with a power output of 6.4253 kilowatts, that is, 6.8% higher than the NACA 2420 blade, which is similar to Equation (21).

6. Effect of Radial Area Variation

In the actuator disc model discussed in [2] and shown in **Figure 1**, the wind velocity in stations 2 and 3, U_2 and U_3, are assumed to be the same and that the area expansion occurring in the streamtube is solely because of U_4 being lesser than U_1. This is not a complete description; hence, the twist distribution of the wind turbine blade calculated using the BEM theory will not be optimal. Analyzing the results from the 3D force analysis code revealed that the expansion of the streamtube is in fact driven in part by the radial forces acting on the blade. It also revealed that the manipulation of these radial forces will lead to further expansion of the streamtube; therefore, reducing the wind velocity just upstream of the blade. This velocity just upstream of the blade is U_2, as shown in the actuator disc model, which is related to the axial induction factor.

$$U_2 = U_1(1-a) \tag{24}$$

The radial momentum equation can be expressed in terms of the streamline curvature, using the streamline curvature approach as outlined in [12], as follows.

$$\frac{1}{\rho}\frac{\partial p}{\partial r} = v_m^2 C + \frac{v_\theta^2}{r} \tag{25}$$

L.H. Smith, Jr. derived an equation to describe the radial variation of circumferentially averaged flow properties inside a turbomachinery blade row [14]. Circumeferential averaging of a flow property "x" is defined as

$$(\bar{x}) = \frac{1}{\theta_s - \theta_p}\int_{\theta_p}^{\theta_s}(x)\,\mathrm{d}\theta \tag{26}$$

where θ is the circumferential angle in radians and subscript "p" stands for the pressure side of a blade and subscript "s" stands for the suction side of the next blade. Smith derived an equation for the circumferentially averaged radial pressure distribution in [14].

$$\frac{\overline{\partial p}}{\partial r} = \frac{\partial \overline{p}}{\partial r} + \left(\frac{p_s - p_p}{\theta_s - \theta_p}\right)\frac{\tan \varepsilon_m}{r} + \left(\overline{p} - \frac{p_p + p_s}{2}\right)\frac{1}{\Lambda}\frac{\partial \Lambda}{\partial r} \tag{27}$$

Λ is the ratio of the open circumference to the total circumference and ε_m is the blade lean angle in the meridional direction.

$$\Lambda = \frac{B\Delta\theta}{2\pi} \tag{28}$$

$$\tan \varepsilon_m = -r\frac{\partial \theta_m}{\partial r} \tag{29}$$

The first term on the right hand side of Equation (27) is the desired radial gradient of the average pressure. The second term is related to the radial component of the blade force on the fluid and the last term is attributed to a blockage effect. Smith [14] goes on to state that the last term vanishes when the pressure varies linearly in the θ direction and drops it from the equation. This is a good approximation for many turbomachinery flows where high velocities are involved such as jet engines. Since wind turbines operate at a mere fraction of the velocities involved with conventional axial turbomachinery and since pressure does not vary linearly in the θ direction, this term cannot be neglected. Substituting Equation (27) in a circumferentially averaged Equation (25) we obtain

$$\left(\frac{1}{\rho}\right)\frac{\overline{\partial p}}{\partial r} = \left(\frac{1}{\rho}\right)\left[\frac{\partial \overline{p}}{\partial r} + \left(\frac{p_s - p_p}{\theta_s - \theta_p}\right)\frac{\tan \varepsilon_m}{r} + \left(\overline{p} - \frac{p_p + p_s}{2}\right)\frac{1}{\Lambda}\frac{\partial \Lambda}{\partial r}\right] = \overline{v}_m^2\overline{C} + \frac{v_\theta^2}{r} \tag{30}$$

When the lean angle $\varepsilon_m \approx 0$ then the above equation becomes

$$\left(\frac{1}{\rho}\right)\frac{\overline{\partial p}}{\partial r} = \left(\frac{1}{\rho}\right)\left[\frac{\partial \overline{p}}{\partial r} + \left(\overline{p} - \frac{p_p + p_s}{2}\right)\frac{1}{\Lambda}\frac{\partial \Lambda}{\partial r}\right] = \overline{v}_m^2\overline{C} + \frac{v_\theta^2}{r} \tag{31}$$

Figure 20 shows the comparison of the first term in Equation (27), the radial gradient of the circumferentially averaged pressure, between the NACA 2420 and NACA 2420-2404 blades. The $\frac{\partial p}{\partial r}$ term for the NACA 2420-2404 blade is lower than the NACA 2420 blade over most of the span, especially from 40% - 60% of the span and from 82% - 100% span.

The third term in Equation (27), the blockage effect term, is compared between the NACA 2420 and the NACA 2420-2404 blades in **Figure 21**. It is slightly higher for the NACA 2420-2404 blade than the NACA 2420 blade at mid-span. From 62% - 100% span, it is lower for the NACA 2420-2404 blade than the NACA 2420 blade.

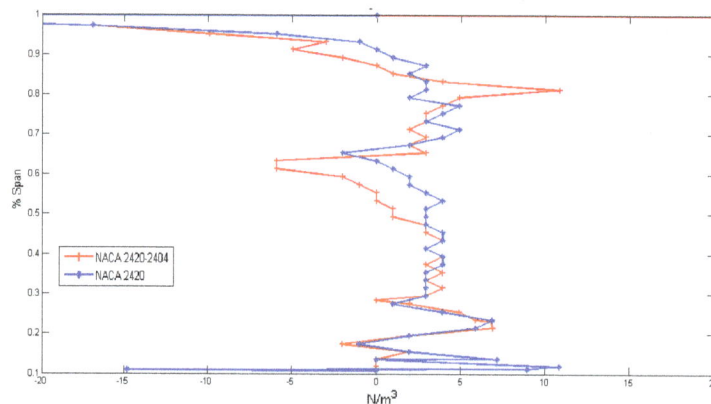

Figure 20. Radial pressure gradient term in Equation (9.14) $\frac{\partial \overline{p}}{\partial r}$ vs. span comparison-NACA 2420 and NACA 2420-2404.

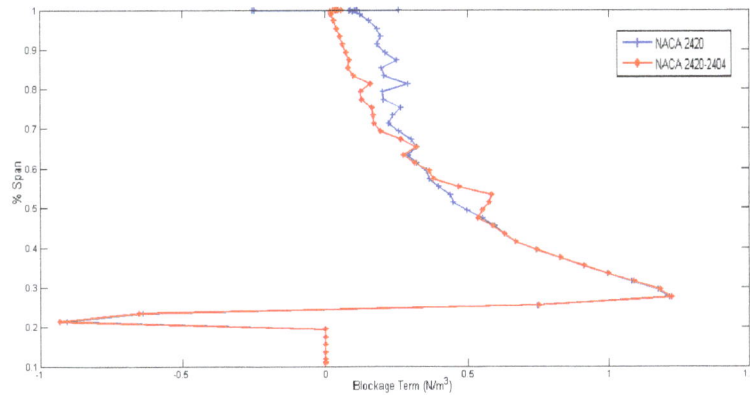

Figure 21. Third term in Equation (9.14) $\left(\overline{p} - \dfrac{p_p + p_s}{2}\right)\dfrac{1}{\Lambda}\dfrac{\partial \Lambda}{\partial r}$ vs. span comparison - NACA 2420 and NACA 2420-2404.

The difference between the two cases is the cause for the difference in the circumferencial averaged radial pressure gradient. The streamline curvature is directly related to the circumferentially averaged radial pressure gradient; hence, the NACA 2420 and NACA 2420-2404 blades have different streamline curvatures. The NACA 2420-2404 blade is able to bend the streamlines more than the NACA 2420 blade, thereby increasing the streamtube area expansion and reducing the wind speed ahead of the blade more effectively. This enables the NACA 2420-2404 blade to extract more energy from the wind and perform better than the NACA 2420 blade.

In order to get the circumferential average streamtube radius at the tip, thirty streamlines were spawned, equi-distant from each other, on a line between the leading edge and trailing edge of the two blades for the NACA 2420 and NACA 2420-2404 cases using Asgard. A comparison of the thirty streamtubes between the NACA 2420 and NACA 2420-2404 blades can be seen in **Figure 22**. The average streamtube radius for the NACA 2420 blade and NACA 2420-2404 blade are shown in **Figure 23** and **Figure 24**. They are embedded on their respective azimuthal averaged static pressure contours. It can be seen how the streamlines are bent more for the NACA 2420-2404 blade at the tip than the NACA 2420 blade. The area ratio is the streamtube area of the NACA 2420-2404 blade over the streamtube area of the NACA 2420 blade. **Figure 25** and **Figure 26** show the streamtube area and area ratio respectively.

The blade lies between Z = 47.7 meters and Z = 48 meters. The upstream streamtube area of the tip streamlines for the NACA 2420-2404 blade is slightly lower than the NACA 2420 blade. When the streamlines reach the blade tip, the difference in area grows large. The NACA 2420-2404 blade is able to reduce the wind velocity and increase the streamtube area at the blade more than the NACA 2420 blade; thereby, extracting more kinetic energy from the wind.

The average axial induction factor of the NACA 2420 blade is 0.1637 and the average axial induction factor of the NACA 2420-2404 blade is 0.1798. Theoretically, the maximum possible axial induction factor is 0.3333 according to Betz [13]. From **Figure 26**, it can be seen that the area ratio far upstream of the blade is approximately 0.98. This area ratio can be related to the average axial induction factor through Equation (16) and the results are very similar. The axial induction factor depends on the streamtube area which, in turn, depends on how much the streamlines are bent in front of the blade. Higher radial area variation on the blade facilitates this bending of the streamlines by creating a higher radial pressure gradient. The lowering of Vz in front of the blade improves the effective angles of attack over the span; thereby increasing the lift coefficients over the span. As a result, the tangential forces acting on the blade are improved; hence, the power output is higher. As shown in Equation (17) and Equation (21), a 1.96% decrease in the streamtube area far upstream of the blade causes the power to increase by 5.661%. The power output results from CFD analysis show that the NACA 2420-2404 blade has an output that is 6.8690% higher than that of the NACA 2420 blade.

7. Conclusion and Future Work

Methods to improve the performance of wind turbines have been explored. 3D CFD analysis was carried out on

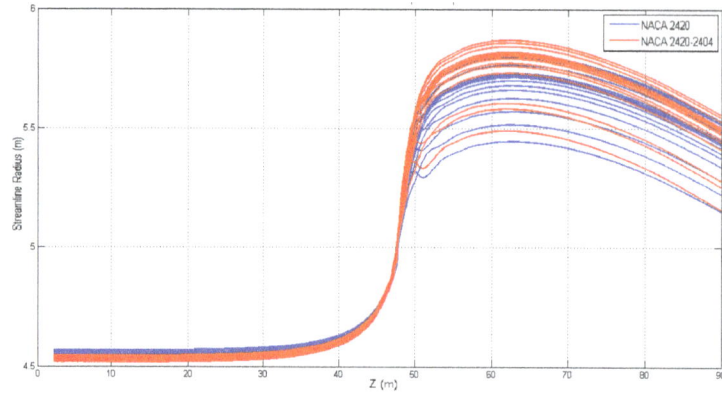

Figure 22. Streamtube radius vs. Z at tip-naca 2420 and naca 2420-2404-30 streamlines.

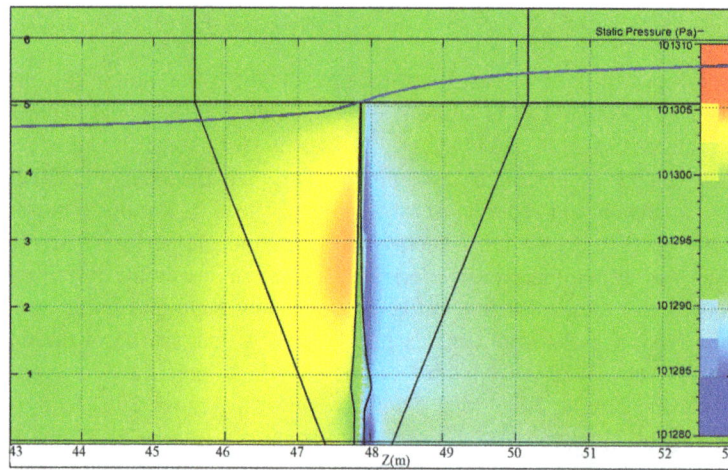

Figure 23. Average streamtube radius vs. Z at tip-naca 2420.

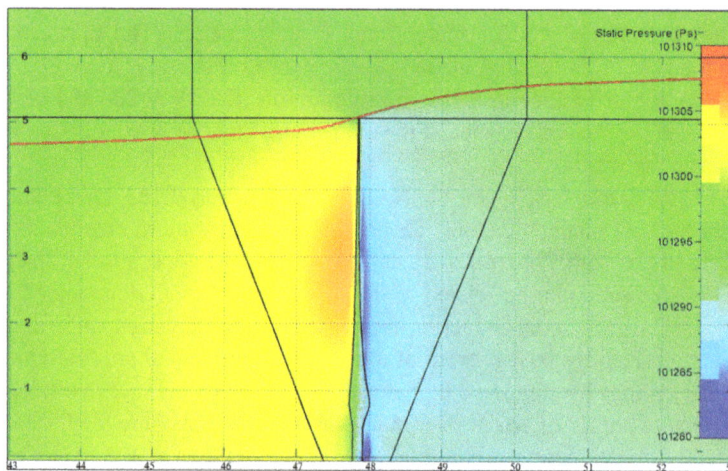

Figure 24. Average streamtube radius vs. Z at tip-naca 2420-2404.

an NREL phase VI wind turbine. 3D analysis of the forces acting on the blade established the magnitude of the radial force acting on the blade to be much larger than the tangential and axial forces. The impact of radial area variation on the performance of the blade was explored using the NACA four digit airfoil series. The thickness to chord ratio was manipulated over the span of the blade to explore its effect.

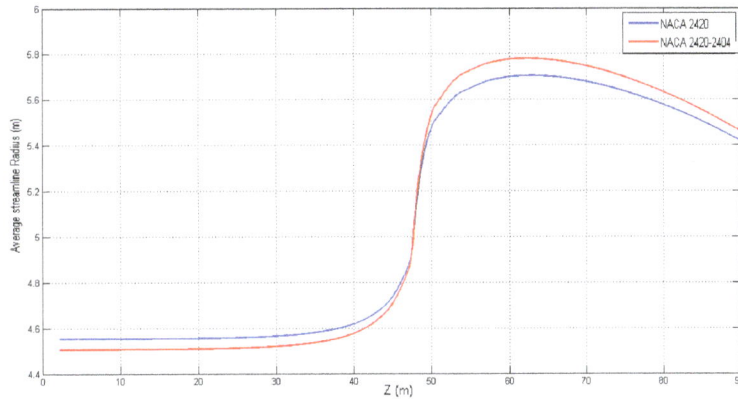

Figure 25. Average streamtube radius vs. Z at tip-NACA 2420 and NACA 2420-2404.

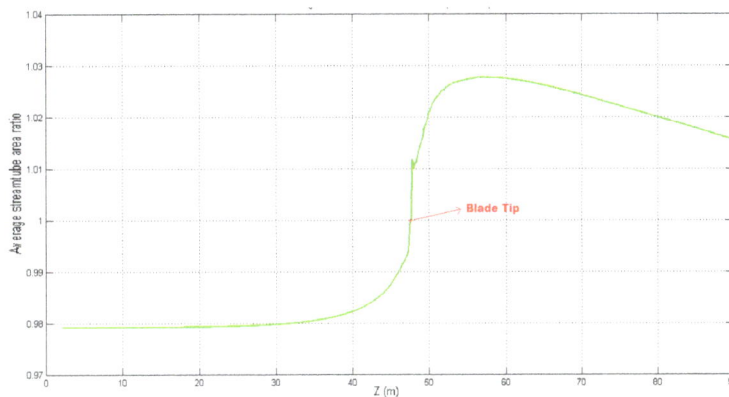

Figure 26. Average streamtube area ratio vs. Z at tip-NACA 2420 and NACA 2420-2404.

The NACA 2420-2404 blade has been found to have the best performance. The key findings of this research work are as follows:

• *The radial component of the surface area of the blade is a contributor to the magnitude of the radial forces acting on the blade.

• *The radial pressure gradient is responsible for bending the streamlines.

• *For wind turbines which operate at low wind speeds, this circumferentially averaged radial pressure gradient term depends on the blockage effect.

• *The axial induction factor depends on the streamtube area which, in turn, depends on how much the streamlines are bent by the blade.

• *The axial induction factor influences the effective angle of attack seen by the blade sections.

• *An increase in the streamtube area by 1.96% results in an increase in power output by 5.661%.

• *The NACA 2420-2404 blade has a power output that is 6.8690% higher than the NACA 2420 blade.

Improved models need to be developed between the radial variation of the cross-sectional area and the axial induction factor. This will aid in designing wind turbine blade geometries with high axial induction factors to extract the maximum possible power from the wind. The 3D force analysis code developed for this thesis can be used for a wide range of applications. The findings presented in this paper are not only restricted to wind turbines but can also be applied to propellers, transonic fans, open rotors and other turbomachinery applications. A highly detailed version of this research can be found in [15].

Acknowledgements

A special thanks to Kiran Siddapaji and Soumitr Dey for their help and brainstorming sessions for my research.

Thanks to Rob Ogden for helping me with many license and Linux issues. I would like to thank Robert Knapke, Marshall Galbraith and Andy Schroder for their support and help. I would also like to thank the NUMECA support team (Roque Lopez and Autumn Fjeld) for their timely help with Fine Turbo trouble-shooting.

References

[1] Appalachian State University-North Carolina Wind Energy. www.wind.appstate.edu

[2] Manwell, J.F. and McGowan, J.G. and Rogers, A.L. (2002) Wind Energy Explained: Theory, Design and Application of Wind Energy. 1st Edition, John Wiley and Sons, Inc., Hoboken.

[3] Dey, S. (2011) Wind Turbine Blade Design System-Aerodynamic and Structural Analysis. Master's thesis, University of Cincinnati, Cincinnati.

[4] Drela, M. and Youngren, H. XFOIL-Subsonic Airfoil Development System. http://web.mit.edu/drela/Public/web/xfoil/

[5] Park, K., Turner, M.G., Siddappaji, K. and Dey, S. and Merchant, A. (2011) Optimization of a 3-Stage Booster Part 1: The Axisymmetric Multi-Disciplinary Optimization Approach to Compressor Design. ASME Proceedings, Paper No. GT2011-46569, 1413-1422.

[6] Siddappaji, K. (2012) Parametric 3D Blade Geometry Modeliing Tool for Turbomachinery Systems. Master's Thesis, University of Cincinnati, Cincinnati.

[7] Siddappaji, K., Turner, M.G., Dey, S., Park, K. and Merchant, A. (2011) Optimization of a 3-Stage Booster-Part 2: The Parametric 3D Blade Geometry Modeling Tool. ASME Proceedings, Paper No. GT2011-46664, 1431-1443.

[8] Giguere, P. and Selig, M.S. (1999) Design of a Tapered and Twisted Blade for the NREL Combined Experiment Rotor. Technical report, NREL/SR-500-26173. http://dx.doi.org/10.2172/750919

[9] Numeca. www.numeca.com

[10] Numeca User Manuals. www.numeca.com

[11] http://www.mathworks.com/matlabcentral/fileexchange/319-polygeom-m

[12] Novak, R.A. (1967) Streamline Curvature Computing Procedures for Fluid-Flow Problems. *Journal of Engineering for Gas Turbines and Power*, **89**, 478-490.

[13] Betz, A. (1926) Windenergie und ihre ausnutzung durch wind-muhlen: Vandenhoeckund ruprecht.

[14] Smith Jr., L.H. (1966) The Radial-Equilibrium Equation of Turbomachinery. *Journal of Engineering for Gas Turbines and Power*, **88**, 1-12.

[15] Sairam, K. (2013) The Influence of Radial Area Variation on Wind Turbines to the Axial Induction Factor. Master's Thesis, University of Cincinnati, Cincinnati.

Nomenclature

a	Axial induction factor
c	Chord
$\dot{m}$	Mass flow rate
p	Pressure
r	Radius
z	Axial direction
A	Cross-sectional area
B	Number of blades
C	Streamline curvature
C_l	Coefficient of lift
C_P	Coefficient of Power
P	Rated Power
Q	Torque
R	Blade radius
T	Thrust
U	Rotor wheel speed

7

Parallel Correlative Study of the Heat Balance Test Process for Testing the Efficiency of Turbo-Generators

Luqman Muhammed Audu[1], Olugbenga Olanrewaju Noah[2], Ogaga Kenneth Ajaino[3], Friday Chukwuyem Igbesi[4]

[1]Department of Mechanical Engineering, Auchi Polytechnic, Auchi, Nigeria
[2]Department of Mechanical Engineering, University of Lagos, Lagos, Nigeria
[3]Department of Mechanical Engineering, Delta State Polytechnic, Ogwashuku, Nigeria
[4]Department of Mechanical Engineering, Delta State Polytechnic, Otefe-Oghara, Nigeria
Email: luqslama@gmail.com, onoah@unilag.edu.ng

Abstract

In order to prolong the life span of a turbo-generator plant and sustain its performance at high efficiency, it is subjected periodically to regular test to monitor the operational profile and efficiency of power conversion from mechanical energy to electrical energy. Analysis of these test data serves as a measure to indicate deviation from normal operation profile and deterioration of plant performance. This present work implemented the heat balance tests process to three turbo-generator units in order to assess the harmony, consistency, and accuracy of results to establish parallel correlation for the test process. The test process involves carrying out a heat balance for the turbo-generators at 50%, 75% and 100% load respectively through the determination of the heat losses through the hydrogen coolers, bearing oil, seal oil and radiation and convention to the atmosphere. Some important results were presented in the paper.

Keywords

Efficiency, Turbo-Generator, Testing, Hydrogen-Cooled, Heat Balance

1. Introduction

Turbo-generators are usually used to produce electricity by using a turbine to drive an alternative current (a.c) generator. The choice of machinery for high power generation currently comes down to either hydrogen cooled generators providing high efficiency or air cooled generator characterized by easy operation and maintenance at low cost [1]. Consequently, the world powered engineering has experienced an increase in the demand for high powered turbo-generators with hydrogen and air cooling [2], for the generation of electric power for the sustenance of vital national production processes and services.

Turbo-generators convert mechanical energy of rotation of the turbine shaft into electricity. The a.c generator consists principally of a magnetic circuit, direct-current (d.c) field winding, a-c armature winding and a mechanical structure including a cooling and lubricating system. The magnetic circuit and field winding are arranged so that as the machine rotates, the magnetic flux linking the armature windings change cyclically, thereby inducing alternating voltage in the armature winding.

Heat is produced in a generator as a result of resistive losses caused by current flow in the stator and field windings, stator core magnetic losses and windage losses. This heat generation if not controlled and regulated through a cooling system can cause breakdown in insulation and other damages to vital component of the machine. By providing forced cooling of the rotating and stationary components, the generator ratings may be increased and the physical components may be made smaller [3]. To ensure reliable operation of turbo-generators, the heating of any electrical part of the machine must be kept within certain limits [4].

Turbo-generators were designed exclusively for air-cooling in the early stages of development. Hydrogen gas was later used as the cooling medium for turbo-generators of high rating, although recent development of a high performance and high efficiency inner air cooling ventilation system generator with rating of 250 MVA has been achieved [1]. This generator was upgraded by the optimization of the stator core duct and ventilation structure, stator coil structure, and the modification of the ventilation structure of the rotor end, pushing the efficiency to 98.80% which is close to that of hydrogen cooled generators [5]. Further modifications were made to significantly reduce the noise level of the turbo-generator [6]; this development put air cooled turbo-generators at par with their hydrogen counterpart in terms of power ratings. **Figure 1** represents the schematic of cooling method for inner cooler cooling system. Air cooled turbo-generators, in general, tend to generate more noise compared to hydrogen-cooled generators.

Inspection, testing and regular checks are vital aspects of maintenance procedures that assist in sustaining the operation, performance of the plants and provide appropriate information for timely maintenance to provide the needed availability and reliability for plant operation and satisfactory performance [7]. The following research works on turbo-generator's operation and performance have been carried out. [8] studied the temperature generation of the elementary conductors of the stator windings of a 225 MW turbo-generator with fiber optics sensors. Suitable parameters for the design specification of the thermal conductivity and heat transfer coefficient of air cooled turbo-generators were obtained and characterized. [5] discussed the development of 250 MVA, 60 HZ air cooled generator with the optimization of the ventilation structure and stator coil strand arrangements. The generator's performance achieved a high efficiency of 98.80% with output characteristics that competes with hydrogen cooled generators. [9] investigated the air gap ventilation, stator strand and rotor winding temperature

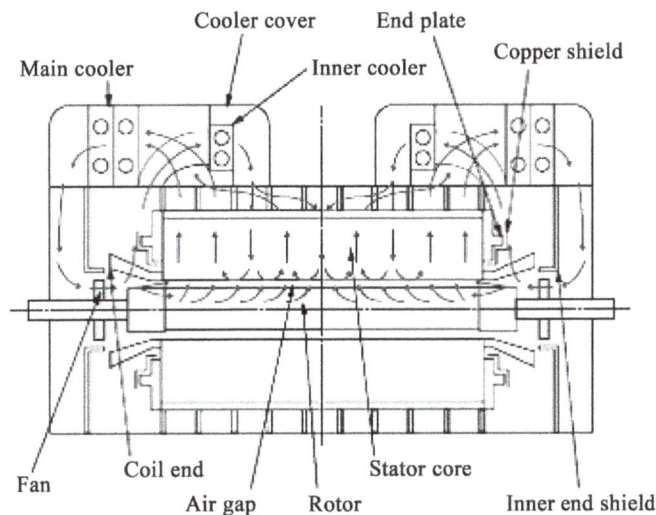

Figure 1. Schematic of cooling method for Inner cooler cooling system.

distribution in a 250 MVA air cooled turbo-generator. The temperature measured under full load conditions was under the limits of class B insulation, while the measured air flow and temperature distribution in the stator, rotor and air gap conformed to the values obtained by the system design programs. [10] evaluated the reliability of a 250 MVA generator by connection to a high capacity load machine with the application of a rated field current to reproduce field conditions. The maximum temperature rise was observed to be accommodated between the limits of class B insulation. [1] evaluated the performance of a 250 MVA (50/60 Hz) large air cooled turbo-generator with inner cooler ventilation system. The results obtained were used to develop a database for the design and construction of hydrogen and air cooled turbo-generator with class B temperature rise standard.

Two general types of test are carried out on turbo-generator in thermal power stations. These are the guaranteed performance test or acceptance test and routine test. The guaranteed performance test is carried out immediately after commissioning to ensure that the performance of the machine meets the specification given by the manufacturer. This performance test is carried out using the Work Test Process. The routine test is carried out during normal operation of the machine at regular interval to ascertain the generator performance and to provide information on generator efficiency, causes of depreciation in generator performance and an effective guide to maintenance of the generators.

The Work Test Process is carried out after generator has attained operational stability. In this process six losses are evaluated for the generator. Two of the losses are considered to be fixed losses. They are fixed with input to load for the generator. The fixed losses are the windage and friction losses and core loss. The rest losses are variable and they include field copper (rotor) losses, armature copper loss, stray load loss and excitation loss [11].

The Work Test Process could not be used to carry out the routine test at Egbin Thermal Station because it entails disengaging the turbo-generator from the national power grid which will result to the loss of at least 220 MW out of an actual national power output that fluctuates between 2600 MW - 3600 MW per day [12]. Statistics from the Federal Ministry of Energy indicates the peak power generation as at October, 2014 to be 3513.5 MW [13]. Therefore, in the face of the present power predicament there is the dire need to develop a routine test process for evaluating the efficiency of turbo-generators without disrupting their operation and power generation to the power grid by the Power Holding Company of Nigeria (PHCN). This is achieved by carrying out a heat balance on the turbo-generator system. The heat balance test process had earlier been used to evaluate the efficiency of one turbo-generator unit at Egbin Thermal Station [14]. In other to establish parallel correlation for the heat balance test process, there is the essential need to extend the test process to three turbo-generator units to assess the harmony and consistency of results between the three units.

The objective of the study is the implementation of the heat balance testing process to three turbo-generator units in order to assess the harmony, consistency and accuracy of results to establish parallel correlation for the test process. The results will also be used to determine if there has been any significant deterioration in the efficiency of the plant by a comparative analysis with the results of the work test process used to commission the plants.

2. Methodology

A heat balance for the turbo-generator plant was carried out by taking readings on embedded and supervisory equipment on the turbo-generator on 100%, 75% and 50% operational mode respectively. The reading were used to calculate the heat losses though the hydrogen coolers, bearing oil, shaft seal oil and by radiation and convention through the generator hood to the atmosphere.

The summary of the process is as follows:

1) Reading of the mass flow rate and temperature elevation of the flow of hydrogen through the rotor.

2) Reading of the water flow rate and temperature elevation of the flow of water through the stator.

3) Reading of the mass flow rate and temperature elevation of the bearing oil.

4) Reading of the mass flow rate and temperature elevation of the shaft seal oil.

5) Measuring of the heat losses through radiation and convention to the atmosphere through the generator hood.

The result of the Heat Balance Test Process was compared to that of the Work Test Process to establish the reliability of the test results using the Hitachi result as a reference point. The result was also used to deduce if there has been any significant deterioration in the efficiency of the plants.

2.1. Temperature Measurement of Generator Hood Surface

The temperature of the generator hood was measured by an infra-red thermometer to determine the loss of heat by radiation and convention to the atmosphere through the generator's hood surface. The generator hood which is semi-circular in shape was divided into nineteen equal parts and the temperature taken at each point. The average temperature was then taken for the generator at the various loads.

2.2. Procedure for Calculating Heat Losses

1) The Heat losses in hydrogen coolers

a) Heat losses in hydrogen cooler $A = Q_1 = M_{WA} \times CP_{WA} \times \theta_{WA}$ (kW) $\qquad$ (1)

b) Heat losses in hydrogen cooler $B = Q_2 = M_{WB} \times CP_{WB} \times \theta_{WB}$ (kW) $\qquad$ (2)

c) Heat losses in hydrogen cooler $C = Q_3 = M_{WC} \times CP_{WC} \times \theta_{WC}$ (kW) $\qquad$ (3)

d) Heat losses in hydrogen cooler $D = Q_4 = M_{WD} \times CP_{WD} \times \theta_{WD}$ (kW) $\qquad$ (4)

where M_W is mass flow rate (kg/s) of water through the cooler, CP_W is the specific heat capacity of water (kJ/kg·K), $\theta_W \mathbb{F}$ is the temperature difference across the cooler (°C).

2) Heat losses through bearing oil

a) Heat losses in bearing oil (turbine side)

$$Q_5 = M_T \times CP \times \theta_T \ (\text{kW}) \qquad (5)$$

b) Heat losses in bearing oil (collector ring side) $Q_6 = M_C \times CP_T \times \theta_c$ (kW) $\qquad$ (6)

3) Heat losses through shaft seal oil (turbine Side) $Q_7 = M_{ST} \times CP_T \times \theta_{st}$ (kW) $\qquad$ (7)

Heat losses through seal oil (collector ring side) $Q_8 = M_{SC} \times CP_T \times \theta_{sc}$ (kW) $\qquad$ (8)

where M_T, M_C, are the mass flow rate (kg/s) of bearing oil at the turbine and the collector ring side respectively, M_{ST}, M_{SC} are the mass flow rate (kg/s) of seal oil at the turbine and collector ring side respectively.

4) Heat losses by convention and radiation $Q_9 = h \times A \times (t_s - t_a)$ (kW) $\qquad$ (9)

where h is the combined coefficient of heat transfer from generator surface, A is the area of generator surface in contact with air (m^2), t_s is the generator surface temperature (°C), t_a is the ambient air temperature (°C).

Generator shaft power input = power generated + total losses

$$\text{Efficiency} = \frac{\text{Power generated} \times 100}{\text{Power input}} \qquad (10)$$

3. Test Data

3.1. Balance Test Data

Various readings taken from the turbo-generator supervisory and monitoring equipment areindicated in the tables below (see **Tables 1-6**) for the various operating conditions. The readings include flow rates and temperature readings for bearing oil, seal oil and water flow in hydrogen coolers.

Unit I (Turbo-Generating Set 1)
Unit 4 (Turbo-Generating Set 4)
Unit 6 (Turbo-Generating Set 6)

3.2. Guaranteed Test Data

Table 5 indicates the calculated losses for the guaranteed test at the operating temperature for generator operating load condition of 221.2 MW at Maximum Continuous Rating (MCR), which is approximately 100% load.

3.3. Material Properties

1) Specific heat capacity of water at the mean temperature = 4.179 kJ/kg·K.
2) Specific heat capacity of bearing oil = 1.881 kJ/kg·K.
3) Combined coefficient of heat transfer from generator surface to air = 12.40 W/m^2.
4) Area of contact of generator surface with air = 60.7085 m^2.
5) Ambient air temperature = 29°C.

Table 1. Flow rates and temperature reading on unit 1.

S/N	Item	Unit	Operation condition		
	Load	MW	219	163	111
i.	Bearing oil inlet temp.	°C	47.50	47.50	47.00
ii.	Bearing oil outlet temp	°C	5		
	Turbine side (TBS)	°C	54.80	54.00	53.00
iii.	Collector ring sides (CLRS)	°C	54.00	53.50	52.50
iv	Bearing oil flow rate (TBS)	kg/s	3.83	3.20	2.62
iv	Bearing out flow rate (CLRS)	kg/s	3.83	3.20	2.62
B	Hydrogen coolers (water flow rate)	kg/s			
i.	Cooler A	kg/s	22.22	19.03	15.57
Ii	Cooler B	kg/s	22.22	19.02	15.56
iii.	Cooler C	kg/s	22.22	19.02	15.56
iv.	Cooler D	kg/s	22.20	19.00	15.43
v	Water inlet temp. (Cooler A-D)	kg/s	34	34	34
vi	Water outlet temp.	°C			
	Cooler A	°C	42.30	41.50	41.00
	Cooler B	°C	41.00	40.00	39.50
	Cooler C	°C	42.00	41.00	40.50
	Cooler D	°C	42.00	41.34	40.50
C	Shaft seal oil	°C			
i	Seal oil inlet temp.	°C	44.00	44:00	44.00
ii	Seal oil outlet temp (TBS)	°C	66.00	63.50	61.50
iii	Seal oil outlet temp(CLRS)	°C	64.50	62.00	60.50
iv	Seal oil flow rate (TBS)	kg/s	1.12	0.972	0.764
v	Seal oil flow rate (CLRS)	kg/s	1.11	0.972	0.764
T_M	Mean generator surface temperature	°C	54	53	49

Table 2. Flow rates and temperature reading on unit 4.

	Item	Unit	Operation condition		
	Load	MW	219	164	110
A	Bearing oil				
i	Bearing oil inlet temp	°C	47.00	47.00	47.00
ii	Bearing oil outlet temp (CLRS)	°C	54.50	54.00	52.00
iii	Bearing oil outlet temp (TBS)	°C	56.00	55.00	53.00
iv	Bearing oil flow rate (TBS)	kg/s	3.83	3.20	2.62
v	Bearing oil flow rate (CLRS)	kg/s	3.83	3.20	2.62
B					
	Water flow rates	kg/s			
i	Cooler A	kg/s	22.21	19.04	15.55
	Cooler B	kg/s	22.21	19.02	15.54
	Cooler C	kg/s	22.22	19.03	15.56
	Cooler D	kg/s	22.21	19.01	15.47
ii	Water inlet temperature Cooler A-D	°C	34	34	34
iii	Water outlet temp Cooler A	°C	42.50	40.00	40.00
	Cooler B	°C	41.50	40.00	39.00
	Cooler C	°C	42.50	41.00	40.50
	Cooler D	°C	41.50	41.00	40.00

Continued

C	Shaft seal oil				
i	Seal oil inlet temperature	˚C	43.50	44.50	44.00
ii	Seal oil outlet temp (TBS)	˚C	66.00	64.00	62.50
iii	Seal oil outlet temp (CLRS)	˚C	65.00	63.00	61.50
iv	Seal oil flow rate (TBS)	kg/s	1.13	0.975	0.764
v	Seal flow rate (CLRS)	kg/s	1.12	0.975	0.764
T_M	Mean generator surface temperature	˚C	53.00	50.00	47.85

Table 3. Flow rates and temperature reading for unit 6.

	Item	Unit	Operation condition		
	Load	MW	220	166	110
A	Bearing oil				
i	Bearing oil inlet temperature	˚C	48.00	48.00	48.50
ii	Bearing oil outlet temperature (TBS)	˚C	56.50	55.00	53.50
iii	Bearing oil outlet temperature (CLRS)	˚C	55.50	54.00	52.50
iv	Bearing oil flowrate (TBS)	kg/s	3.84	3.21	2.63
v	Bearing oil flowrate (CLRS)	kg/s	3.84	3.21	2.63
B					
	Water flow rates	kg/s			
	Cooler A	kg/s	22.24	19.05	15.44
i	Cooler B	kg/s	22.24	19.04	15.45
	Cooler C	kg/s	22.24	19.05	15.48
	Cooler D	kg/s	22.24	19.01	15.20
ii	Water inlet temperature A-D	˚C	34.00	34.00	34.00
	Water outlet temp.				
	Cooler A	˚C	42.00	41.00	40.50
iii	Cooler B	˚C	42.00	41.50	40.50
	Cooler C	˚C	43.00	41.00	40.00
	Cooler D	˚C	42.50	41.00	40.00
C	Shaft seal oil				
i	Seal oil inlet temperature	˚C	44.50	44.00	44.00
ii	Seal oil outlet temp (TBS)	˚C	66.50	64.00	61.50
iii	Seal oil outlet temp (CLRS)	˚C	65.00	63.50	60.00
iv	Seal oil flowrate (TBS)	kg/s	1.103	0.981	0.792
v	Seal flowrate (CLRS)	kg/s	1.100	0.981	0.792
T_M	Mean generator surface Temperature	˚C	53.00	50.47	47.21

Table 4. Total water flow rates for unit 1, unit 4 and unit 6.

Load	Total water flowrates (kg/s)		
	Unit 1	Unit 4	Unit 6
50%	62.12	62.12	61.57
75%	76.07	76.10	76.15
100%	88.86	88.85	88.96

3.4. Heat Balance

Tables 7-9 indicate the heat losses through the bearing oil, seal oil, hydrogen coolers and losses through radiation and convention to the atmosphere. The operating conditions of the turbo-generators are also indicated in percentage of their installed capacity to specify their output in correlation with the losses from the plants.

The efficiencies of the plant at various rated condition are also indicated.

Table 5. Generator efficiency at rated conditions.

Description	Operating temp (°C)	Power loss (kW)
1. Armature copper loss	95	329
2. Rotor copper loss	95	758
3. Core loss		501
4. Stray load loss		665
5. Windage and friction loss at rated hydrogen pressure		502
6. Excitation loss		76
7. Total generator loss		2831
8. Efficiency (%)	98.73	

Source: Egbin Thermal Station Operation Manual by PHCN [15].

Table 6. Summary of generator losses and efficiency.

	Unit				
1. Operating condition (MCR)	%	100	75	50	25
2. Total generator loss	KW	2831	2156.3	1629.8	1289.0
3. Generator efficiency	%	98.73	98.71	98.54	97.71

Table 7. Heat balance for unit 1.

	Load	Unit	Operating conditions (losses)		
1	Load	MW	219	163	111
	Load	%	99.55	74.09	50.45
A	Bearing oil losses				
I	Turbine side (TBS)	kW	57.63	39.125	29.569
II	Collector ring side(CLRS)	kW	46.83	36.115	27.105
	Total bearing oil losses	kW	104.46	75.240	56.674
B	Losses through H₂ coolers				
I	Cooler A	kW	770.72	596.448	455.469
II	Cooler B	kW	649.710	476.907	357.639
III	Cooler C	kW	789.290	556.392	422.664
IV	Cooler D	kW	742.190	555.807	419.133
	Total cooler losses	kW	2951.910	2185.554	1654.905
	Losses through seal oil				
C	Turbine side (TBS)	kW	46.350	35.652	25.149
	Collector ring side (CLRS)	kW	42.800	32.910	23.719
	Total seal oil losses	kW	89.150	68.562	48.868
D	Losses by radiation and convention	kW	18.82	18.069	15.075
E	Total losses	kW	3164.34	2308.335	1777.522
F	Turbo-generator shaft power input		222,164.34	165,308.335	112,775.522
G	Generator efficiency	%	98.60	98.60	98.42

Table 8. Heat balance for unit 4.

S/N	Items	Unit	Operating conditions (losses)		
1	Load	MW	219	164	110
	Load	%	99.55	74.54	50
A	Bearing oil losses				
i	Turbine side (TBS)	kW	64.838	64.838	29.56
ii	Collector ring side (CLRS)	kW	54.032	54.032	24.641
	Total bearing oil losses	kW	118.87	118.870	54.210
B	Losses through H_2 coolers				
i	Cooler A	kW	788.933	596.761	389.901
ii	Cooler B	kW	696.117	596.134	324.708
iii	Cooler C	kW	789.286	596.444	390.151
iv	Cooler D	kW	696.117	556.100	387.895
v	Total cooler losses	kW	2970.453	2345.442	1492.656
C	Losses through seal oil				
i	Turbine side (TBS)	kW	47.82	35.765	26.238
ii	Collector ring side (CLRS)	kW	45.294	33.929	25.149
iii	Total seal oil losses	kW	93.118	69.694	51.387
D	Losses by radiation and convention	kW	18.069	15.808	14.190
E	Total losses	kW	3200.51	2479.448	1612.442
F	Turbo-generator shaft power input	kW	2,222,000.51	166,479.448	111,612.442
G	Generator efficiency	%	98.56	98.56	98.55

Source: PHCN, Egbin Design Manual [16].

Table 9. Heat balance for unit 6.

S/N	Item	Unit	Operating conditions (losses)		
1	Load	MW	220	166	110
	Load	%	100	75.45	50
A	Bearing oil losses				
i	Turbine side (TBS)	kW	61.39	42.266	27.209
ii	Collector ring side (CLRS)	kW	54.173	36.228	22.262
	Total bearing losses	kW	115.569	78.494	49.147
B	Losses through H_2 coolers				
i	Cooler A	kW	789.998	557.269	419.404
ii	Cooler B	kW	743.193	596.761	419.676
iii	Cooler C	kW	788.933	557.270	452.836
iv	Cooler D	kW	742.524	556.069	381.125
	Total cooler losses	kW	3064.649	2267.340	1673.041
C	Losses through seal oil				
	Turbine side (TBS)	kW	45.644	36.905	26.071
	Collector ring side (CLRS)	kW	43.959	35.983	23.836
	Total losses in seal oil	kW	89.603	72.888	49.907
D	Losses by radiation and convention	kW	18.067	16.162	12.955
E	Total losses		3287.888	2434.888	1785.05
F	Turbo-generator shaft power input	kW	223,287.88	168,434.940	111,785.374
G	Generator efficiency	%	98.57	98.55	98.40

3.5. Summary of Results

Table 10 contains the summary of the results of the efficiencies of the turbo-generators at various rated load conditions for the heat balance test process and the guaranteed test.

4. Analysis of Results

Tables 7-9 indicate the test result of the efficiencies of the turbo-generators. For unit 1, the efficiency is 98.60%, 98.60% and 98.42% at 99.55%, 74.09% and50.45% percent load respectively. For unit 4 the efficiency is 98.56%, 98.56% and 98.55% at 99.55%, 74.54% and 50% load respectively. While for unit 6 the efficiency is 98.57%, 98.55% and 98.40% at 100%, 75.45% and 50% load respectively.

In comparison, the result of the present test shows a very good correlation with the guaranteed efficiency test result of the turbo generator using the work test process. This is represented in **Figure 2**, which is a plot of efficiency versus load for the various units and test processes. This agreement in the test result was established in spite of the fact that different approaches were used in both test processes. The slight difference in the test result for the various units may be due to the difference in procedure.

Tables 7-9 and **Figure 2** show that there is consistency of result between the efficiency values for the various units using the heat balance test process in the efficiency test. Thus, there is no significant variation of the efficiency value between the units as they fall within the same range. The plots of efficiency versus load for the various units follow the same pattern. Plots of load versus losses for the various plants follow the same pattern. This is represented in **Figure 3**.

Consequently, it is established that the test result are accurate and reliable. It could also be inferred that with good maintenance culture the test result from one unit could positively reflect the test results for other units. A close examination of **Figure 2** and **Figure 3** show that both the efficiency and losses from the turbo-generator cooling system increases with increase in load. From **Figure 4** it can be deduced that the efficiency of the cooling system depend to a large extent on the effectiveness of the hydrogen cooling system since over 93% of the

Table 10. Summary of test results.

		Unit 1	Unit 4	Unit 6	
Heat balance test process	Eff. (%)	98.6	98.6	98.42	
	Load (kW)	219	163	111	
	Eff. (%)	98.56	98.56	98.55	
	Load (kW)	219	164	110	
	Eff. (%)	98.57	98.55	98.4	
	Load (kW)	220	166	110	
Guaranteed test	Eff. (%)	98.73	98.71	98.54	97.71
	Load (kW)	220	165	110	55

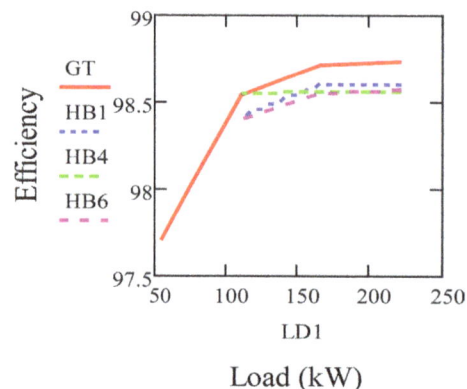

GT: Guaranteed Test, HB: Heat Balance Test, LDI: Load

Figure 2. Plot of efficiency versus load for heat balance and guaranteed test results.

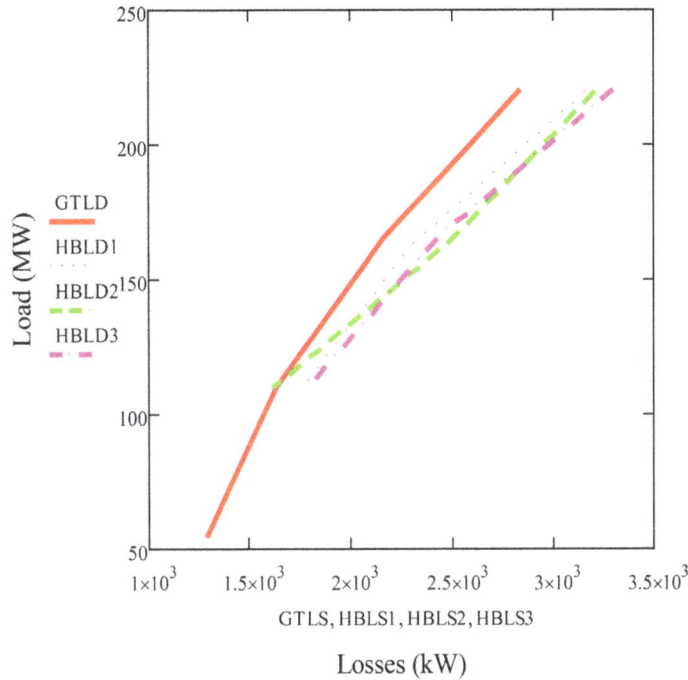

GTLD: Guaranteed Test Load, HBLD: Heat Balance Test Load,
GTLS: Guaranteed Test Losses, HBLS: Heat Balance Test Losses.

Figure 3. Plots of load versus losses for heat balance and guaranteed test results.

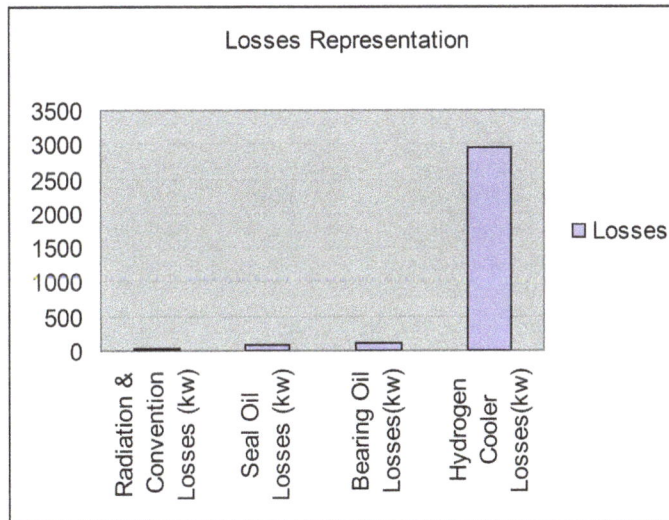

Figure 4. Chart of losses for unit 1.

heat losses in the turbo-generator is removed through this medium. An ineffective hydrogen cooling system will lead to the accumulation of heat in the turbo-generator, raising the temperature of parts and component to levels which could easily lead to their deterioration, breakdown and distortion of shapes. Thus, the efficiency, durability and availability depend on the efficiency of the hydrogen cooling system.

From the tables (**Tables 7-9**) of flow rates and temperatures readings on the supervisory equipment on the turbo-generators for the same item for corresponding loads, the range of the values of the readings are the same and show slight difference to two places of decimal. This is a pointer to the uniformity of the readings and the accuracy of the test process.

Table 11 shows a summary of losses for turbo-generator unit 1.

Table 11. Summary of losses for unit 1.

Radiation & convention losses (kW)	18.82
Seal oil losses (kW)	89.15
Bearing oil losses (kW)	104.46
Hydrogen cooler losses (kW)	2951.91

5. Conclusions

The heat balance test to determine the efficiency of the turbo-generators was done by carrying out an energy balance on the machine systems and the efficiency determined from the ratio of output power to input power to the generator shaft. The efficiency results for unit I were 98.60% at 100% load, 98.60% at 75% load, 98.42% at 50%, unit 4 had 98.56% at 100% load, 98.56% at 75% load and 98.55% at 50%. While those of unit 6 were 98.57% at 100% load, 98.55% at 75% load and 98.40% at 50% load.

Careful observation of the results shows that there exist very good harmony and consistency between the test results using the heat balance test process for the three units. Thus, there is no significant variation of the efficiency value between the units as they fall within the same range. The plots of efficiency versus load for the various units follow the same pattern. Plots of load versus losses for the various plants follow the same pattern also. This high degree of harmony and consistency is also established between the results of the heat balance test results and guaranteed test results. Thus, parallel correlation is hereby established for the three units of turbo-generators in the determination of their efficiencies using the heat balance test process. The heat balance test process is an effective tool to monitor turbo-generator's performance and efficiency profile which serve as a good basis to indicate deviation and deterioration in plants performance.

Acknowledgements

The authors acknowledged the invaluable assistance of the management of Egbin Thermal Station, Lagos and Mechanical Engineering Department, University of Benin, Benin City for their cooperation during the study. Special appreciation goes to Prof L, A. Salami for his support and counseling in the study.

References

[1]   Hattori, K., Ide, K., Goto, F., Semba, A. and Watanabe, T. (2002) Sophisticated Design of Turbine Generator with Inner Cooler Ventilation System. *Hitachi Review*, **51**, 148-152.

[2]   Andreev, A.M., Azizov, A.S., Bezborodov, A.V., Papkov, A.A. and Pak, V.M. (2011) Investigation of Possibility of Developing High Efficiency Conducting System of Electrical Insulation for Turbogenerators with Air and Hydrogen Cooling. *Russian Journals of Electrical Engineers*, **82**, 180-183. http://dx.doi.org/10.3103/S106837121104002X

[3]   Nag, P.K. (2001) Power Plant Engineering. 2nd Edition, McGraw Hill, New Delhi.

[4]   Kostenko, M. and Piotrovsky, L. (1977) Electrical Machines: Alternating Current Machines. MIR Publisher, Moscow.

[5]   Muramatsu, S., Hattori, K., Takahashi, K., Nakahara, A. and Iwashige, K. (2005) Development of 250-MVA Air-Cooled Turbine Generator. *Hitachi Review*, **54**, 121-125.

[6]   Kakimoto, T., Hattori, K., Fugane, K., Mae, H., Kazuhiko, T.K. and Iwashige, K. (2008) High Efficiency and Low Noise Air-Cooled Turbine Generator "GH1550A". *Hitachi Review*, **57**, 285-291.

[7]   Higgins, R.L. (1995) Maintenance Engineering Handbook. 5th Edition, McGraw Hill, New York.

[8]   Gurevic, E.I., Lyamin, A.A. and Shelemba, I.S. (2010) Measurement of the Temperature of a Stator Winding with Fibre Optics Sensors Temperature in Bench Test of a Turbogenerator. *Power Technology and Engineering*, **44**, 249-245.

[9]   Hattori, K., Okabe, H., Ide, K., Kobashi, K. and Watambe, T. (2004) Performance Evaluation of a 250 MVA Class Air Cooled Turbogenerator. CIGRE Session, 1-6.

[10] Okabe, H., Onoda, M., Hatori, K., Watambe, T., Morooka, H. and Higashimura, Y. (2003) Development and Performance Evaluation of a High-Reliability Turbine Generator. *Hitachi Review*, **52**, 89-95.

[11] Marrin, H. and Klayto, P.E. (1997) Fundamental Electrical Technology. Addison-Wesley, London.

[12] Zubair, A.M. and Olarewaju, S.O. (2014) Production Index of Electricity Generation and Consumption in Nigeria. *International Journal of Engineering and Science*, **3**, 11-17.

[13] Odungide, M. (2014) Power Statistics in Nigeria. Academia. http://www.academia.com

[14] Audu, M.L., Zubair, J., Anakhuagbon, A. and Ajaino, O.K. (2012) Efficiency Testing of Turbo-Generators: The Heat Balance Approach. *International Journal of Engineering Innovations*, **4**, 94-102.

[15] PHCN (1978) Egbin Power Station Operation. Manual Vol. 4, Generator and Auxiliaries.

[16] PHCN (1978) Egbin Thermal Station Design Manual.

8

Energy Integration in South America Region and the Energy Sustainability of the Nations

Miguel Edgar Morales Udaeta, Antonio Gomes dos Reis, José Aquiles Baesso Grimoni, Antonio Celso de Abreu Junior

GEPEA/EPUSP, Energy Group of the Department of the Electrical Energy and Automation Engineering/ Polytechnic School of the University of São Paulo, São Paulo, Brazil
Email: udaeta@pea.usp.br

Abstract

The objective of this manuscript is to analyze relation involving the energy sector and socioeconomic growth and, then, contextualize the process of energy integration within the development policies in South America. The methodology considers data related to the world's economy and energy consumption and energy integration policy in countries and regions; and, South America's energy potential and the energy integration process. Results show that despite the political and institutional difficulties involving the process, energy integration can bring a lot of benefits for countries development. The process of energy integration in South America is divided in three moments, but in both periods the transnational energy projects were restricted, mostly, by a bilateral plan and the creation of physical links in a region. In the 21th century's context, it should be noted Brazil's participation which has been consolidated as a lead country in this process, and, also the IIRSA (Initiative for the Integration of Regional Infrastructure in South America, nowadays renamed as COSIPLAN) like the main initiative in energy integration in the continent, in a context where the projects are no longer limited to traditional economic blocs. Finally, we note a lack of consensus in defining a comprehensive model of integration and solving asymmetries both within countries and between them.

Keywords

Energy Integration, Energy Planning, Energy Resources, Regional Geo-Energy, South America, Energy Policy, Development

1. Introduction

Despite the economic integration process arising in Europe (including the energy integration), the related discussions soon spread worldwide, leading initiatives in other regions, including South America. From the second half of the 20th century, some economic integration mechanisms have been developed in the South-American region such as the creation of the Andean Community of Nations (CAN), the Southern Common Market (Mercosur) and the Union of South American Nations (UNASUR), plus some bilateral initiatives geared to the use of shared energy resources or trade them.

In this last, century has noticed a significant increase in the number of energy projects in South America, largely associated with the Initiative for the Integration of Regional Infrastructure in South America (IIRSA), the resulting economic growth in the region and, thus, the increase of demand energy. Indeed, studies of the International Energy Agency [1] and the World Energy Council [2], show that the energy demand of developing countries has increased significantly—due to the considerable growth of their economies—a phenomenon which also includes America South and specifically Brazil, which in 2011 occupied the 6th position in the world ranking of Gross Domestic Product (GDP) [3] and was seventh country to consume more energy in the world [4].

Energy and Development

The availability of energy is necessary for human development throughout history base, so that the use of different energy sources is the thread of history man stuff having made possible the two major changes in their relationship with nature: Neolithic Revolution and the Industrial Revolution [5].

In fact, when looking at **Figure 1**, below, one sees that the energy needs of society monitor the evolution and development of mankind. From an energy consumption of about 2000 kcal per day, which characterized the primitive man, energy consumption increased by 1 million years to 230,000 kcal per day, taking into account the consumption pattern of the so-called technological man [6].

The invention of the steam engine (a mark of the Industrial Revolution, which started in the 18th century in England) created the technical basis for the development of new forms and sources of exploitation and use of energy, replacing human labor with machines and subsequently developing means of transport [7]. By enabling the large-scale production and fast shipping of goods, the Industrial Revolution spurred the development of capitalism, specifically the economic development of industrialized nations. Moreover, this process resulted in the gradual increase in world energy demand (especially in developed countries) and, thus, for new sources and ways to harness and convert energy. This historical process shows a dialectical relationship between energy and development, in which the ability to use and power is at the same time, therefore the level of technical and eco-

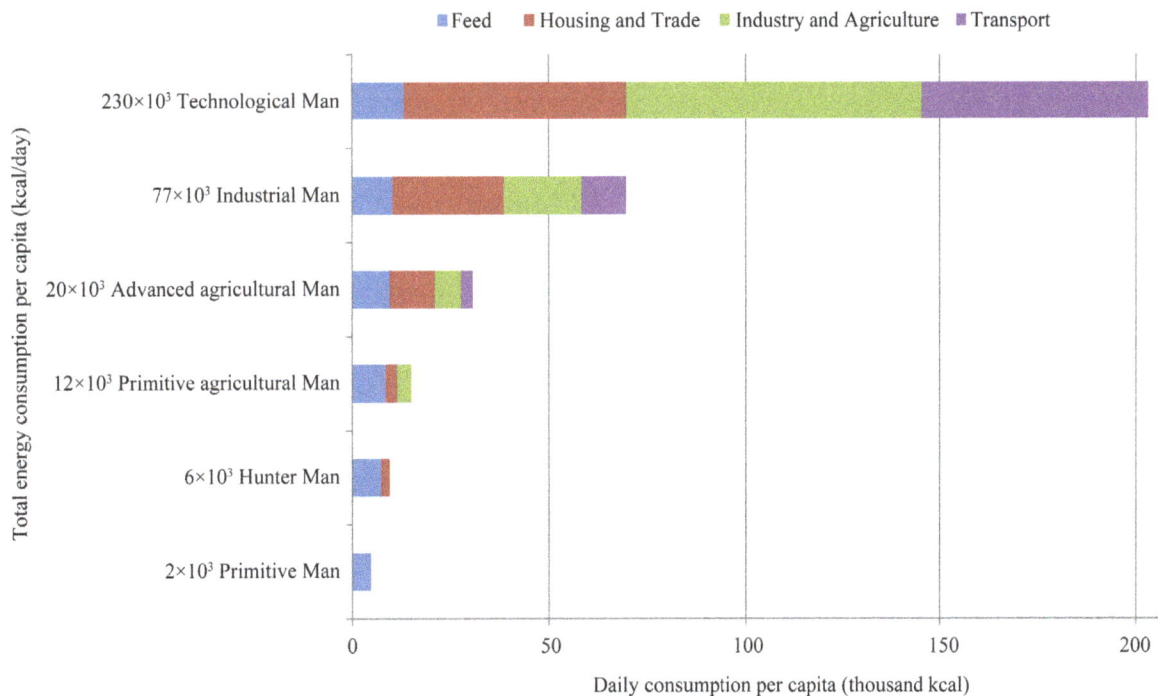

Figure 1. Evolution of human energy consumption troughout history (based on [6]).

nomic development of a society, but also the catalyst of this same development.

The relationship between energy and development is evidenced in **Table 1** and **Table 2**.

Indeed, by analyzing countries' data, related to GDP and energy consumption, we notice that countries that have higher energy consumption tend to be those with higher GDP. Of the 10 countries with the highest GDP in the world, only Italy and the UK do not appear among the 10 largest energy consumers—showing the relationship between economic development and energy consumption.

However, this relation between GDP and energy consumption, measured by calculating energy intensity—where it calculates the energy required to produce one unit of GDP (in Dollar case)—has been changing in recent times. Thus, OECD countries show a GDP growth without corresponding to a proportional increase in energy consumption, reducing their energy intensities. According to the World Energy Council (2004), this phenomenon is due to energy efficiency policies adopted in these countries, as well as changes in their economic structures or the energy matrices [2].

On the other hand, for non-OECD countries energy demand has increased considerably as a consequence of population, economic, urban and industrial growth. According to studies by the IEA (2010) [1], between the years 2008 and 2035 energy consumption in non-OECD countries is expected to increase 64%, while in OECD the growth is only 3%. Thus, it is expected that the share of developed countries in world energy consumption countries, which declined from 61% in 1973 to 44% in 2008, decrease to only 33% in 2035.

The same studies show that the OECD countries have, on average, an annual growth of primary energy demand of 0.1% (and in some cases, like Japan, rates are negative), while China and India have an annual increase of over 2.0%. These two countries are those that contribute most to the increase in global energy consumption. According to the IEA studies, the increase in primary energy demand in China in the period 2008-2035 is expected to be 75%, corresponding to 36% of global growth and resulting, therefore, in 22% of the total demand

Table 1. Ranking of the energy comsumption by countries (2011) [4].

Country	Consumption (mtoe)
1° China	2648
2° USA	2225
3° India	759
4° Russia	725
5° Japan	469
6° Germany	317
7° Brazil	268
8° Canada	266
9° South Korea	257
10° France	257

Table 2. Ranking of global GDP (2011) [3].

Country	GDP (millions USD)
1° USA	2648
2° China	2225
3° Japan	759
4° Germany	725
5° France	469
6° Brazil	317
7° United Kingdom	268
8° Italy	266
9° Russia	257
10° Canada	257

on the planet. India is the second largest contributor to the increase of energy consumption, accounting for 18% of it and having the highest annual growth rate (over 3.0%). Not coincidentally, these two countries are those whose studies point to have the highest rates of growth of GDP (forecast 5.7% and 6.4%, respectively). Brazil is expected to be the third country in the world regarding the growth rates of GDP and energy demand in the period 2008-2035 settling with one of the largest economies in the world and configured as one of the largest energy consumers in the world.

Still, besides being associated with the issue of economic development, we must not forget the social function involving energy production, since the primary purpose of energy services is to meet the needs of human beings, which means production and consumption, means to achieve it [8]. Thus, countries with higher HDI (Human Development Index) tend to be those with higher energy consumption [9], with an income factor that most influences this consumption (among the indicators that make up the HDI) [6]. Still, it must be emphasized that the low power consumption is not the only cause of poverty and development, however, a good indicator of its causes—as, for example, low levels of education and poor health systems.

Once established the relationship between energy and human development, it is important to highlight that the use of energy, by being associated with the technical development and the appropriation of space, notably affects the spatial configuration and, therefore, the environment, resulting in socio-cultural changes and significant demographic. Thus, projects involving energy production should be considered in all stages of the production chain, the social impacts of these types of development [9]. Otherwise, it will be creating a great contradiction, since above all, energy production is primarily intended to provide the satisfaction of human needs.

2. Essentials of Energy Integration

In the current context of increasing demand for natural resources, and the prospect of depletion of many of these, energy planning becomes very important and includes the research and development of alternative sources of energy production technologies that are associated with renewable resources and causing an environmental impact and minimum share.

As noted in the introduction, the energy planning involves access to energy resources. Due to the fact that they are unevenly distributed across the planet, access to them is the subject of disputes filed by various interests, being a matter of great geopolitical importance to a state.

Discussing the dependence of developed for natural resources located in other regions of the world, Hobsbawm (2007) [10] points out that one of the motivations of European imperialism over other regions of the world, occurred between the years 1875 and 1914 was the technological development that generated the need for "raw materials, due to weather or geological chance would be found exclusively or deep in remote places" (p. 96) supply. In many cases, even having plenty of resources in their territories, developed countries seek to exploit them in other regions. A clear example would be the American experience with oil after the Second World War, when the country encouraged their companies to exploit the oil fields of the Middle East, preserving located in its territory, economic, national security strategy [11]. Harvey (2011) [12] discusses the US interests in the Middle East associating with the fact that the region has the largest oil reserves in the world. Thus, by controlling the region, the United States also controls the access to this feature and with it, the global oil Market [12].

Several other authors discuss the relationship between energy security and military action in countries outside their national territories. Triola (2008) [13], an employee of the US Navy, maintains Harvey's assertions (2011) by arguing that the energy supply of the United States is a matter of national security and, therefore, also involves military interests. Nagy (2009) [14], in turn, suggests the militarization of energy security as a responsibility of the Organization of the North Atlantic Treaty (NATO), leaving her secure supply of energy resources.

These kinds of assertions show us that the access to energy resources involves different kinds of interests between countries that can be conflicting. Thus, as a reasonable alternative, it is considered that policies aimed at energy integration can meet harmoniously interests involved. The central idea of the energy integration is noted the contribution that economic and energy sectors in each country can the social and economic development process within the framework of regional integration [9] [15]. By enabling the commercialization of energy resources, or electricity itself, based on multilateral agreements, energy integration can provide a more reliable and efficient supply to large consumers of energy, also bringing economic gains for countries that sell their energy resources and its surplus electricity. In the long term, is optimized energy production, while taking advantage of the diversity resulting from connection to energy sources from neighboring countries, eliminating the dependence on a single source of energy and reducing supply costs. Also, the creation of economic blocs and energy strengthens the integrated region, leveraging the commercial, political, social and cultural relations between its members.

Moreover, despite the potential benefits related to cross-border energy integration, there are many elements that hinder their achievement, they order being political, technical, economic and environmental.

One of the main difficulties associated with the implementation of integration projects refers to the articulation of rules and congruent with the stimulus to investment and energy interdependence policies. It involves a number of agreements, targets and regulations that involve complex legal issues facing opening markets and thereby enabling the creation of rules to facilitate transactions and equity investments (state, private and national is required private multinational). This process involves the countries internal political issues—related to the approval and acceptance of laws and internal projects involving diverse interests within the nation—as well as elements associated with the foreign policy of each state and its geopolitical interests in the region.

With regard to differences of interests among countries in the South American case, one can use the question as an example of Bolivian gas, in which Bolivia nationalized refineries belonging to Petrobras, claiming that the contracts had been established the wounded interests of the Bolivian nation. Another example relates to historical differences between Chile and Bolivia, involving Bolivia access to the Pacific Ocean as a barrier to agreements between the two countries [16].

Obviously, the larger the number of agents involved in the process, the greater the difficulty in establishing policies of interest to everyone. That's why the most successful experiments were those made them bilaterally arising from projects with strong participation of national states.

From a technical standpoint, the interconnections require an infrastructure with bi-reaching goals—or multi—that includes the participation of all involved and interested. So that the integration process is done in a cohesive manner, it is essential to studies that provide adequate planning be made—with regard to the generation, transmission and distribution of energy as well as the interests and economic returns for the various agents involved in the issue. The greater the need for infrastructure and technical complexity related to the projects become more expensive the same—which implies the need for large investments of money (and, most often, in various financing). In the case of South America, for example, infrastructure integration projects to sizable proportions by both distances, as the natural difficulties imposed by the environment.

Resourcefulness along the World of Energy Integration

A) European Union

Throughout the twentieth century a number of policy initiatives and energy integration have been deployed worldwide, the most successful being developed within the European Union—that is, in a larger context of economic and political integration. Unlike what happened historically in most cases, the European experience has been guided by multilateralism and the creation of supranational regulators.

The first step in this process occurred in 1951, with the signing of the ECSC Treaty, which established the creation of the European Coal and Steel in a context in which the countries of the continent sought to economically rebuild the region after the Second World War. With this treaty, signed by France, Germany, Italy, Netherlands, Belgium and Luxembourg, we sought to integrate the Franco-German production of coal and steel—raw materials essential to industrial activity and the local economy at the time—through the creation of a common market aimed at economic development, job creation and improved quality of life.

In 1957, were signed the Treaties of Rome establishing the European Economic Community (EEC) and the European Atomic Energy Community (Euratom), establishing the creation of a common market on the continent and recognizing the importance and need to develop common energy policies the member countries in the context of regional economic and social development. To overcome the uncertainties related to traditional energy sources, the Member States of Euratom sought on nuclear energy a means to ensure energy security and independence. Thus, according to documents of the European Union itself (2013), "as the cost of investing in energy beyond the means of individual States, the founding members joined together to form the Euratom".

Throughout the following decades, the process of integration into the European continent was deepening, also encompassing the energy sector—seen as crucial to regional socioeconomic development. The main frame of this integration initiative came in 1992 with the Maastricht Treaty, which created the European Union and in which it commits to the creation and development of trans-European networks in the sectors of transport infrastructure, telecommunications and energy. Thus, it is for the authority to "promote the interconnection and interoperability of national networks as well as access to such networks" through the actions of its supranational political bodies [17].

From the descriptions above, it can be seen that the integration of the energy sector is part of European policies since the mid-twentieth century, are subordinate, in turn, to the initiatives of economic and political integration—and therefore cannot be analyzed outside this context. Similarly, we note that, over time, these policies are no longer focused on specific energy sources (coal and nuclear), for, after the Maastricht Treaty, extended to the whole European energy system and further increase use of sources clean and renewable energy—increasingly promoted by the European Union's energy policy, with the aim of reducing emissions of greenhouse gases on the continent—and also the integration into the natural gas supply.

The strategy of integration of renewable sources resulted in a less centralized and diversified system, strengthening the European integrated network. The Policy of 2009 set by the European Parliament and the Council of the European Union, concerning common rules for the internal European market and the unequal terms of trade of electricity between the member states must be overcome by the right of free choice suppliers reassured consumers.

The transmission of electricity on the continent is through the ENTSO-E network, established in July 2009, according to Policy 2009 and composed of 42 operators in 34 countries, with 305.000 km of transmission and 828 GW of generation, to supply and demand of 3400 TWh/year, serving more than 525 million citizens. The purpose of this initiative was to integrate the different operators of the system to European legislation, promoting development through reliable operation, technical and administrative support, and security in meeting the demand of the system. In this context, the intelligent transmission networks (smart grids) send electricity from points of generation to consumers using a monitoring system with digital technology, allowing the integrated use of decentralized energy sources—like solar and wind—by assimilating its entry in periods of wind and sun.

Naturally, the process of integration of the European energy sector also faces obstacles, but in the context of this work concerns us understand how is the process of multilateralism and integration with respect to interconnections and to political and economic technical aspects involved—so compared with other experiences around the world.

B) Africa

Apart from the European experience, energy integration initiatives were implemented in other continents, so that, in most cases, they gave bilaterally and were much less extensive than in the European Union.

In Sub-Saharan Africa, for example, the first cross-border electricity interconnections were deployed around large hydroelectric projects. The first interconnection was built in 1958, with the transmission line of 132 kV connecting the hydroelectric plant of Owens Falls, in Uganda, the capital of Kenya (Nairobi). Built in the Democratic Republic of Congo and interconnected with neighboring countries such as Zambia, Congo and others—later binational hydroelectric, as the Kariba North (on the border between Zambia and Zimbabwe) and the development of the Inga hydroelectric complex fall were built the southern region of the continent [18]. Parts of such linkages were incorporated into the Southern African Power Pool (SAAP), one regional system operating on the continent until 2008, when it began operating the West African Power Pool. A Power Pool is an interconnected system obtained by joining two or more interconnected electric systems—managed as if it were a system—by reallocating the demand and supply of energy and generation capacity, in order to operate more efficiently and secure.

The SAPP was created in August 1995 as an association of electric power enterprises vertically integrated, representing the 12 nations of the Commonwealth of the Southern African Development Community (SADC, for its acronym in English): Angola, Botswana, DRC, Lesotho, Malawi, Mozambique, Namibia, South Africa, Swaziland, Tanzania, Zambia and Zimbabwe [19].

C) North America

Energy integration of North America shall be considered in the context of the Free Trade Agreement (NAFTA)—an agenda of neoliberal policies designed to achieve deeper levels of regulatory, institutional and policy integration that enable the integration of markets the continent [20]. But despite NAFTA establish a "trilateral landmark" for trade in energy and electricity [21] features the experiences of energy integration in the region are, paradoxically, based on bilateral relations, in which the United States appear as large buyer power and energy resources (such as oil and natural gas) from Mexico, and especially in Canada—energy trade between Mexico and Canada is almost non-existent [22].

In fact, 99% of all Canadian energy exports are destined for the United States [23], indicating the absence of trade in energy resources and electricity from Canada to Mexico. Even Mexico is a major oil supplier to the

United States, its energy market is poorly integrated into the US (compared to Canada)—due to political and economic and internal [22] issues.

With respect to the electric integration, specifically, it is less advanced than in the natural gas and oil industries. Even so, there is a trade of electricity in the region, made possible by interconnections linking the United States to Mexico, and especially in Canada. In this context, the United States are characterized as major importers of electricity, while also exporting (to a lesser extent) to its neighbors.

Finally, it is important to note that there are differing opinions regarding the benefits of the integration process in North America. While Doucet (2007) [22] argues that this process is extremely beneficial for Canada, as the United States are the major buyers, Campbell (2007) [20] believes that these relationships undermine the energy security of the country. To justify his point of view, this author takes as argument the fact that, by the rules of NAFTA, Canada can only reduce its exports of oil and gas to the United States in the same proportion that reduce their production to market internal, with this, the country loses the right to reduce its exports to the US, even aiming to prolong the durability of their internal reserves or reduce its imports. For Campbell (2007), the fact that Mexico to impose certain restrictions on the integration model imposed by the United States represents a defense of local energy security , and is therefore beneficial for the country.

3. Energy Integration in South America

3.1. South American Energetic Potential

Because it is a region rich in energy resources, especially oil and water resources, various integration projects in terms of energy already deployed or are in process in South America is worth mentioning that many of these features occur so border (making it complex to manage and operate), highlighting the Andes and the Amazon River.

The Andes is a large mountain range spanning five countries in South America—as well as being a natural border between them—and constitutes a major mineral area in the world: gold, silver, copper, zinc, nickel, iron granite among others. The Amazon River, its waters represent much of the available freshwater in the world. This river originates in Peru, crossing the border with Colombia until you reach the tri-border involving these two countries and Brazil. Not coincidentally, these three countries are those with the potential hydroelectric May in South America, as shown in **Figure 2**.

By observing the above data, it is seen that the total hydropower potential of the American South is 590 GW, so that Brazil is by far the country with the greatest potential, exceeding 250 GW. No wonder that this country is one that has the highest hydroelectric power generation on the continent (403 TWh in 2010), corresponding to

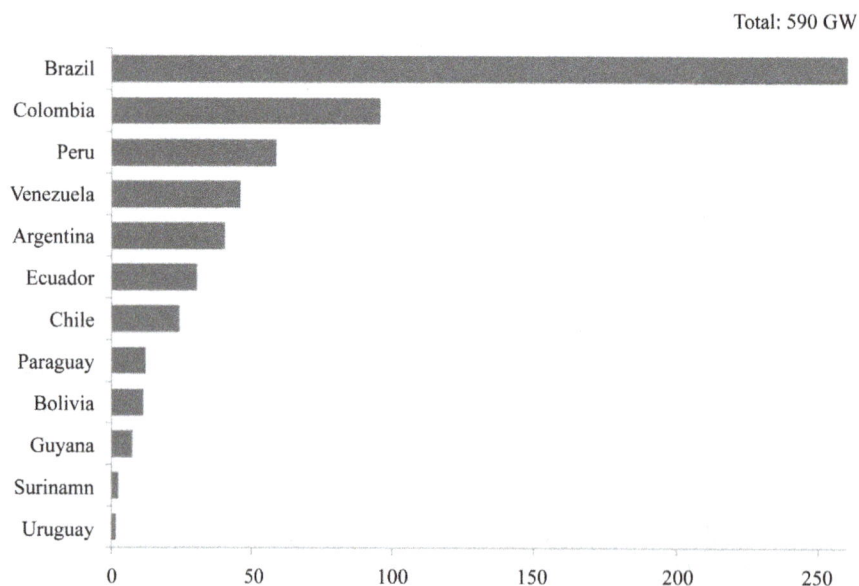

Figure 2. Hydroelectric potential in South America (2010).

11.5% of the world's hydroelectric production, according to **Table 3**. Likewise, it should be highlighted Venezuela, since this country occupies the 9th place ranking producer of hydroelectricity in the world and is the second largest South American producer.

Regarding hydrocarbons should be highlighted reserves of oil and natural gas on the continent. South America has about 4% of global natural gas reserves, so that the majority is located in Venezuela. **Figure 3** shows the distribution of proved reserves of natural gas in the region.

Oil cannot be overlooked when it comes to meeting the South American energy potential. The region has large proven reserves of this resource, highlighting, once again, to Venezuela. According to data from the US Energy Information Administration—an agency linked to the US government—in Venezuelan territory is the second largest proven oil reserves in the world (211.2, billion barrels of oil) smaller only than that of Saudi Arabia. Brazil also deserves mention, being in 15th place in the global ranking of countries with the largest oil reserves. **Figure 4** shows that, after Venezuela and Brazil, Ecuador has the third largest proven oil reserves in South America.

3.2. Regional Integration Process in South America

The process of South American energy integration starts from the mid-twentieth century, by means of natural gas, having two axes of the main action, oriented from the actions of the CAN (Andean Community of Nations)

Table 3. The biggest hydroelectricity producers in the world (IEA, 2010).

Producers	TWh	% of the global installed capacity
China	722	20.5
Brazil	403	11.5
Canada	352	10
United States	286	8.1
Russia	168	4.8
Norway	118	3.4
India	114	3.3
Japan	91	2.6
Venezuela	77	2.2
France	67	1.9
Other Countries	1118	31.7
World	**3516**	**100**

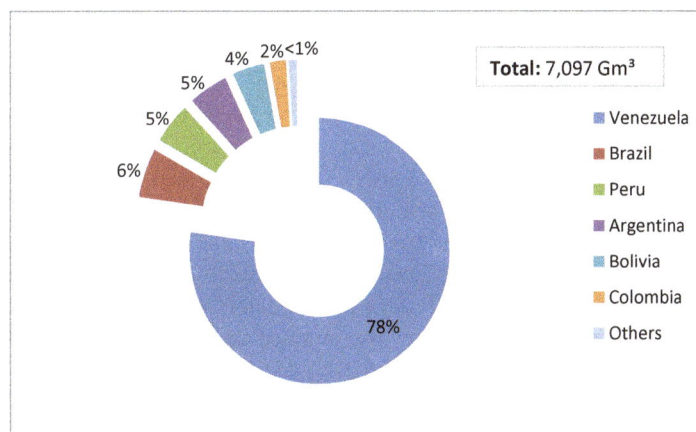

Figure 3. Regional distribution of proved reserves of natural gas in South America in 2010.

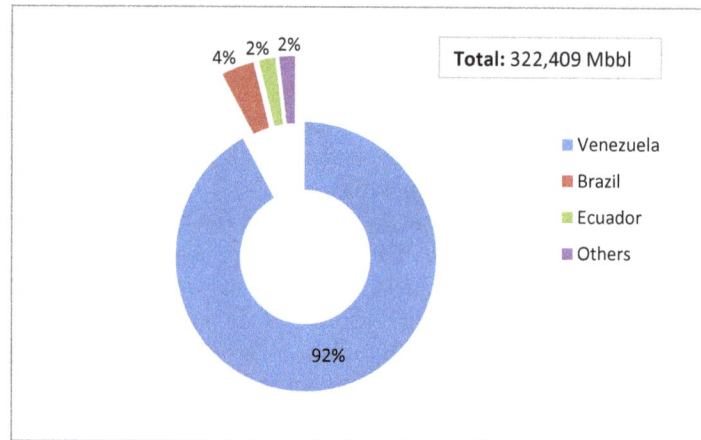

Figure 4. Distribution of proved oil reserves in South America (2010)[1].

and Mercosur [7]. It should be noted that these economic blocks have economic complementation agreements among themselves and with Chile—a country that is not a member of both.

The process of energy integration occurred in the twentieth century can be divided into two stages.

The first, between the early 1970s and the late 1980s, was marked by a great performance of National States in building binational projects. These must be highlighted, the Bolivia - Argentina, hydroelectric plants Salto Grande (Argentina and Uruguay), Itaipu (Brazil/Paraguay) and Yacyretá (Paraguay/Argentina), and transmission lines associated with these binational plants.

The period that occurred after the end of the 1980s was marked by economic and political reforms in neoliberal, resulting in decreased performance of states and the increased private participation in the South American country's economy. Thus, has begun the implementation of projects with varying degrees of participation of private, mixed and public enterprises—involving mainly the hydrocarbons sector. In this period many binational pipelines were constructed, demonstrating the importance of this resource within the energy integration projects in South America[2].

Despite the differences involving the form of state action, it is noticed that, in both periods mentioned above, projects were restricted to the bilateral framework, demonstrating the absence of a regional integration policy indeed.

The political and economic changes on the continent, in the 21th century turned, once again, the landscape of energy integration in the region. First, the election of heads of state of the Left parties made to gather strength in the region's anti-neoliberal and anti-imperialist discourse, changing the logic of financing projects in the region and strengthening the participation of States in the economies of the countries again. Meanwhile, economic growth achieved by regional countries, especially Brazil, resulted in increased energy demand of us. With this, the first decade of this century, there was a tendency to turn that vision bilaterally in order to give a more regionalist and multilateral integration projects to character. It was in this context that was created in 2000, the Initiative for the Integration of Regional Infrastructure in South America (IIRSA), in order to promote the interconnection of transport, telecommunications , energy, oil and gas pipelines.

With the construction of infrastructure related to IIRSA project, and with funding from financial institutions such as the Inter-American Development Bank (IDB), the World Bank, the Andean Development Corporation (CAF) and the Financial Fund for the Development of PrataBasin (FONPLATA), was intended to establish conditions conducive to the development of trade agreements and energy integration. Thus was deposited much confidence in the success of IIRSA, despite the logical difficulties inherent to their success—such as: history of binational conflicts; disparity markets; privilege certain actors, and difficulty of regulatory consistency between countries [9]. But in 2009, IIRSA has been discontinued and its projects were linked to the South American Council of Infrastructure and Planning (COSIPLAN)—linked to the Union of South American Nations (UNASUR). Also, COSIPLAN along with UNASUR, retains the idea of strengthening multilateral relations in South

---

[1]Source: Based on OLADE.

[2]In fact, natural gas is one of the vectors of energy integration on the continent, representing most of the energy matrix of some South American countries such as Argentina, Bolivia and Chile.

America, in order to give greater political support infrastructure integration projects.

The following is a brief discussion of the process of energy integration in South America, with a focus on initiatives both within the economic bloc (Mercosur and CAN), and outside them.

A) Mercosur Region

Mercosur is an economic bloc, created in 1991, consisting of Brazil, Argentina, Paraguay, Uruguay and Venezuela, whose members Bolivia, Chile, Ecuador, Colombia and Peru.

Despite the creation of this economic bloc be linked to the need for expansion of domestic markets and stimulate the circulation of goods and services in the region—without being tied directly to the energy issue—the binational relations among member countries prior to its creation. Since the creation of Mercosur, the energy sector of member countries showed considerable changes. Among them stand out the reform of the role of the state (acting more as a regulator than as an entrepreneur) and consolidation of natural gas as an integrating feature of the region—as all countries in the region have construction projects in the pipeline [9].

Among the transnational projects of power generation worth mentioning the construction of Itaipu, Salto Grande and Yacyretá (already mentioned above) plants, besides the Central Salta—a combined cycle power station, built by a Chilean company to generate electricity from gas natural from Argentina, but without providing energy for this country. We cannot forget that Chile is considered a poor country in terms of ownership of energy resources.

In the case of Itaipu (binational venture involving Brazil and Paraguay, aiming harness the hydroelectric potential of the Paraná River), Brazil was responsible for the setup project and investments for the construction of the plant—in addition to financing the part that would fit Paraguay [7]. In 2008, the plant secured supply 87.3% of all electricity consumed in Paraguay and 19.3% of the demand of the Brazilian Interconnected System. Importantly, Paraguay does not consume all of your energy generated, since it takes only 20% of what is produced to meet its domestic demand for energy, causing the remaining 80% are sold to Brazil.

B) Andean Community of Nations Region

The Andean Community of Nations (CAN) is formed by Peru, Bolivia, Ecuador and Colombia. In the past, Chile and Venezuela also integrated organization but abandoned (at different times) for political and economic mismatches. The region covered by the CAN has a great potential energy—both in terms of hydrocarbons (oil and gas) as regards the hydro, among others. With regard to the oil market, Venezuela, Colombia and Ecuador are configured as suppliers to countries like Brazil, Chile and Peru—since they have oil consumption that exceeds their production, in contrast with Bolivia and Venezuela that have resources that exceed their local demands. You need to highlight the situation of Chile, due to its lack of energy resources, is much interested in integration projects that enable the supply of its domestic energy demand.

According to Castro, Dassie and Delgado (2009) [24], the process of energy integration in the Andean region began in 1969 with the construction of the transmission line Zulia - La Fria, connecting Colombia and Venezuela. Although the energy exchange was not significant due to what they call security problems in the power supply, the authors argue that this project was the first step for the energy integration occurred in the region. The same authors state that the evolution of the process of electrical interconnection between the Andean countries has enabled advances, such as the prediction of building an interconnection between Bolivia and SIEPAC. Udaeta, Burani, Fagá and Oliva (2006) [9] also highlight the role of interconnections in the integration process in CAN, noting that Bolivia appears as a "hinge", because of its possibilities of interconnection with Brazil, Chile, Argentina and Peru.

Besides the physical integration, one cannot ignore the advances that have occurred in political and legal terms to enable the marketing and access to transmission networks between countries. Thanks to that, better expectations around integration projects in the region. Even so, it is of great political, economic and technical complexity to implementation of transnational projects—so that the more countries involved, and the larger the area covered, the greater the difficulty. Antunes (2007) [16] cites the proposal for energy integration taken by Chile in 2007, and that would involve this country along with Bolivia, Ecuador, Peru and Colombia, based on an Andean multilateral agreement aimed at "prevent interference geopolitical discussions on exit to the sea for Bolivia lost"(p. 5).

C) Other Energetic Integration Projects in South America

Obviously, one cannot reduce the process energy integration in South America only to economic blocs in the region. In this sense, the creation of IIRSA, and later the COSIPLAN makes integration projects that go beyond the limits of economic blocs, giving greater geopolitical cohesion to the region are encouraged. In the case of

IIRSA, a portfolio of projects divided into eight sub-regions was taken, clearly transcending the boundaries of economic blocs: the Hub; Capricorn Hub, Amazon Hub; Guianese Shield Hub, Southern Hub, Central Interoceanic Hub; MERCOSUR-Chile, Brazil and Peru-Bolivia.

Throughout its existence, the IIRSA delivered 524 projects, of which 451 (about 86%) belonging to the transport sector and whose investments were received from more than 55 billion dollars. The energy sector received 64 designs, but although they represent 12% of the total, his works surpassed the cost of 44 billion dollars (about 42% of the investment portfolio of IIRSA), due to the complexity and magnitude of the projects. Have the communications sector received only 9 projects, which together cost less than $50 million [25]. By analyzing these data, it is clear that the IIRSA projects favored the transport sector, to the detriment of integration of energy and communication sectors. Thus, we can say that IIRSA fostered a process that is configured much like physical interconnection, than actual integration [26]. This phenomenon is mainly related to building a road network connecting the Pacific and Atlantic interests, in order to facilitate the flow of goods across the continent.

IIRSA-related projects are targets for some criticism. Despite advances in regard to strengthening networks of transport, energy and communication sectors were left in the background. Gudynas [26] questions the interests that were behind the IIRSA project by privileging the physical interconnections (such as roads, waterways and pipelines), and seek not intrinsic to strengthen other aspects integration process—as political, productive and cultural ties. Other limitations are related to IIRSA the lack of progress in the harmonization of sectorial policies and relevant regulations and little consideration given to social and environmental aspects.

As regards the pursuit of multilateralism in the South American integration process, it should be noted the role of Bolivia, which has a role of coordination between the countries of the Southern Cone and the Andean Community. Being a country with large reserves of natural gas and has low power consumption; Bolivia is shaping up as a major supplier of this resource to countries that need to import it—like Chile and Brazil. You can use the example of the Bolivia-Brazil pipeline, named by the authors as a major milestone of the energy integration in South America, the importance of consolidating gas sector in the Brazilian energy matrix.

Energy integration also consolidates Brazil as a major buyer of energy, since, despite having large reserves of energy resources, the country has shown a significant increase in its demand. Therefore, the Brazilian government has increasingly sought to stimulate energy ventures outside its territory. However, in addition to seeking to satisfy its energy demand, there is a clear interest in encouraging the participation of Brazilian capital on projects through funding from its National Bank for Economic and Social Development (BNDES) and the participation of Brazilian construction companies civil.

In this context, the process of energy integration between Brazil and Peru should be highlighted. In 2009, was signed by both countries an agreement for the construction of six hydroelectric plants geared to supply the energy markets of both countries and located in the Amazon basin of Peru's territory. Due to the location where it will be deployed, covering areas of high biodiversity and where local communities, projects of this nature are carried live by controversy due to contradictions involving their economic gains and environmental impacts they generate. In fact, despite these initiatives have targeted to meet economic interests in the region, their social and environmental impacts are left in the background.

4. Some Conclusions Related Endogenous Energy-Integration

Despite the difficulties involving energy integration (as questions of sovereignty of countries and divergence of interests among stakeholders—be they citizens, companies, nation states or other agents), the process of energy integration can bring a lot of benefits for developing countries, since the production and consumption of energy are factors that directly influence the economic development process. Likewise, energy resources are not distributed evenly around the planet, so that trade in energy resources and electricity can benefit both the importing country and the exporting—depending on the way the procedure is conducted and interests involved.

Second, one cannot forget that the energy integration policies should be analyzed in a broader context of economic integration. Without economic integration in fact, the process of energy integration weakens or worse it become unviable, since it depends on complex and coherent political, legal and economic foundations established by supranational institutions respected by all. Often, historical political conflicts between countries are an obstacle to the process of economic and political integration, undermining thereby the energy integration. The South American case shows that, a context in which the process of economic integration is a little-advanced stage makes energy integration slow down.

By analyzing the process of historical energy integration in South America, it appears that this has gone through several changes over the past decades—following the political and economic changes occurring in the continent. In the 21th century, for example, strengthening the anti-imperialist and anti-neoliberal discourse has resulted in increased participation of states and their companies in making decisions and profits relating to the exploitation of energy resources, and reducing reliance on funding coming from the IDB and the World Bank—that guided the economic policies in the region. Also, the economic growth of countries resulted in an increase in their energy demands, requiring new developments in the energy sector. It should be noted Brazil, which has been consolidated as a lead in this process—stimulating new projects and the participation of the Brazilian capital, through BNDES financing and the involvement of construction companies—due to their economic power—and thereby politician in the region—as well as their energy requirements, which have been increasing in recent times.

Comment is also likely the fact that transnational energy projects are restricted, mostly, to bilaterally. This shows that the process of energy integration in South America was not an end in itself, but a means to meet the energy and economic needs of certain countries and agents in certain recent historical periods. In other words, projects like Itaipu and Bolivia Brazil Gas Pipeline had main objective to serve the economic interests of the parties involved—and not energy integration itself. Even in more recent initiatives, such as IIRSA, the integration of the energy sectors of the South American countries occurs in a larger plan: the integration of infrastructure (which also involves networks of movement and communication)—is therefore a means to achieve greater integration project. Even so, IIRSA—and now COSIPLAN—is the main initiative of energy integration in South America in the beginning of the 21th century, in a context where projects happen beyond the traditional economic blocs.

Moreover, the integration process is not limited to the creation of physical links in a region. It requires a series of policies and regulations to facilitate harmonic different types of flows inherent in this process as well as the narrowing of political, productive and cultural ties. In this sense, the results were still unsatisfactory, since we focused on the transportation industry related projects, at the expense of others—that cater to specific economic interests, but they do not contribute to reducing social and economic inequalities present within the southern countries Americans, and the continent in general. So far it needs to advance in South America—if we want a restricted integration only to facilitating the flow of goods across the continent

Finally, the various changes occurring in the energy integration process in the region show a lack of consensus in defining a comprehensive model of integration and satisfying interests of all States and involved actors in the region. Something that is directly related to existing internal asymmetries within countries, but also is between them. Thus, for this to be overcome, it is necessary that these asymmetries are resolved both within countries and between them.

References

[1] Agência Internacional de Energia (IEA) (2010) World Energy Outlook 2010. Paris.

[2] World Energy Council (WEC) - Comitê Brasileiro (2004) Eficiência Energética: Uma análise mundial. Rio de Janeiro. Disponível no www.worldenergy.org

[3] Fundo Monetário Internacional (FMI) (2012) Dados disponíveis no www.imf.org

[4] ENERDATA (2012) Dados disponíveis no www.enerdata.net

[5] Cipolla, C.M. (1961) Sources d'énergieethistoire de l'humanité. In: *Annales. Économies, Sociétés, Civilisations.* 16e année, N. 3, 521-534.

[6] Goldemberg, J. and Lucon, O. (2008) Energia, Meio Ambiente e Desenvolvimento. São Paulo, Edusp.

[7] D'ÁVALOS, Victorio Enrique Oxília. *Raízes Socioeconômicas da Integração Energética na América do Sul: análise dos projetos Itaipu Binacional, Gasbol e Gasandes.* Tese de Doutorado. PPGE-USP. São Paulo, 2009.

[8] Udaeta, M.E.M. (1997) Planejamento Integrado de Recursos Energéticos para o Setor Elétrico—PIR (Pensando o Desenvolvimento Sustentável). Epusp, tese de doutorado, São Paulo.

[9] Udaeta, M.E.M., Burani, G.F., Fagá, M.T.W. and Oliva, C.R.R. (2006) Ponderação analítica para da integração energética na América do Sul. *Revista Brasileira de Energia,* **12.**

[10] Hobsbawm, E. (2007) A era dos impérios 1875-1914. Ed. Paz e Terra, São Paulo.

[11] Scarlato, F.C. (2008) O Espaço Industrial Brasileiro: Sociedade, Industrialização e Regionalização do Brasil. In: Ross, J.L.S., Org., *Geografia do Brasil*, 5th edição, Edusp, São Paulo.

[12] Harvey, D. (2011) O novo imperialismo. Edições Loyola, São Paulo.

[13] Triola, L.C. (2008) Energy and National Security: An Exploration of Threats, Solutions and Alternative Futures. *IEEE Energy* 2030, Atlanta, 17-18 November 2008, 13-35.

[14] Nagy, K. (2009) The Additional Benefits of Setting up an Energy Security Centre. *Energy*, **34**, 1715-1720.

[15] Suárez, L.P.L. (2006) O Papel das Petrolíferas para o Desenvolvimento da Integração Energética: A formação do Mercado de Gás Natural na América do Sul. Unicamp, dissertação de mestrado, Campinas.

[16] Antunes, J.C.A. (2007) Infraestrutura na América do Sul: Situação atual, necessidades e complementaridades possíveis com o Brasil. Comissão Econômica para a América Latina e o Caribe (CEPAL).

[17] Tratado De Maastricht. União Europeia, 1992.

[18] WEC—World Energy Council (2005) Regional Energy Integration in Africa. Londres.

[19] Southern African Power Pool (SAAP), 2013. http://www.sapp.co.zw/

[20] Campbell, B. (2007) Una perspectiva nacional de la integración continental del sector canadiense del petróleo y el gás. In: Vargas, R. and Ugalde, J.L.V., Org., *Dos modelos de Integración Energética—América del Norte/América del Sur*, CISAN, Ciudad de Mexico.

[21] Márquez, D. (2007) El TLCAN plus: La homologación de estándares y sus implicaciones legales para México. In: Vargas, R. and Ugalde, J.L.V., Org., *Dos modelos de Integración Energética—América del Norte/América del Sur*, CISAN, Ciudad de Mexico.

[22] Doucet, J. (2007) La integración energética norteamericana. Una perspectiva canadiense. In: Vargas, R. and Ugalde, J.L.V., Org., *Dos modelos de Integración Energética—América del Norte/América del Sur*, CISAN, Ciudad de Mexico.

[23] Us Energy Information Administration, 2012. http://www.eia.gov/countries/analysisbriefs/Canada/canada.pdf

[24] Castro, N.J., Dassie, A.M. and Delgado, D. (2009) Indicadores Mundiais do Setor Elétrico—As Experiências Latino-Americanas de Integração Energética. GESEL/UFRJ, Rio de Janeiro.

[25] IIRSA (2010) IIRSA diez años después: Sus logros y desafíos. BID-INTAL, Buenos Aires.

[26] Gudynas, E. (2008) As instituições financeiras e a integração na América do Sul. In: Verdum, R., Org., *Financiamento e megaprojetos*: *Uma interpretação da dinâmica regional Sul-Americana*, Inesc, Brasília.

9

Experimental and Numerical Investigation on the Flow Characteristics around a Cross-Flow Wind Turbine

Takaaki Kono[1*], Akira Yamagishi[2], Takahiro Kiwata[1], Shigeo Kimura[2], Nobuyoshi Komatsu[2]

[1]Research Center for Sustainable Energy & Technology, Kanazawa University, Kanazawa, Japan
[2]Division of Mechanical Science and Engineering, Kanazawa University, Kanazawa, Japan
Email: *t-kono@se.kanazawa-u.ac.jp

Abstract

This study investigated the flow characteristics around a cross-flow wind turbine. A wind tunnel experiment (WTE) was performed to measure the flow characteristics past the wind turbine when operating at the optimal tip-speed ratio of $\lambda = 0.4$. In addition, computational fluid dynamics (CFD) simulations were performed for the flow field around the wind turbine that was operating at tip-speed ratios of $\lambda = 0.1$, 0.4, and 0.7. The CFD approach was validated against the WTE measurements. CFD results confirmed that with an increase in λ, the velocity deficit was generally increased in the leeward of the return side of the wind turbine, while it was generally decreased in the leeward of the drive side of the wind turbine. It was also confirmed that with an increase in λ, the turbulence kinetic energy was generally increased in the leeward of the return side of the wind turbine, while it generally decreased in the leeward of the drive side of the wind turbine.

Keywords

Cross-Flow Wind Turbine, Wind Tunnel Experiment, CFD

1. Introduction

A small wind turbine with a cross-flow runner (hereafter referred to as "cross-flow wind turbine") has a high starting torque and is quiet. Thus, it is suitable for it to be introduced in urban areas where wind speed is generally low and careful attention to noise reduction is required. However, it has a drawback in that its maximum

---
*Corresponding author.

power coefficient is extremely low (about 10%) when compared with that of other small wind turbines. To date, several studies have addressed ways to improve the efficiency of cross-flow wind turbines [1]-[4]. However, knowledge of flow characteristics around a cross-flow wind turbine is extremely limited. Therefore, there is great potential to improve cross-flow wind turbine's efficiency by taking flow characteristics into account.

In this study, we conduct computational fluid dynamics (CFD) simulations and a wind tunnel experiment (WTE) to clarify the flow characteristics around a cross-flow wind turbine.

2. Experimental Method

Figure 1 shows the schematic of the experimental setup. The experiment was conducted using a closed circuit wind tunnel with an open test section. The size of the cross section of the wind tunnel outlet was 1250 mm × 1250 mm.

The cross-flow wind turbine being tested, which is shown in **Figure 2**, had an outer diameter of $D = 80$ mm, an inner diameter of $d = 65$ mm, and a lateral length of $L = 400$ mm. The shape of the blades was an arc of a circle with radius of $r = 11.5$ mm, angle of $\theta = 114°$, and chord length of $l_c = 10.5$ mm. The inlet angle of the blades was $\beta = 40°$, and the number of the blades was $N = 15$.

The wind turbine was placed at 800 mm downwind from the wind tunnel outlet and 525 mm above the floor. The wind turbine then was connected to a torque meter and a direct current motor that controls the wind turbine's number of rotations.

The free stream velocity was set to $U = 7$ m/s, and the turbulence intensity was less than 0.5%. Assuming that the origin lay at the center of the wind turbine, wind velocity distribution was measured using an X-array hotwire probe (KANOMAX, 0252R-T5) at $z/D = 0$ in the lateral direction, at the range of $x/D = -0.5$ to 2.0 in the streamwise direction, and at the range of $y/D = -1.5$ to 1.5 in the vertical direction, as shown in **Figure 3**. The spacing of the measurement points was 5 mm in both the x and y directions.

3. Numerical Analysis Method

3.1. Abbreviations and Acronyms

Two-dimensional CFD simulation was performed for the flow field around the wind turbine on the x-y plane at $z/D = 0$. The governing equations are the Reynolds-averaged continuity equation

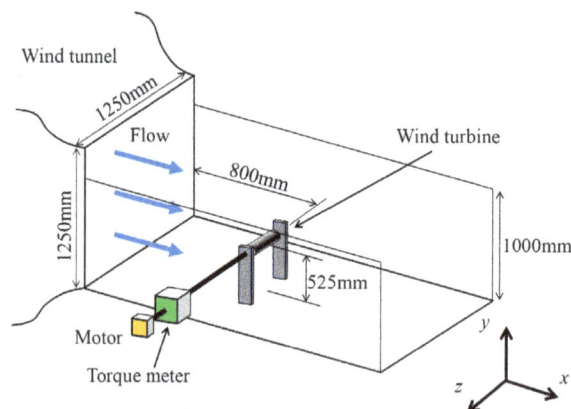

Figure 1. Experimental setup.

Figure 2. Cross-flow wind turbine.

$$\frac{\partial \langle u_i \rangle_R}{\partial x_i} = 0 \tag{1}$$

and the Reynolds-averaged Navier-Stokes (RANS) equations

$$\frac{\partial \langle u_i \rangle_R}{\partial t} + \frac{\partial \langle u_i \rangle_R \langle u_j \rangle_R}{\partial x_j} = -\frac{1}{\rho}\frac{\partial \langle p \rangle_R}{\partial x_i} + \nu \frac{\partial^2 \langle u_i \rangle_R}{\partial x_j^2} - \frac{\partial}{\partial x_j}\langle u_i'' u_j'' \rangle_R \tag{2}$$

where u_i is the velocity component in the x_i direction, t is the time, ρ is the density of air, p is the pressure, ν is the kinetic viscosity, and $\langle \phi \rangle_R$ is the Reynolds-average of a flow variable ϕ. The Reynolds stresses $\langle u_i'' u_j'' \rangle_R$ were computed using the k-ω shear-stress transport (SST) turbulence model [5]. The advection term was discretized by the second-order upwind scheme. Other spatial derivatives were discretized via the second-order central difference scheme. The Pressure-Implicit with Splitting of Operators (PISO) algorithm was used for velocity-pressure coupling.

3.2. Computational Conditions

The detail of the computational grid is shown in **Figure 4**. The computational domain consists of a rotational area, which includes the wind turbine, and a stationary area that surrounds it. A sliding mesh technique was used

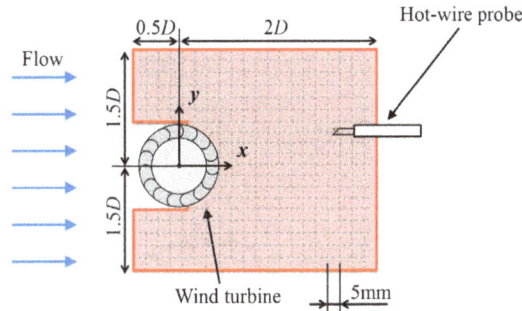

Figure 3. Measurement range.

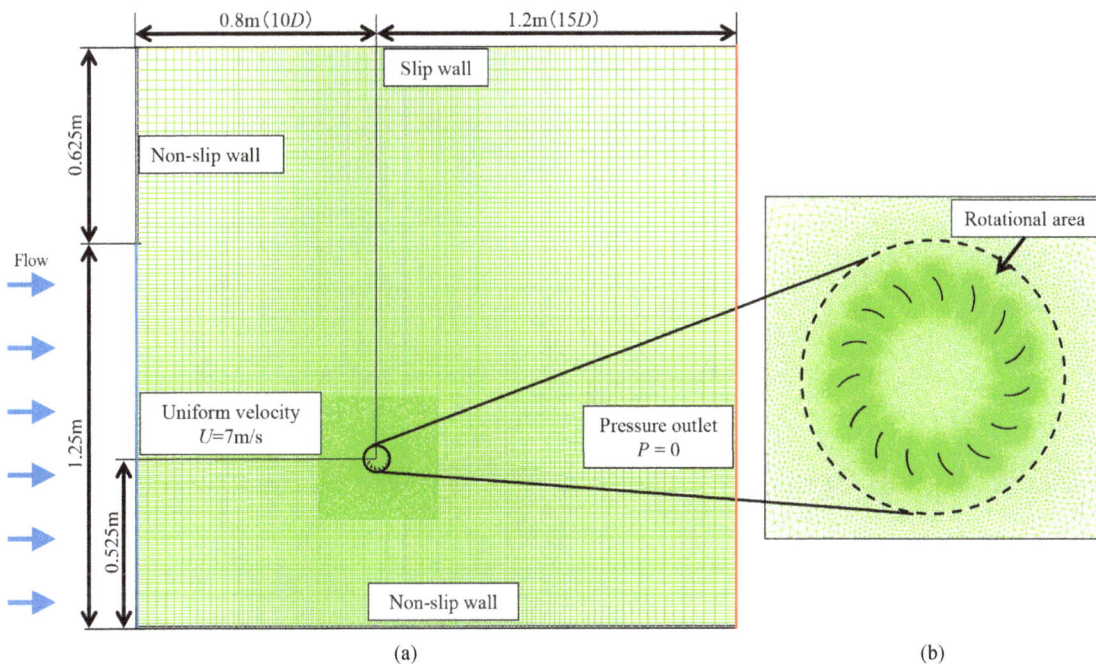

(a)   (b)

Figure 4. Computational grid. (a) Entire domain, (b) enlarged view.

to couple the rotational grid and stationary grid, as described in literature [6]. The total number of grid points was approximately 250,000.

On the inflow boundary, a stream-wise wind speed of $U = 7$ m/s with the turbulent intensity of 0.5% was implemented. On the outlet boundary, the pressure outlet boundary condition was imposed. On the bottom boundary of the computational domain and blade surface, the non-slip boundary condition was set. On the top boundary of the computational domain, the free-slip boundary condition was implemented.

The tip-speed ratio λ ($= D\omega/U$) of the wind turbine was set with U being constant and ω, the turbine's angular velocity, being changed. The time step Δt ($= \pi/360\omega$) is the period of time for the wind turbine to rotate by 0.5 degrees. The statics were summed up from $3600\Delta t$ to $7200\Delta t$.

4. Results and Discussion

4.1. Power Coefficient

Figure 5 shows the dependence of the power coefficient C_P ($= 2T\omega/LD\rho U^3$) on the tip-speed ratio λ, where T is the turbine's torque. It is observed that the CFD results match well.

4.2. Flow Characteristics around Wind Turbine at Optimal Tip-Speed Ratio

In this section, we focus on the flow characteristics around the wind turbine when operating at $\lambda = 0.4$.

Figure 6 shows the time-averaged velocity profiles of the flow past the wind turbine. The profiles of the CFD and WTE results appear to be in good agreement except for the streamwise velocities in the range of $y/D = -0.5$ to 0.0. The large discrepancies between the CFD and WTE results for the streamwise velocity likely due to the difference in the generation frequency of the strong counter-clockwise vortices, one of which is indicated by the dashed oval in **Figure 7**, shed from the return side of the wind turbine. Based on **Figure 8**, which shows the

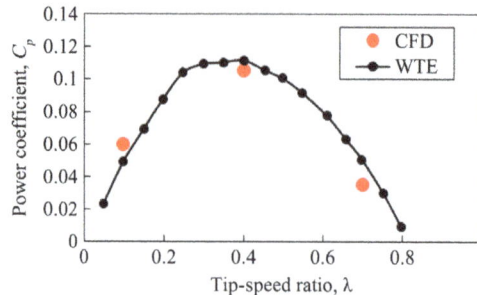

Figure 5. Dependence of the power coefficient C_P on the tip-speed ratio λ.

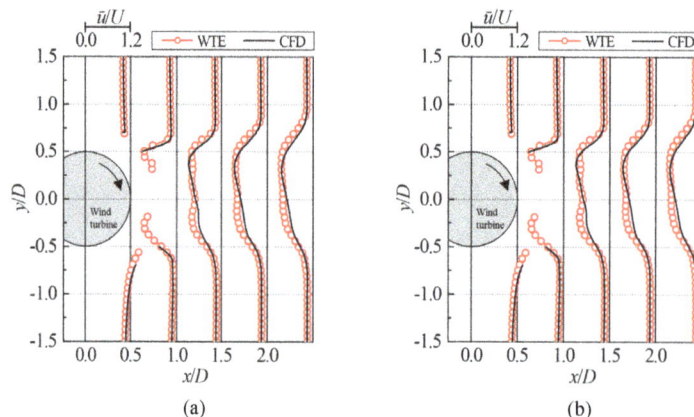

Figure 6. Time-averaged velocity profiles of the flow past the wind turbine when operating at $\lambda = 0.4$. (a) Streamwise velocity, (b) vertical velocity.

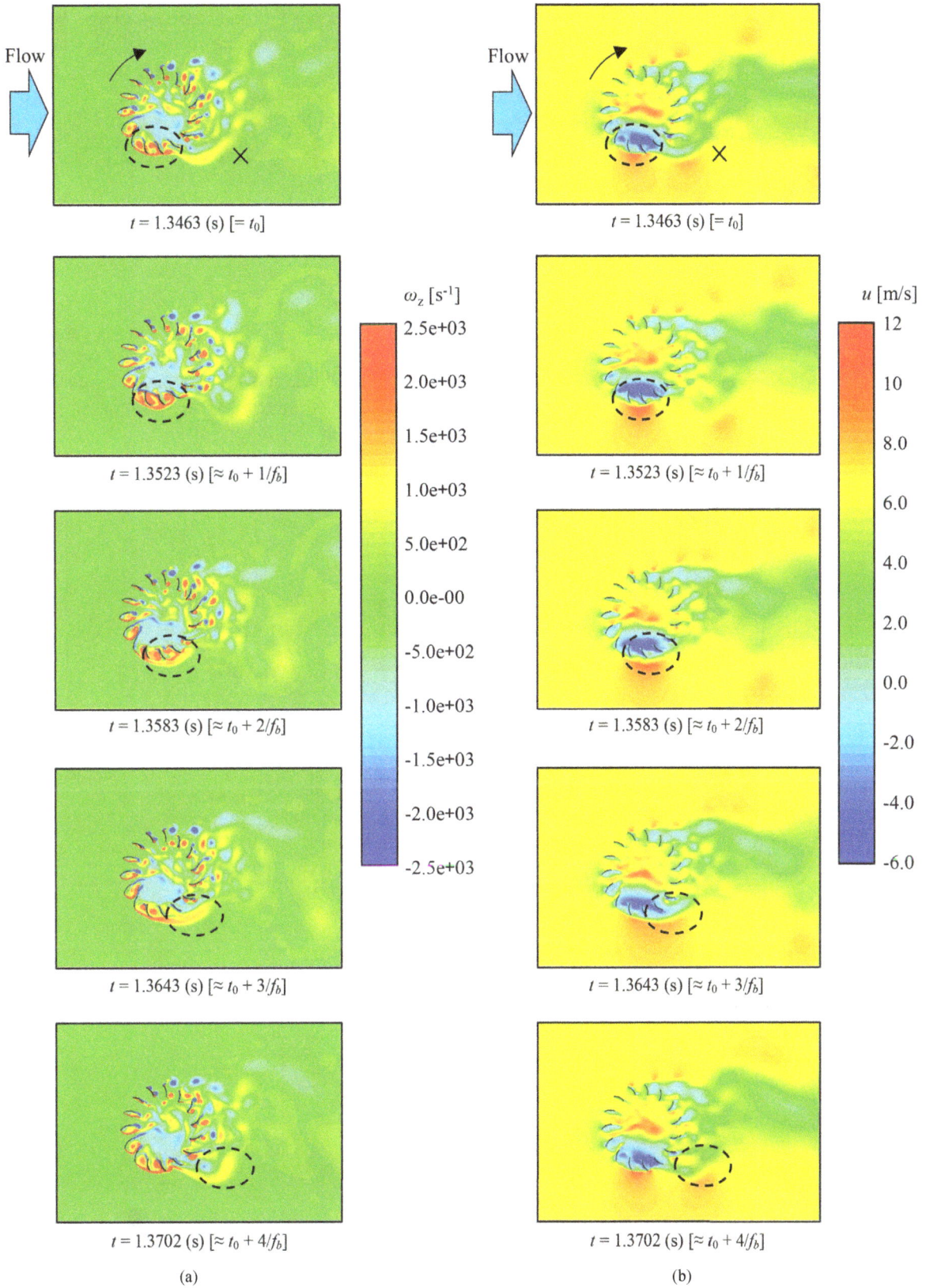

Figure 7. Contours of axial vorticity and streamwise velocity around the wind turbine when operating at $\lambda = 0.4$. (a) Axial vorticity, (b) streamwise velocity.

frequency spectra of the streamwise velocity fluctuations at the position indicated by a cross mark in **Figure 7(a)**, the strong counter-clockwise vortices are periodically shed from the return side of the wind turbine at a frequency of approximately $f_b/4$ Hz in CFD and $f_b/3$ Hz in WTE. Here, f_b is the blade passing frequency (the product of the rotor rotation frequency and the number of blades) shown in **Table 1**.

Figure 9 shows the turbulence kinetic energy (TKE) profiles of the flow past the wind turbine. Here, the values of TKE of WTE are calculated by

$$\text{TKE}_{\text{WTE}} = \frac{1}{2}\left(\overline{u'^2} + \overline{v'^2} + \overline{w'^2}\right),$$ (3)

and the values of TKE of CFD are calculated using the following approximate equations, (see **Appendix**):

$$\text{TKE}_{\text{CFD}} = \frac{1}{2}\left(\overline{u'''^2} + \overline{v'''^2} + \overline{w'''^2}\right) + \overline{k},$$ (4)

where $\overline{\phi}$ is the time average of a flow variable ϕ, ϕ' is $\phi - \overline{\phi}$, ϕ''' is $\phi' - \phi''$, and k is the TKE of the SST k-ω turbulence model. **Figure 9** confirms that the profiles of the WTE and CFD results match qualitatively well, having two peaks around $y/D \approx -0.5$ and $y/D \approx 0.5$ to 0.9. It is considered that these two peaks were generated by the vortices released from the wind turbine, as shown in **Figure 7(a)**.

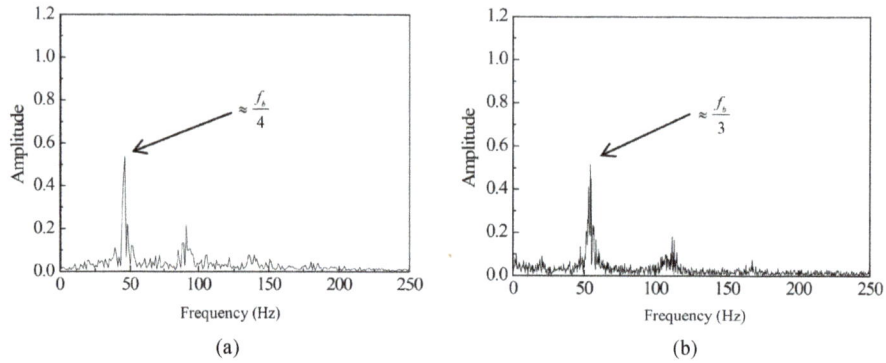

(a)   (b)

Figure 8. Frequency spectra of the streamwise velocity fluctuations at the position indicated by a cross mark in **Figure 7(b)**. (a) CFD, (b) WTE.

Table 1. Blade passing frequency.

Tip speed ratio, λ	0.1	0.4	0.7
Blade pasing frequency, f_b (Hz)	41.5	167.25	291.75

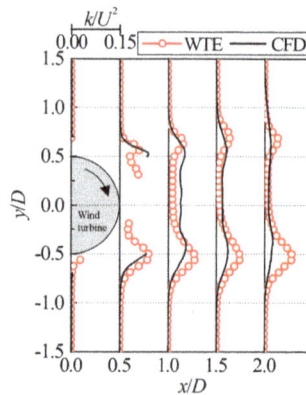

Figure 9. Turbulence kinetic energy profiles of the flow past the wind turbine when operating at $\lambda = 0.4$.

4.3. Dependency of Flow Characteristics on the Tip-Speed Ratio of the Wind Turbine

In this section, based on the CFD results, we discuss the dependency of the flow characteristics on the tip-speed ratio of the wind turbine.

Figure 10(a) compares the time-averaged streamwise velocity profiles of the flow past the wind turbine when operating at various tip-speed ratios. With an increase in λ, the velocity deficit is generally increased in the leeward of the return side of the wind turbine ($y/D < 0$), and in the leeward of the drive side of the wind turbine ($y/D > 0$), the velocity deficit is generally decreased.

Figure 10(b) compares the TKE profiles of the flow past the wind turbine when operating at various tip-speed ratios. With an increase in λ, the values of TKE are generally increased in the leeward of the return side of the wind turbine, and in the leeward of the drive side of the wind turbine, the values of TKE are generally decreased.

With regard to the causes of the dependencies of the velocity deficit and TKE on λ, the characteristics of the vortices shed from the drive side and from the return side of the wind turbine vary depending on λ. **Figure 11** shows the contours of the axial vorticity around the wind turbine at $\lambda = 0.1$ and 0.7, respectively. According to **Figure 7(a)** and **Figure 11**, the vortices shed from the return side of the wind turbine become larger and stronger as λ increases. As a result, the velocity deficit and TKE in the leeward of the return side increases with an increase in λ. The formation of larger and stronger vortices with an increase in λ is considered to stem from the fact that the shear stress near the blades on the return side of the wind turbine increases due to an increase in the speed of the blades that move in the opposite direction of the flow around the wind turbine. Moreover, based on **Figure 7(a)** and **Figure 11**, the vortices shed from the drive side of the wind turbine become smaller and weaker as λ increases. As a result, the velocity deficit and TKE in the leeward of the drive side decreases with an increase in λ. The formation of smaller and weaker vortices with an increase in λ is considered to occur because the interaction between the opposite sign of the vortices from the edges of a blade becomes less frequent due to an increase in the frequency of the reattachment of the vortices from the inner edge of a blade.

5. Conclusions

To clarify the flow characteristics around a cross-flow wind turbine, a wind tunnel experiment (WTE) and computational fluid dynamics (CFD) simulations were conducted. The CFD simulations were performed for the cases in which the wind turbine was operating at tip-speed ratios of $\lambda = 0.1$, 0.4, and 0.7. The validity of the CFD approach was confirmed through the comparison with the WTE results for the optimal tip-speed ratio of $\lambda = 0.4$. The main findings are summarized as follows.

1) With an increase in λ, the velocity deficit is generally increased in the leeward of the return side of the wind turbine, while it is generally decreased in the leeward of the drive side of the wind turbine.

2) With an increase in λ, the turbulence kinetic energy is generally increased in the leeward of the return side of the wind turbine, while it is generally decreased in the leeward of the drive side of the wind turbine.

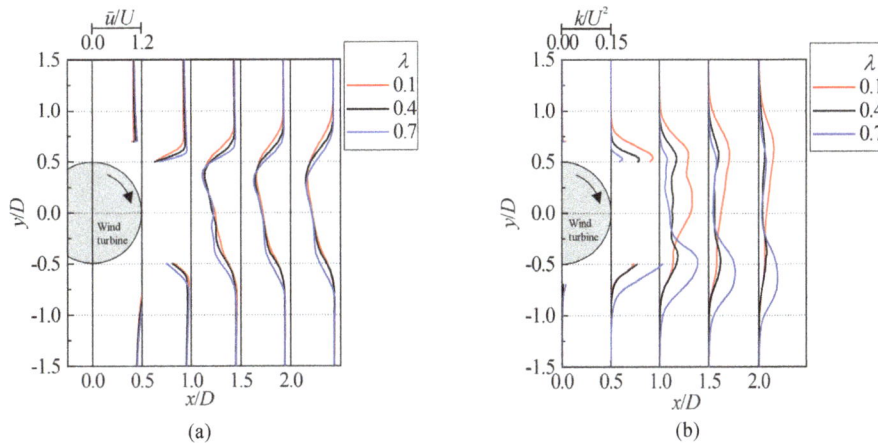

Figure 10. Time-averaged streamwise velocity and turbulence kinetic energy profiles of the flow past the wind turbine. (a) Streamwise velocity, (b) turbulence kinetic energy.

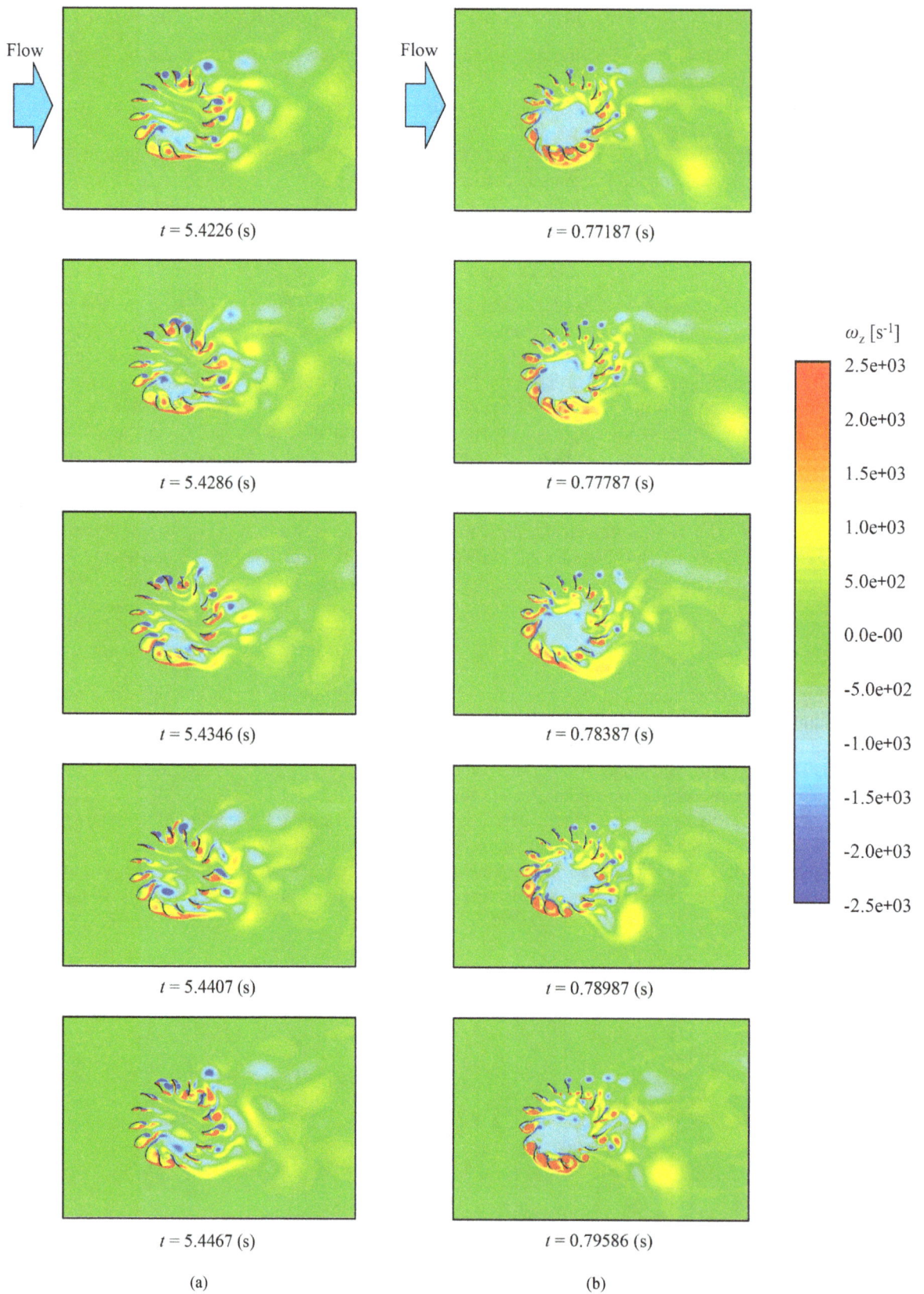

Figure 11. Contours of axial vorticity around the wind turbine when operating at $\lambda = 0.1$ and 0.7. (a) $\lambda = 0.1$, (b) $\lambda = 0.7$.

3) With an increase in λ, the vortices shed from the return side of the wind turbine tend to be larger and stronger.

4) With an increase in λ, the vortices shed from the drive side of the wind turbine tend to be smaller and weaker.

References

[1]   Al-Maaitah, A.A. (1993) The Design of the Banki Wind Turbine and Its Testing in Real Wind Conditions. *Renewable Energy*, **3**, 781-786. http://dx.doi.org/10.1016/0960-1481(93)90085-U

[2]   Tanino, T. and Nakao, S. (2005) Improving Ambient Wind Environments of a Cross-Flow Wind Turbine near a Structure by Using an Inlet Guide Structure and a Flow Deflector. *Journal of Thermal Science,* **14**, 242-248. http://dx.doi.org/10.1007/s11630-005-0008-0

[3]   Shigemitsu, T., Fukutomi, J. and Takeyama, Y. (2009) Study on Performance Improvement of Cross-Flow Wind Turbine with Symmetrical Casing. *Journal of Environment and Engineering*, **4**, 490-501. http://dx.doi.org/10.1299/jee.4.490

[4]   Dragomirescu, A. (2011) Performance Assessment of a Small Wind Turbine with Crossflow Runner by Numerical Simulations. *Renewable Energy*, **36**, 957-965. http://dx.doi.org/10.1016/j.renene.2010.07.028

[5]   Menter, F.R., Langtry, R.B., Likki, S.R., Suzen, Y.B., Huang, P.G. and Volker, S. (2006) A Correlation-Based Transition Model Using Local Variables—Part I: Model Formulation. *Journal of Turbomachinery*, **128**, 413-422. http://dx.doi.org/doi:10.1115/1.2184352

[6]   ANSYS Inc. (2009) Fluent 12.0 User's Guide.

Appendix

The approximate Equation (4) is derived as follows.

With time averaging and Reynolds averaging, an instantaneous velocity is decomposed as

$$u_i = \overline{u}_i + u_i', \tag{A.1}$$

$$u_i = \langle u_i \rangle_R + u_i'' = \overline{u}_i + u_i''' + u_i'', \tag{A.2}$$

The turbulence kinetic energy based on time-averaging operation is defined as

$$\text{TKE} = \frac{1}{2}\left(\overline{u'^2} + \overline{v'^2} + \overline{w'^2}\right), \tag{A.3}$$

Using the relationship in Equation (A.2), Equation (A.3) is expressed as

$$\text{TKE} = \frac{1}{2}\left(\overline{\left(u'' + u'''\right)^2} + \overline{\left(v'' + v'''\right)^2} + \overline{\left(w'' + w'''\right)^2}\right), \tag{A.4}$$

Because using the present CFD approach is not possible for computing the values of w'^2 and $\overline{u_i'' u_j'''}$ in Equation (A.4), the present study assumes these values to be zero. As a result, Equation (4) is obtained.

10

Bootstrapped Multi-Model Neural-Network Super-Ensembles for Wind Speed and Power Forecasting

Zhongxian Men[1,2], Eugene Yee[2,3], Fue-Sang Lien[1,2], Hua Ji[1], Yongqian Liu[4]

[1]Waterloo CFD Engineering Consulting Inc., Waterloo, Ontario, Canada
[2]Department of Mechanical & Mechatronics Engineering, University of Waterloo, Waterloo, Onatrio, Canada
[3]Defence Research and Development Canada, Suffield Research Centre, Medicine Hat, Alberta, Canada
[4]School of Renewable Energy, North China Electric Power University, Beijing, China
Email: eyee0309@gmail.com

Abstract

The bootstrap resampling method is applied to an ensemble artificial neural network (ANN) approach (which combines machine learning with physical data obtained from a numerical weather prediction model) to provide a multi-ANN model super-ensemble for application to multi-step-ahead forecasting of wind speed and of the associated power generated from a wind turbine. A statistical combination of the individual forecasts from the various ANNs of the super-ensemble is used to construct the best deterministic forecast, as well as the prediction uncertainty interval associated with this forecast. The bootstrapped neural-network methodology is validated using measured wind speed and power data acquired from a wind turbine in an operational wind farm located in northern China.

Keywords

Artificial Neural Network, Bootstrap Resampling, Numerical Weather Prediction, Super-Ensemble, Wind Speed, Power Forecasting

1. Introduction

There has been an increasing emphasis towards a greater use of renewable energy (e.g., solar, wind, geothermal) as a strategy to reduce greenhouse gas emissions and to mitigate climate change. In this context, one of the fastest growing sources of renewable energy for the generation of "green electricity" is the power obtained from wind turbines. The ever increasing use of wind power poses new challenges. One important challenge is how to accommodate the unpredictable fluctuations in wind speed and direction which lead to variability and uncertainty in the wind power generation. The latter has significant implications for unit commitment and determination of scheduling and dispatch decisions (economic dispatch) needed for the optimal utilization of wind energy within a mixed power system. In this regard, wind power forecasting has become a critical component in the efficient management of a green electrical power system (required by generation companies and utilities) and in electrical market operations (required by energy market analysts and traders).

The development of wind power forecasting models for improving the efficiency and reliability of mixed electrical power systems and for supporting electrical market operations has been reviewed by Costa et al. [1], Ma et al. [2] and Foley et al. [3]. Methodologies for wind power forecasting can be categorized into three broad classes as follows: statistical, physical, and machine learning. Because time series of wind speed and power are frequently measured in the vicinity of wind farms, statistical approaches based on time series analysis and forecasting have been applied for the prediction of wind speed and power. Towards this purpose, time series forecasting based on the popular Box-Jenkins methodology [4], as applied to autoregressive (AR) models, moving average (MA) models, and autoregressive moving average (ARMA) models, has been utilized for wind speed and power forecasting using historical time series of wind speed and power. As an example, Erdem and Shi [5] demonstrated the application of an ARMA model for wind speed and direction forecasting. More sophisticated nonlinear time series models that accommodate the temporal evolution of the variance (heteroskedasticity) such as the autoregressive conditional heteroskedasticity (ARCH) model by Engle [6] and the generalized autoregressive conditional heteroskedasticity (GARCH) models by Bollerslev [7] have been applied to describe the intrinsic variability in the wind speed and the associated generated power. For instance, the ARMA-GARCH and the GARCH-in-mean (GARCH-M) models proposed by Liu et al. [8] have been applied to model the mean and volatility of the wind speed.

The second general class of models is physically-based models for wind speed prediction based on numerical weather prediction (NWP) or computational fluid dynamics (CFD). Utilizing equations of physics such as the conservation principles of mass, momentum, and energy in conjunction with various parameterizations for sub-grid scale physical processes that cannot be resolved explicitly by the necessarily finite number of grid points that are used to represent the atmospheric flow, NWP and/or CFD models provide hydrodynamic and thermo-dynamic models of the atmosphere that can be used to furnish a prediction of the flow field in a prescribed region. The prediction of the wind velocity field can be used in conjunction with the power curve for a wind turbine to provide a generated power forecast. Numerical weather prediction models have a number of limitations, including limited spatial resolution resulting in a coarse representation of the local terrain [9]. To overcome the latter problem, Liu et al. [10] considered the possibility of coupling a synoptic scale flow model to a large-eddy simulation model for wind energy applications and Li et al. [11] recently introduced a short-term wind forecasting methodology based on the use of CFD pre-calculated flow fields.

The third general class of models for wind power forecasting is based on machine learning approaches such as artificial neural networks, fuzzy systems, and support vector machines [3] [12] [13]. Unlike the parametric models used for time series forecasting, machine learning uses either a "gray" or "black" box (essentially non-parametric) representation for the underlying physical processes (defining a nonlinear mapping from an input to an output), and then utilizes various learning algorithms and historical time series of wind speed or power to "train" the gray (black) box. The black box trained in this manner can be applied subsequently to make predictions of the future wind speed or generated power (from a wind turbine).

In this paper, we propose to use the bootstrap resampling method in conjunction with an ensemble artificial neural network (ANN) approach for the multi-step-ahead forecasting of wind speed and generated power. The artificial neural network combines machine learning with physical modeling by using NWP wind speed data from a physical model as the exogenous input to the network. The purpose of the bootstrap resampling method is to reduce the bias in prediction of the wind speed and power and to obtain more accurate estimates for the standard deviation (uncertainty) of these predictions. More importantly, the confidence bands in these predictions can be determined, which can be used to provide a more rigorous uncertainty assessment in wind speed and power forecasting.

2. Bootstrapping Ensembles of Artificial Neural Networks

As discussed in the previous section, a major concern of wind energy management is the uncertainty quantification of multi-step-ahead predictions of the wind speed (at the turbine hub height) and the corresponding power generated by the turbine. Instead of choosing a single best ANN for forecasting, we propose instead to use the bootstrap resampling method in the context of an ensemble of ANNs for predictive uncertainty analysis.

Let $y \equiv (y_1, y_2, \cdots, y_T)'$ be a sample of realizations of a (scalar) random variable y, where the positive integer T represents the sample size and the $'$ denotes the transpose of a vector (or matrix). In this paper, y represents either the measured wind speed (at turbine hub height) or the power generated by a wind turbine. Similarly, let $x \equiv (x_1, x_2, \cdots, x_T)'$ to be a sample of the corresponding predicted wind speed (at turbine hub height) obtained from a numerical weather prediction model. Finally, let $\zeta \equiv (z_1, z_2, \cdots, z_T)'$ with $z_t = (y_t, x_t)$ ($t = 1, 2, \cdots, T$ where t is the time step). Note that ζ consists of the sample of collections of the measured wind speed (or power) y and the corresponding modeled wind speed x.

We want to first represent (model) the functional relationship between y and x based on the training set ζ, by using an ensemble of artificial neural networks, and then employ all of the trained ANNs in the ensemble to forecast y when new values of x become available. In this sense, the forecasting of wind speed and generated power is obtained by conditioning not only on a single best ANN (model), but on an entire ensemble of plausible ANNs (models).

The nonlinear parameterized mapping f from an input x to an output y can be described generally by an ANN given by

$$y = f(x; \theta). \tag{1}$$

The output of the ANN is a continuous function of the input and θ is a collection of weights and biases (parameters) that determine the architecture of the neural network. By virtue of the universal approximation theorem demonstrated by Hornik *et al.* [14], it is known that an ANN with three or more layers can approximate any continuous function provided the activation function is a locally bounded, piecewise continuous function. In view of this, in our application, we use a three-layer neural network structure consisting of an input layer, a hidden layer, and an output layer. More specifically, the mapping for the ANN for our application has the following explicit form:

$$v_j^{(1)} = \sum_l w_{jl}^{(1)} x_l + b_j^{(1)}; \quad h_j = f^{(1)}\left(v_j^{(1)}\right) \tag{2}$$

for the hidden layer and

$$v_i^{(2)} = \sum_j w_{ij}^{(2)} h_j + b_i^{(2)}; \quad y_i = f^{(2)}\left(v_i^{(2)}\right) \tag{3}$$

for the output layer. In Equations (2) and (3), the index l varies over the input, the index j varies over the hidden units, and the index i varies over the output. Furthermore, $f^{(1)}(\cdot)$ and $f^{(2)}(\cdot)$ are activation functions for the hidden and output layers of the network, respectively. For our application, the activation functions used to define the neural network architecture are $f^{(1)}(v) = 1/(1 + \exp(-v))$ (logistic sigmoidal function) and $f^{(2)}(v) = v$ (simple linear function).

Note that each neuron in the network is a unit that combines and processes all the data coming into the layer and then passes the transformed data (output of the activation function) to all the neurons of the successive layer. Specifically, the input of a neuron is a weighted sum of the outputs of all the neurons in the previous layer plus a bias. The weights w and the biases b collectively make up the parameter vector θ. The ANN is trained using the data set ζ by selecting the parameter vector θ so as to minimize some error function which measures how close the modeled output $f(x; \theta)$ is to the measured output y of the training set. Two specific forms of the error function that will be used for this purpose are the root-mean-square error (RMSE)

$$E_{\text{RMSE}}(\theta) = \left(T^{-1} \sum_{t=1}^{T} \left(y_t - f_t(x; \theta) \right)^2 \right)^{1/2} \tag{4}$$

and the mean absolute error (MAE)

$$E_{\text{MAE}}(\theta) = T^{-1} \sum_{t=1}^{T} \left| y_t - f_t(x; \theta) \right|. \tag{5}$$

The minimization of these error functions is achieved using the particle swarm optimization algorithm [15]. Once an ANN has been trained, it can be used for the out-of-sample forecasting of the dependent (output) variable y.

To apply the bootstrap resampling procedure [16] to the ANN, we have to impose a statistical distribution on the sample data ζ. To this purpose, we follow the standard bootstrap method and assume that the empirical distribution function $\hat{F}$ of z_t $(t = 1, 2, \cdots, T)$ is a uniform distribution (viz., one assigns equal probability to each sample value z_t). More specifically, we assign to each z_t a probability of T^{-1} and realize the distribution $\hat{F}$ such that $\hat{p}(z = z_t) = T^{-1}$ for $t = 1, 2, \cdots, T$. This distribution implies that any statistic of the observed data (viz., any functional of the data) is invariant under all permutations of the components in the sample ζ. The nonparametric simulation of bootstrap data sets based on the empirical distribution function $\hat{F}$ is simple: we sample with replacement from the components of ζ (viz., because $\hat{F}$ places equal probability on each of the original data values in ζ, each sample is obtained by independently sampling at random from these data values). By so doing, we can obtain a bootstrap sample with a sample size of T. We can repeat this process, say N times. Then, we will have N bootstrap samples drawn from the data values in ζ, which we will denote by $\zeta^{(1)}, \zeta^{(2)}, \cdots, \zeta^{(N)}$.

For each of these bootstrap samples, we can train an ensemble of ANNs with the same network architecture, but with each member of the ensemble having different numbers of neurons in the hidden layer (recall that the number of neurons in the input and output layers are determined *a priori* by the dimension of the input and output vectors, respectively). To be more specific, assume that the number of neurons in the hidden layer of the network architecture varies from j to q $(j \le q)$ inclusive. Furthermore, in order to explicitly treat the initialization uncertainty (viz., the uncertainty arising from the initialization of the weights used in training an ANN), we will train each ANN model structure (with a fixed number of hidden nodes) starting from m different sets of initial weights θ. In consequence, we have a super-ensemble of ANNs consisting of $Nm(q - j + 1)$ members [viz., $(q - j + 1)$ different ANN model structures, each of which is trained starting from m different random initializations of the weights on a particular bootstrap sample $\zeta^{(n)}$ with $n = 1, 2, \cdots, N$]. Each member of this super-ensemble of ANNs can be used to conduct a multi-step-ahead prediction of y.

The procedure for bootstrapping an ensemble of neural networks is summarized as follows:

1. Assign the nonparametric distribution $\hat{F}$ to the observed data $\zeta \equiv (z_1, z_2, \cdots, z_T)'$,

$$\hat{p}(z = z_t) = \frac{1}{T}, \quad t = 1, 2, \cdots, T. \tag{6}$$

2. Draw a (nonparametric) bootstrap sample (with replacement) from the empirical distribution function $\hat{F}$,

$$z_1^{(*)}, z_2^{(*)}, \cdots, z_T^{(*)} \overset{\text{iid}}{\sim} \hat{F}, \tag{7}$$

and train m ANNs starting from m different (random) initializations for the weight vector θ for a fixed ANN model structure (viz., an ANN with a fixed number of nodes in the hidden layer). Repeat the process for different ANN model structures with the number of nodes in the hidden layer varying from j to q $(j \le q)$ inclusive. The multi-step-ahead forecasting of the wind speed and generated power $\tilde{y}$ is carried out for each of the trained ANNs in the ensemble.

3. Repeat Step 2 N times to obtain bootstrap replications for the forecasted wind speed and power; namely, $\tilde{y}^{(1)}, \tilde{y}^{(2)}, \cdots, \tilde{y}^{(K)}$ with $K = Nm(q - j + 1)$. Calculate the mean and standard deviations of the bootstrap predictions of $\tilde{y}$ using the procedure described below. These quantities can be used in conjunction with a bootstrap-t method to obtain confidence intervals for the forecasts of the wind speed and generated power.

We use a two-stage weighted averaging method to provide the predictive uncertainty assessment for the wind speed and power. For *each* bootstrap sample, we calculate the predictions (forecasts) of the multi-step-ahead wind speed and power using the $L \equiv m(q - j + 1)$ members of the ensemble of ANNs trained using the bootstrap sample. A statistical combination of these forecasts is used to obtain the best forecast based on the current ensemble whose members have been trained using the *given* bootstrap sample. This procedure is repeated for each bootstrap sample, and the optimum forecast is calculated as a weighted average (statistical combination) of these N best forecasts obtained from the N bootstrap samples in the super-ensemble. The information can be used also to determine the confidence intervals in the forecasted quantities.

For each bootstrap sample, $\zeta^{(n)}$ for $n = 1, 2, \cdots, N$, we have $L \equiv m(q-p+1)$ in-sample predictions of y, which we denote as $\hat{y}^l \equiv \left(\hat{y}_1^l, \hat{y}_2^l, \cdots, \hat{y}_T^l\right)'$ $(l = 1, 2, \cdots, L)$. Similarly, we have L out-of-sample predictions $y^l \equiv \left(\tilde{y}_{T+1}^l, \tilde{y}_{T+2}^l, \cdots, \tilde{y}_{T+d}^l\right)'$ for $l = 1, 2, \cdots, L$, where $d \geq 1$ is the prediction horizon. Define the weight vector $w^{(n)} \equiv \left(w_1^{(n)}, w_2^{(n)}, \cdots, w_L^{(n)}\right)'$ with $w_l^{(n)} \geq 0$ and $\sum_{l=1}^L w_l^{(n)} = 1$. For each bootstrap sample (fixed n), the weights $w^{(n)}$ can be chosen to define that statistical combination of the forecasts from the various (L) members of the ensemble that provide the best forecast. To this purpose, the statistical weights $w^{(n)}$ for each model are determined from the training data set by a constrained[1] minimization of the following quadratic (objective) function:

$$\mathcal{E}_L\left(w^{(n)}\right) = \frac{1}{L}\left\|\hat{E}w^{(n)}\right\|^2 = w^{(n)'}H_L w^{(n)}, \tag{8}$$

where $\hat{E} \equiv \left(\hat{e}^1, \hat{e}^2, \cdots, \hat{e}^L\right)$ is the error matrix, with $\hat{e}^l \equiv y - \hat{y}^l$ $(l = 1, 2, \cdots, L)$. Here, $\|\cdot\|$ denotes the Euclidean norm and $H_L \equiv \frac{1}{L}\hat{E}'\hat{E}$ is an $(L \times L)$ matrix.

Once the weights $w_l^{(n)}$ for $l = 1, 2, \cdots, L$ have been obtained for each bootstrap sample (fixed n), we can use them to perform the best in-sample predictions $\hat{y}$ and best multi-step-ahead predictions $\tilde{y}$ of the output variable y. More specifically, for a given ensemble of ANNs that has been trained for a fixed bootstrap sample, the in-sample predictions of y for the l-th member of the ensemble are denoted explicitly by $\hat{y}^l \equiv \left(\hat{y}_1^l, \hat{y}_2^l, \cdots, \hat{y}_T^l\right)'$ and the multi-step-ahead predictions of y for the l-th member of the ensemble are denoted by $\tilde{y}^l \equiv \left(\tilde{y}_{T+1}^l, \tilde{y}_{T+2}^l, \cdots, \tilde{y}_{T+d}^l\right)'$ [recall $d > 0$ is the prediction horizon]. These best predictions (in-sample and multi-step-ahead) are given by

$$\hat{y}_t^{(n)}\left(w^{(n)}\right) = \sum_{l=1}^L \hat{w}_l^{(n)}\hat{y}_t^l, \quad t = 1, 2, \cdots, T, \tag{9}$$

$$\tilde{y}_t^{(n)}\left(w^{(n)}\right) = \sum_{l=1}^L \hat{w}_l^{(n)}\tilde{y}_t^l, \quad t = T+1, \cdots, T+d, \tag{10}$$

where $w^{(n)} = \left(\hat{w}_1^{(n)}, \hat{w}_2^{(n)}, \cdots, \hat{w}_L^{(n)}\right)$ denotes the solution to the minimization of the quadratic error function $\mathcal{E}_L\left(w^{(n)}\right)$ given in Equation (8).

For a *fixed* bootstrap sample, the standard deviation vector $\tilde{\sigma}^{(n)} \equiv \left(\tilde{\sigma}_{T+1}^{(n)}, \tilde{\sigma}_{T+2}^{(n)}, \cdots, \tilde{\sigma}_{T+d}^{(n)}\right)'$ for the multi-step-ahead prediction of y is estimated as follows (using the L forecasts $\tilde{y}_t^l$ $(l = 1, 2, \cdots, L)$ for y_t obtained from the ANN ensemble at the fixed time index t):

$$\tilde{\sigma}_t^{(n)} = \left(\sum_{l=1}^L \hat{w}_l^{(n)}\left[\tilde{y}_t^l - \overline{y}_t\right]^2\right)^{1/2}, \qquad t = T+1, \cdots, T+d, \tag{11}$$

where $\overline{y}_t \equiv \sum_{l=1}^N \tilde{y}_t^l / L$ is the (ensemble) mean of the samples in the set $\left\{\tilde{y}_t^1, \cdots, \tilde{y}_t^L\right\}$. Because this is a biased estimator of the standard deviation, we use instead the following formula

$$\tilde{\sigma}_t^{(n)} = \left(\left[1 - \sum_{l=1}^L \left(\hat{w}_l^{(n)}\right)^2\right]^{-1} \cdot \sum_{l=1}^L \hat{w}_l^{(n)}\left[\tilde{y}_t^l - \overline{y}_t\right]^2\right)^{1/2}, \qquad t = T+1, \cdots, T+d, \tag{12}$$

---

[1]Recall the weights $w_l^{(n)}$ are non-negative and satisfy $\sum_{l=1}^L w_l^{(n)} = 1$.

to obtain an unbiased estimator of the standard deviation.

At this point, we have N different forecasts for the output variable y (along with the standard deviations in these forecasts) obtained by applying the weighted-averaging schema described above to the ensemble of ANNs trained on each of the N bootstrap samples. The second stage of the process is to apply the weighted-averaging schema again to calculate the bootstrap predictions of $\tilde{y}$ based on these N different forecasts. To this purpose, we define another weight vector $\boldsymbol{\eta} = \left(\eta_1, \cdots, \eta_N \right)'$, whose estimate $\hat{\boldsymbol{\eta}} = \left(\hat{\eta}_1, \cdots, \hat{\eta}_N \right)'$ can be evaluated similarly by minimizing an objective function similar to Equation (8), except now the error matrix is constructed from the residuals between y and $\hat{y}^{(n)}$ $\left(n = 1, 2, \cdots, N \right)$, where $\hat{y}^{(n)} = \left(\hat{y}_1^{(n)}, \hat{y}_2^{(n)}, \cdots, \hat{y}_T^{(n)} \right)'$ is calculated by using Equation (9). Once the weight vector η has been estimated, the weighted bootstrap predictions of y are obtained from

$$\tilde{y}_t^B \left(\hat{\boldsymbol{\eta}} \right) = \sum_{n=1}^{N} \hat{\eta}_n \tilde{y}_t^{(n)}, \qquad t = T+1, \cdots, T+d, \tag{13}$$

and the corresponding bootstrapped standard deviations from

$$\tilde{\sigma}_t^B = \left(\sum_{n=1}^{N} \hat{\eta}_n \left[\tilde{\sigma}_t^{(n)} \right]^2 \right)^{1/2}, \qquad t = T+1, \cdots, T+d, \tag{14}$$

where $\tilde{\sigma}_t^{(n)}$ (standard deviation obtained from the ANN ensemble trained on the n-th bootstrap sample) is computed in accordance to Equation (12).

Confidence intervals for the forecast of the output variable y will be obtained using the bootstrap-t method. Towards this purpose, the confidence intervals of the multi-step-ahead prediction $\tilde{y}^B$ of y can be determined in accordance to the following formula:

$$\left(\tilde{y}_t^B \right)_{\text{est}} = \tilde{y}_t^B \pm \tilde{\sigma}_t^B t_{\alpha, v}, \qquad t = T+1, \cdots, T+d, \tag{15}$$

with $t_{\alpha, v}$ being the α-level critical value of a Student's-t distribution with $v = K-1$ degrees of freedom [recall $K = Nm(q-p+1)$]. Alternatively, $t_{\alpha, v}$ can be replaced by the bootstrap percentiles of the sample $\tilde{y}_t^{(k)}$ $\left(k = 1, 2, \cdots, K \right)$. So, for example, extracting the 2.5% and 97.5% bootstrap percentiles of this sample would allow one to construct a bootstrap-t based confidence interval for the prediction at a coverage level of 95%.

3. Application of the Methodology

3.1. Data Preparation

The two data sets that we analyse were collected from a specific wind turbine, referred to as WT24 hereafter, located in a wind farm in northern China. One of these data sets corresponds to the hourly-averaged wind speeds measured at the turbine hub height and the other corresponds to the associated hourly-averaged power generated by the turbine. The wind speed and generated power time series, consisting of 432 observations each, were measured over a period of 18 days. The measurements collected over the first 15 days (corresponding to 83% of the entire length) of the time series were used as the training set and the remaining 3 days were reserved for the forecast assessment and validation. In addition to these measured time series, wind speed data at turbine hub height obtained from a numerical weather prediction model was available and this information was used as the exogenous input for artificial neural network training. In particular, the modeled wind speed and direction data over the region occupied by the wind farm were obtained from a NWP model executed with a temporal resolution of 1 h on a computational grid with a 3-km spatial resolution centered on the location of the wind farm. We applied a simple bilinear interpolation (BI) on this coarse-resolution NWP wind speed data to obtain the wind speed at the location of the WT24 wind turbine, which was then subsequently used as an exogenous input for ANN training.

As described in the previous section, we bootstrapped (resampled with replacement) the training data set to generate $N = 200$ "phantom" (bootstrap) data sets. We used each of these bootstrap data sets to train three-layer

ANNs with a variable number of nodes in the hidden layer ranging from 5 to 30 nodes inclusive (so, $j = 5$ and $q = 30$).[2] In order to reduce the uncertainty arising from initialization of the network parameter (weight) vector, each network was trained 5 times each time starting from a different randomly chosen initialization of the weight vector (viz., $m = 5$). For each bootstrap sample (training data set), we have an ensemble consisting of $L \equiv m(q - p + 1) = 130$ member ANNs trained on sample. In consequence, the super-ensemble of ANNs trained on the entire set of bootstrapped samples consists of $K \equiv Nm(q - p + 1) = 26,000$ member ANNs. The information embodied in this ensemble can be used for the multi-step-ahead forecasting of the wind speed and generated power, along with a quantitative assessment of the prediction confidence.

3.2. Results

Figure 1 compares the measured wind speed at the hub height for WT24 with the out-of-sample forecasted wind speed obtained from bootstrapping an ensemble of ANNs. The best deterministic forecast for the wind speed based on a statistical combination of the individual forecasts in the super-ensemble [determined in accordance to Equations (10) and (13)] is shown in this figure (dot-dashed line labeled "Best forecast"). A comparison of the best deterministic forecast with the measured wind speed shows that the forecast captures adequately the longer temporal trends in the measured wind speed. In addition, the 95% prediction confidence intervals obtained using the two-stage weighted-averaging method is exhibited in **Figure 1** as the dotted lines demarcated using an open circle. Note that the 95% prediction uncertainty range appears to cover most of the observations, providing an observation coverage that is consistent (approximately or better) with the quoted level of confidence in the uncertainty interval. A quantitative assessment of the forecast performance in this case is summarized in **Table 1**, which summarizes the RMSE and MAE in the wind speed prediction using the bootstrapped neural-network methodology. The performance of this forecast methodology can be compared to that obtained from a simple persistence model forecast which uses the current wind speed to predict the value of the future wind speed.

Next, we consider the forecast performance for the wind power using the bootstrapped neural-network methodology. The forecast for the generated power is more complicated than that for the wind speed owing to

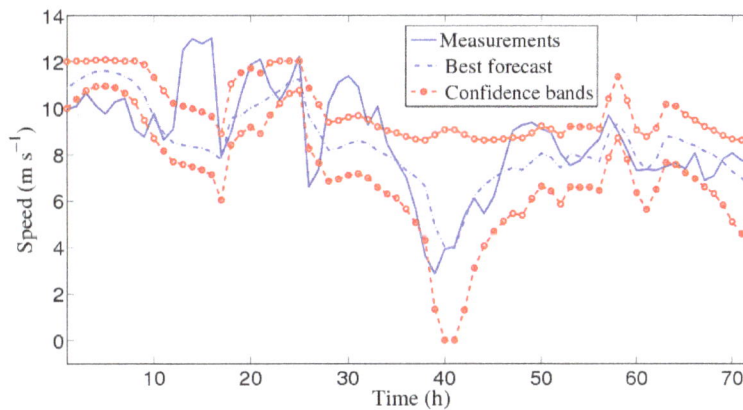

Figure 1. Out-of-sample wind speed forecasting obtained using the bootstrapped neural-network methodology. The 95% confidence interval for the forecast was calculated by using the two-stage weighted-averaging method applied to the ANN members of the super-ensemble.

Table 1. Wind speed forecast assessment of the bootstrapped neural-network methodology.

Criterion	Persistence	Best
RMSE	2.3554	1.6394
MAE	1.8693	1.2145

---

[2]Recall that the number of nodes in the input and output layers of the neural network are determined (fixed) by the dimensionality of the input and output vectors, respectively.

the fact that the wind power is censored from above. More specifically, wind turbines are designed so that when the wind speed exceeds a certain value (referred to as the rated output wind speed), a limit to the power generation is imposed implying that the power generated is censored from above. To account for this maximum limit in the wind power generation by a wind turbine, the bilinear interpolation of the NWP wind speeds to the location of WT24 were censored (to the rated output wind speed of the turbine) before they were used as the exogenous input for the neural network training. Furthermore, the measured wind power used in the training of the ANNs were already censored from above by the maximum limit for power generation by the turbine. Indeed, for the current example, if the modeled wind speeds exceeded 11.5 m·s^{-1} (rated output wind speed for WT24), then the generated wind power associated with this range of wind speeds was limited above to 1550 W (rated output power for WT24).

Figure 2 compares the measured power generated by the turbine WT24 with the forecasted wind power obtained using the bootstrapped neural-network methodology. The best deterministic forecast ("Best forecast") is shown based on a particular statistical combination of the individual forecasts obtained from the various members of the super-ensemble using the two-stage weighted-averaging process. In addition, the 95% confidence intervals for the predicted wind power is superimposed on the plot. **Table 2** summarizes the values of the RMSE and MAE for power forecasts obtained using the persistence methodology and the bootstrapped neural-network methodology. A comparison of the best deterministic forecast of the power (dash-dotted line) with the measured power shows that the broad features in the variation of the power is captured fairly well. From **Table 2**, it is seen that this best deterministic forecast gives misfits (either RSME or MAE) that are roughly a factor of three less than those obtained using the simple naïve persistence forecast. The observation coverage for the 95% prediction uncertainty intervals captures the power measurements fairly well, although qualitatively it is judged that this coverage is not as good as for the case of wind speed prediction. This would indicate that there may be sources of uncertainty in the wind power forecasting that have not been accounted for in the bootstrapping process.

4. Summary and Conclusions

In this paper, we proposed a novel bootstrapped artificial neural-network approach for wind speed and generated power forecasting. The approach provides a multi-ANN model super-ensemble that can be used to provide a best deterministic forecasting for these quantities, as well as to provide a quantitative assessment of the related

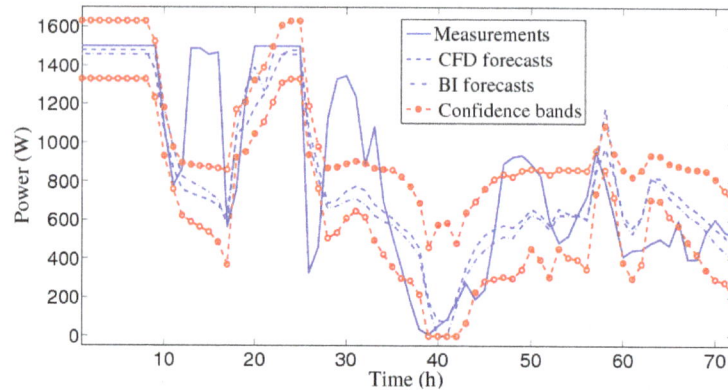

Figure 2. Out-of-sample wind power forecasting using the bootstrapped neural-network methodology. The 95% confidence intervals for the forecast were calculated by using the two-stage weighted-averaging method applied to the ANN members of the super-ensemble.

Table 2. Wind power forecast assessment of the bootstrapped neural-network methodology.

Criterion	Persistence	Best
RMSE	811.652	292.7335
MAE	653.252	210.8805

prediction uncertainty. In this approach, the individual ANNs that comprise the super-ensemble are first trained using a data set of available wind speed and power measurements and wind speed predictions (obtained from a numerical weather prediction model). The training consists of fitting various artificial neural network architectures (model structures) against the observation and exogenous input to determine the optimal statistical weights for each model.

The advantage of this methodology is that the biases in the forecast can be reduced and good predictions (forecasts) can be obtained through a statistical combination of the individual forecasts from the super-ensemble to give the best deterministic forecast. Applications to a wind turbine in northern China show that our proposed method works quite well. Because the method also provides prediction uncertainty bounds in the forecasts, it is anticipated that this approach would be very useful for green electrical system power management. Indeed, with the rapid pace of increases in computational power, it will become easier in the near future for power system managers and energy system traders/analysts to take advantage of the super-ensemble approach, for providing optimal forecasts and quantitative assessments of the uncertainty associated with these forecasts (allowing this information to be used in a more accurate and reliable manner for various applications).

References

[1]   Costa, A., Crspo, A., Navarro, J., Lizano, G., Madsen, H. and Feitosa, E. (2008) A Review on the Young History of the Wind Power Short-Term Prediction. *Renewable and Sustainable Energy Reviews*, **12**, 1725-1744. http://dx.doi.org/10.1016/j.rser.2007.01.015

[2]   Ma, L., Luan, S., Jiang, C., Liu, H. and Zhang, Y. (2009) A Review on the Forecasting of Wind Speed and Generated Power. *Renewable and Sustainable Energy Reviews*, **13**, 915-920. http://dx.doi.org/10.1016/j.rser.2008.02.002

[3]   Foley, A.F., Leahy, P.G., Marvugliaand, A. and McKeogh, E.J. (2012) Current Methods and Advances in Forecasting of Wind Power Generation. *Renewable Energy*, **37**, 1-8. http://dx.doi.org/10.1016/j.renene.2011.05.033

[4]   Box, G.E.P. and Jenkins, G. (1970) Time Series Analysis, Forecasting and Control. Holden-Day, San Francisco.

[5]   Erdem, E. and Shi, J. (2011) ARMA Based Approaches for Forecasting the Tuple of Wind Speed and Direction. *Applied Energy*, **88**, 1405-1414. http://dx.doi.org/10.1016/j.apenergy.2010.10.031

[6]   Engle, R.F. (1982) Autoregressive Conditional Heteroskedasticity with Estimates of the Variance of United Kingdom Inflation. *Econometrica*, **50**, 987-1007. http://dx.doi.org/10.2307/1912773

[7]   Bollerslev, T. (1986) Generalized Autoregressive Conditional Heteroskedasticity. *Journal of Econometrics*, **31**, 307-327. http://dx.doi.org/10.1016/0304-4076(86)90063-1

[8]   Liu, H., Erdem, E. and Shi, J. (2011) Comprehensive Evaluation of ARMA-GARCH(-M) Approaches for Modeling the Mean and Volatility of Wind Speed. *Applied Energy*, **88**, 724-732. http://dx.doi.org/10.1016/j.apenergy.2010.09.028

[9]   Martí Perez, I., Nielsen, T.S., Madsen, H., Navarro, J., Roldán, A., Cabezón, D. and Barquero, C.G. (2001) Prediction Models in Complex Terrain. *Proceedings of the European Wind Energy Conference*, Copenhagen, 2001, 875-878.

[10]   Liu, Y., Warner, T., Liu, Y., Vincent, C., Wu, W., Mahoney, B., Swerdlin, S., Parks, K. and Boehnert, J. (2011) Simultaneous Nested Modeling From the Synoptic Scale to the LES Scale for Wind Energy Applications. *Journal of Wind Engineering and Industrial Aerodynamics*, **99**, 308-319. http://dx.doi.org/10.1016/j.jweia.2011.01.013

[11]   Li, L., Liu, Y., Yang, Y. and Han, S. (2013) Short-Term Wind Speed Forecasting Based on CFD Pre-Calculated Flow Fields. *Proceedings of the Chinese Society of Electrical Engineering*, **33**, 27-32.

[12]   Kariniotakis, G., Stavrakakis, G.S. and Nogaret, E.F. (1996) Wind Power Forecasting Using Advanced Neural Network Models. *IEEE Transactions on Energy Conversion*, **11**, 762-767. http://dx.doi.org/10.1109/60.556376

[13]   Kariniotakis, G. (2003) Forecasting of Wind Parks Production by Dynamic Fuzzy Models with Optimal Generalisation Capacity. *Proceedings of the 12th Intelligent System Application to Power Systems* 03, ISAP 03/032, Lemnos, September 2003.

[14]   Hornik, K., Stinchcombe, M. and White, H. (1989) Multilayer Feedforward Networks are Universal Approximators, *Neural Networks*, **2**, 359-366. http://dx.doi.org/10.1016/0893-6080(89)90020-8

[15]   Kennedy, J. and Eberhart, R.C. (2001) Swarm Intelligence. Morgan Kaufmann Publishers, San-Francisco.

[16]   Efron, B. and Tibshirani, B. (1993) An Introduction to the Bootstrap. Chapman & Hall/CRC, Boca Raton.

11

Transmission and Consumption of Air Power in Pneumatic System

Shengzhi Chen[1], Chongho Youn[2], Toshiharu Kagawa[2], Maolin Cai[3]

[1]Department of Mechano-Micro Engineering, Tokyo Institute of Technology, Kanagawa, Japan
[2]Interdisciplinary Graduate School of Science and Engineering, Tokyo Institute of Technology, Kanagawa, Japan
[3]School of Automation Science and Electrical Engineering, Beihang University, Beijing, China
Email: chen.s.af@m.titech.ac.jp

Abstract

In recent 20 years, energy saving has been done in many projects. However, in pneumatic system, it is not easy to determine or measure the air power flow because of the compressibility of pneumatic system. In this paper, we used air power meter (APM) to measure the energy consumption of flow in pneumatic cylinder actuator system. Meter-in circuit and meter-out circuit of speed control system are used in this research. The model of cylinder system is based on four equations: state equation of air, energy equation, motion equation and flow equation. The model estimates the pressure change in charge and discharge side of cylinder, and also the displacement and velocity of the piston. Furthermore, energy consumption could theoretically be calculated when the change of air state is regarded as isothermal change. Lastly, some data of these two circuits are shown, and the consumption of energy is discussed.

Keywords

Air Power, Meter-In Circuit, Meter-Out Circuit, Energy Consumption

1. Introduction

Pneumatic cylinder actuator is widely used in factory automation field as a driving machine. To compare with the electrical motor, it is commonly used for conveying system because of its advantage of reciprocal linear motion. Furthermore, in meter-out circuit, the most remarkable characteristic is the response of speed control becomes stable easily when adjusting the speed control valve. Recently, PTP (Point To Point) is a representative application in the industry field. In these ten years, the development of pneumatic technology has become better due to the widely use of the pneumatic cylinder.

As we know, the power of compressed air used for mechanical work is based on the electricity consumption of compressor. The energy saving in pneumatic system is important. So many methods had been proposed [1]. In order to achieve the effective use of available energy, the assessment of energy consumption is necessary. Previous research showed that the available energy of air consists of two parts: power transmission energy and expansive energy [2]. It is clarified that over half of supplied available energy is used and remaining energy is lost [3]. However, the measurement of the consumption of available energy in a pneumatic cylinder system has not been conducted. The reason is that there are not any effective methods to measure the energy. To compare with the traditional energy consumption assessment method based on air flow rate, a new method using air power in terms of available energy has been proposed [4]. With air power, mechanism and factors will make objective of energy savings clearly.

In this paper, we use APM (Air Power Meter) to measure the energy consumption of the pneumatic cylinder based on meter-in and meter-out circuit, and the experimental results will be discussed.

2. Circuit of Speed Control

In general, the speed control of the pneumatic cylinder is controlled by a speed-control valve which is consisted of a variable throttle valve and a check valve. In term of the adjustment of the charging or discharging flow when using the speed-control valve, the meter-in and meter-out circuit have the different characteristics. As shown in **Figure 1**, the meter-in circuit could control the speed of the pneumatic cylinder by varying the charging flow. **Figure 2** is a schematic of meter-out circuit which is controlled by varying the discharging flow. To compare with the meter-out circuit, meter-in circuit has two advantages as follows. Assume that these two circuits are driven in the same conditions of the supply pressure and load mass. One advantage is the miniaturization of machine and another one is the less consumption of air.

However, meter-out circuit is more popular than meter-in circuit. Many pneumatic mechanism makers recommend people to use meter-out circuit expect for the particular situation. We consider that the reason is the simplification and stability of the speed control. The simplification indicates that the setting of speed is easily because of the speed is proportional to the regulated size of the speed-control valve. And the stability indicates that the speed response is not relevant to the load mass. That is, the speed response reaching to steady-state is independent on the change of load mass.

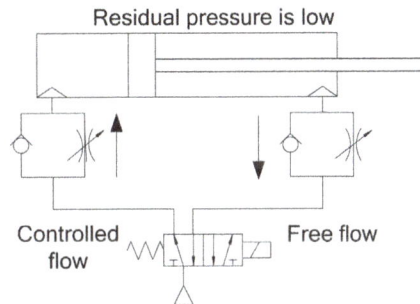

Figure 1. Meter-in circuit.

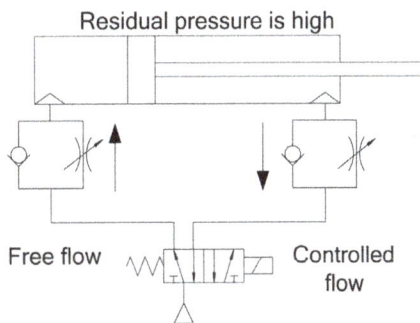

Figure 2. Meter-out circuit.

In addition, two merits of the meter-out circuit are described below.

One is the initial acceleration is very small because of the backpressure in discharge side, so that the piston will not move quickly. Another one is the cushion in the end part of the cylinder could play an important role due to the constant value of the pressure in discharge side.

3. Equations of Pneumatic System

The circuit of pneumatic system is shown in **Figure 3**. Four basic equations are shown as follows.

3.1. State Equation

When we derivative the state equations of air ($PV = mR\theta$) in charge side and discharge side, the following equations are obtained. Here, P is pressure; u is velocity of flow; S is pressured area; θ is temperature. And V represents volume; R represents gas constant of air; G represents mass flow rate. The subscript c and d refer to the charge side and discharge side, respectively.

$$V_c \frac{dP_c}{dt} = -S_c P_c u + R\theta_c G_c + \frac{P_c V_c}{\theta_c} \frac{d\theta_c}{dt} \tag{1}$$

$$V_d \frac{dP_d}{dt} = S_d P_d u + R\theta_d G_d + \frac{P_d V_d}{\theta_d} \frac{d\theta_d}{dt} \tag{2}$$

3.2. Energy Equation

Assuming that the value of heat transfer coefficient is constant, from the conservation of energy and the state equations of air, we obtain the following equations:

$$\frac{C_v P_c V_c}{R\theta_c} \frac{d\theta_c}{dt} = C_v G_u \left(\theta_a - \theta_c\right) + R\theta_a G_c - S_c P_c u + h_c S_{hc} \left(\theta_a - \theta_c\right) \tag{3}$$

$$\frac{C_v P_d V_d}{R\theta_d} \frac{d\theta_d}{dt} = R\theta_d G_d + S_d P_d u + h_d S_{hd} \left(\theta_a - \theta_d\right) \tag{4}$$

where C_v represents the specific heat at constant volume, h represents the heat transfer coefficient and S_h is the heat transfer area.

3.3. Motion Equation

The friction of piston is given by

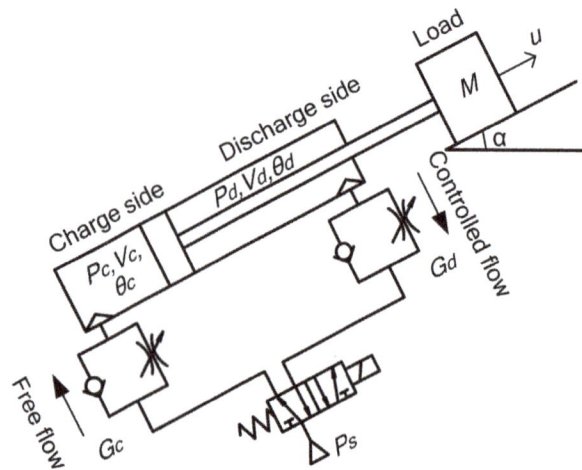

Figure 3. Meter-out circuit when driving a load.

$$F_f = \begin{cases} F_s & u = 0 \\ F_c + Cu & u \neq 0 \end{cases} \tag{5}$$

So the motion equation of the piston is

$$M\frac{du}{dt} = S_c P_c - S_d P_d - P_a(S_c - S_d) - F_f - Mg\sin\alpha \tag{6}$$

The atmosphere pressure is represented by P_a.

3.4. Flow Equation

The air mass flow for the charge and discharge side of the cylinder are expressed as

$$G_c = C_c P_s \rho_0 \sqrt{\frac{\theta_0}{\theta_a}}\phi(P_s, P_c) \tag{7}$$

$$G_d = -C_d P_d \rho_0 \sqrt{\frac{\theta_0}{\theta_d}}\phi(P_d, P_a) \tag{8}$$

where the function ϕ is defined as

$$\phi = \begin{cases} 1 & P_2/P_1 \leq b \\ \sqrt{1 - \left(\dfrac{P_2/P_1 - b}{1 - b}\right)^2} & P_2/P_1 > b \end{cases} \tag{9}$$

C is called as sonic conductance and b represents the critical pressure ratio. Where ρ_o refers to the air density and θ_0 refers to the air temperature (ANR).

Using the equations above, we can calculate the pressure and temperature change in the cylinder chamber, also the displacement and velocity of piston.

3.5. Energy Consumption

As shown in **Figure 4**, we consider a pneumatic cylinder system which is driven in vertical with a load. When the load is lifted by piston, the cylinder chamber is full of the charging air. Assuming that the state change of air in cylinder is an isothermal change, the following equation is used to calculate the approximation of the energy consumption E [5] [6].

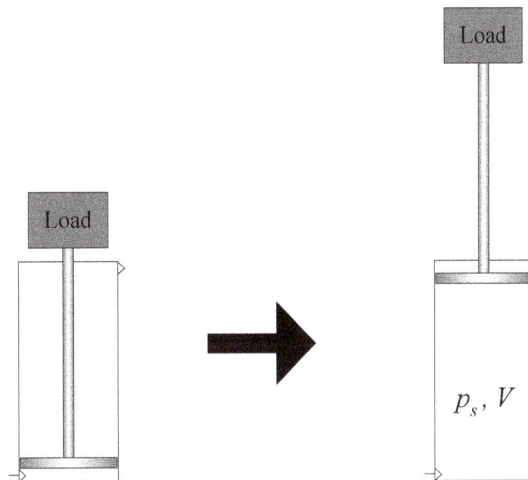

Figure 4. Energy consumption of cylinder.

$$E = P_s V \ln\left(\frac{P_s}{p_a}\right) \qquad (10)$$

From this equation, we can clarify that the energy consumption in this condition is only relevant to the supply pressure.

4. Evaluation Experiment of Meter-Out Circuit

Experimental apparatus is shown in **Figure 5**. A cylinder (MBF 40-200, SMC Co., Ltd.) of inner diameter 40 mm and stroke 200 mm was set up vertically and driven upwards. A load of mass 5 kg/16kg was set in the front head of the piston. And there is an orifice of diameter 0.4 mm at the discharge side of the cylinder.

A solenoid valve was used to control the air flow direction and a regulator was used to keep supply pressure constant and to vary the initial pressure. In order to measure the energy consumption of cylinder, we used air power meter (APM) which could measure the transient flow rate and air power at the charging side.

Before the experiment began, the charge chamber was set up to atmosphere pressure and discharge chamber was set up to supply pressure. Then, we opened the solenoid valve and began the experiment in different supply pressure.

At first, the result of the PQ characteristics of orifice is shown in **Figure 6**. Here, the value of C and b are approximately 0.04 dm^3/(s·bar) and 0.5, respectively.

Secondly, the relationship between supply pressure and equilibrium velocity is shown in **Figure 7**. It can be seen that the equilibrium velocity become quickly with the increase of supply pressure. And then the velocity reaches to a constant value when supply pressure is higher than 350 kPa (abs). This is an important characteristic of meter-out circuit because of the velocity of air is chocked.

Furthermore, the relationship between supply pressure and energy consumption is shown in **Figure 8**. As stated above, energy consumption is only relevant to the supply pressure when the load is driven by cylinder. It

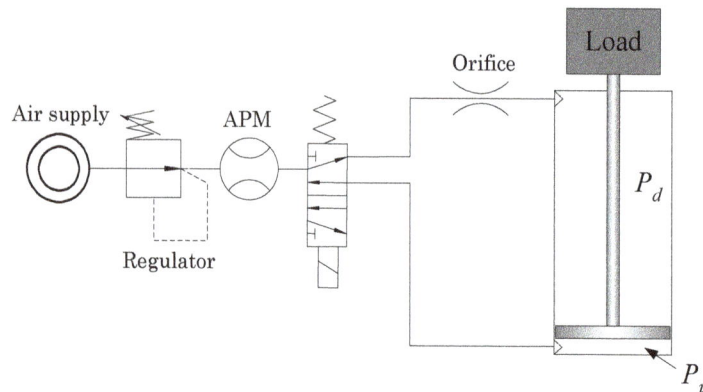

Figure 5. Meter-out circuit.

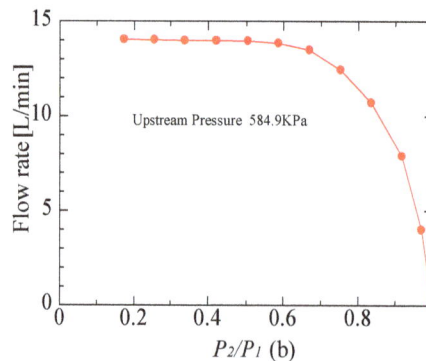

Figure 6. PQ characteristics of orifice.

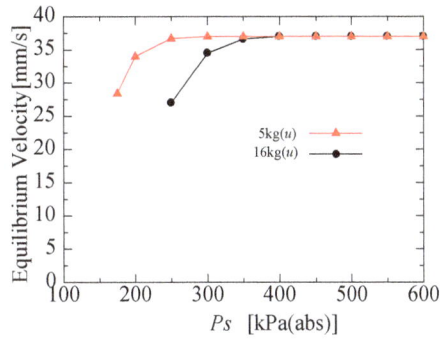

Figure 7. Relationship between supply pressure and equilibrium velocity.

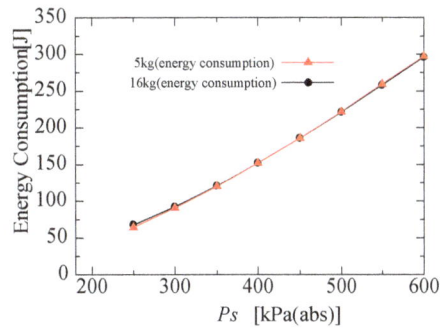

Figure 8. Relationship between supply pressure and energy consumption.

can be seen that the two curves of results in different condition are almost the same. It can be inferred that the energy consumption is increasing along with the supply pressure increased. In addition, we show the other experimental results in **Figure 9** as a reference.

The result showed that when the supply pressure is set up to 600 kPa (abs), the experimental result of energy consumption becomes 300 J during 6 seconds. On the other hand, from Equation (10) we obtained the theoretical result of energy consumption is 176 J. That means, approximately 40% of energy is lost without being used in this experiment. We considered that the reasons are orifice, piston friction and acceleration.

Furthermore, when the supply pressure is set up to 300 kPa (abs), the red line shows the power is approximately 15 W. Then we used velocity of flow and flow rate to calculate the power used in the process of charging is $W = PQ = 13.23$ W. Here, the velocity of flow is 35 mm/s. That means, over 80% of power is used for the control of velocity in meter-out circuit.

5. Evaluation Experiment of Meter-In Circuit

Experimental apparatus of meter-in circuit is shown in **Figure 10**. In this experiment, orifice was set up to the charge side. We use the same state equations and expression of energy consumption as in the meter-out circuit experiment. Experimental results in different conditions are shown in **Figure 11**. If supply pressure is low, the motion of piston reaches to an equilibrium velocity when the load is lifted. However, if supply pressure is too high, piston will move quickly due to the high-speed. We could not confirm that whether the motion of piston has reached to an equilibrium velocity or not. So we use the reaching time here instead of the velocity, and the relationship between reaching time and supply pressure is shown in **Figure 12**. It can be seen that the higher the supply pressure is, the shorter the reaching time is. In addition, **Figure 13** shows the relationship between supply pressure and energy consumption, the trends of curves are the same as the results of meter-out circuit. We can use the expression (10) to calculate the approximation of the energy consumption E. Here, according to the same supply pressure and equation, we obtained the same theoretical result of energy consumption is 176 J. The experimental result of energy consumption is 260 J. That means, the ratio of the energy loss is approximately 32%.

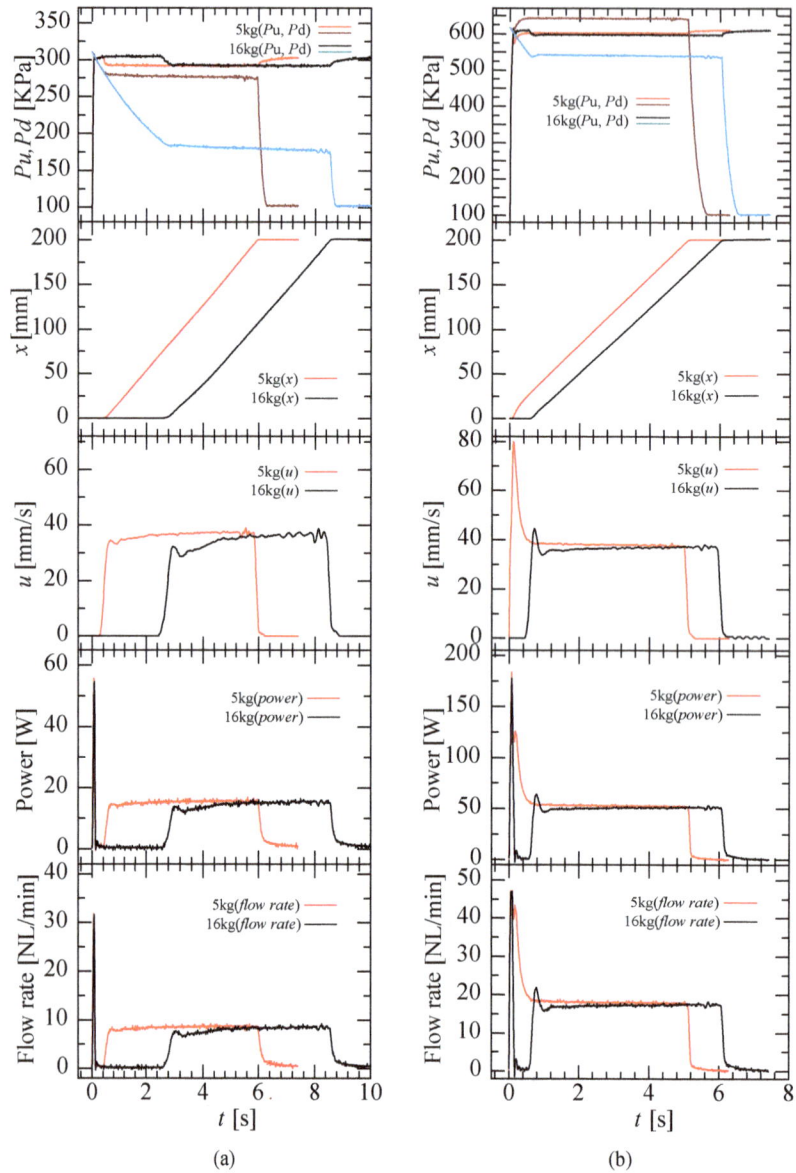

Figure 9. Experimental results of meter-out circuit. (a) P_s = 300 kPa (abs); (b) P_s = 600 kPa (abs).

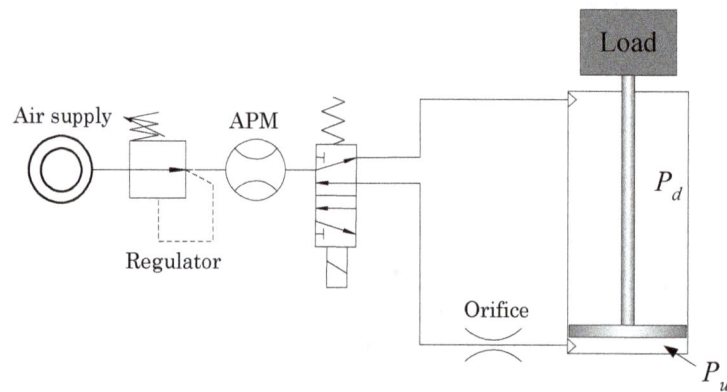

Figure 10. Meter-in circuit.

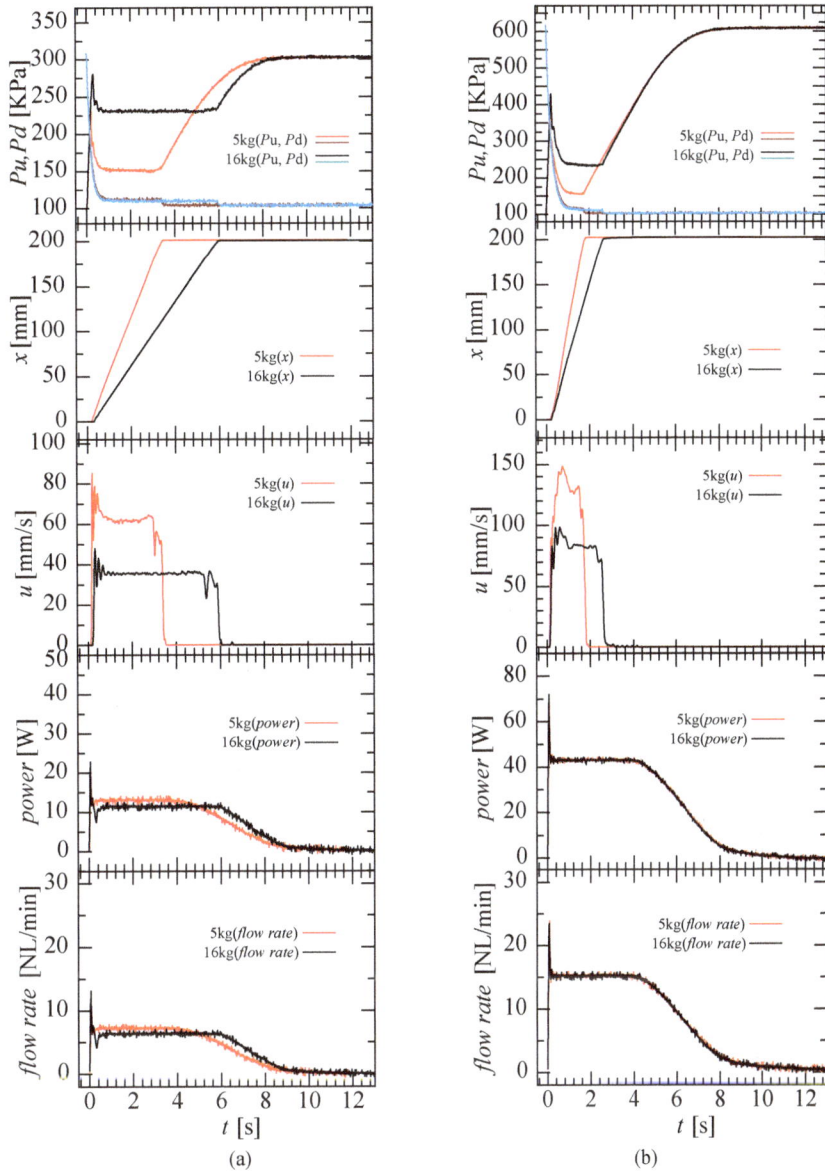

Figure 11. Experimental results of meter-in circuit. (a) P_s = 300 kPa (abs); (b) P_s = 600 kPa (abs).

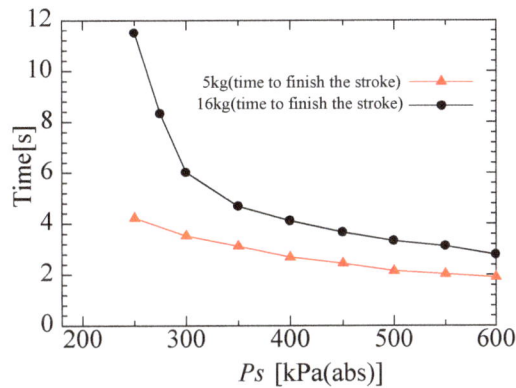

Figure 12. Relationship between supply pressure and reaching time.

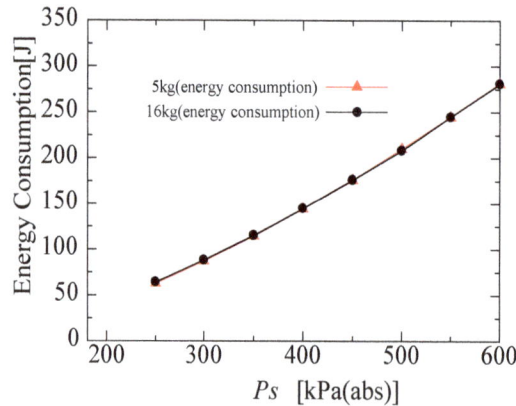

Figure 13. Relationship between supply pressure and energy consumption.

6. Conclusions

This paper showed experimental results of air transmission and energy consumption in pneumatic cylinder system by using air power meter. Experimental results showed:

1) Energy consumption is not relevant to the mass of load. It is mostly relevant to the supply air pressure.

2) The energy consumption in meter-out circuit is almost the same as in meter-in circuit.

3) There is approximately 30% - 40% energy loss occurs in this experiment. It is considered that the reasons are orifice, piston friction and acceleration.

4) In meter-out circuit, over 80% of power is used for the control of velocity.

With the assessment of energy consumption established, the quantification of energy transmission in pneumatic cylinder system will be realized. It is expected to be useful for energy saving research.

References

[1]   Shi, Y.X., Li, X.N. and Teng, Y. (2005) Research on Pneumatic Cylinder's Exhausted-Air Reclaiming Control Devices. *Proceedings of the JFPS International Symposium on Fluid Power*, Tsukuba, 7-10 November 2005, 558-563.

[2]   Cai, M.L., Fujita, T. and Kagawa, T. (2001) Energy Consumption and Assessment of Pneumatic Actuating Systems. *Journal of The Japan Fluid Power System Society*, **32**, 118-123.

[3]   Cai, M.L., Fujita, T. and Kagawa, T. (2002) Distribution of Air Available Energy in Pneumatic Cylinder Actuation. *Journal of the Japan Fluid Power System Society*, **33**, 91-98. http://dx.doi.org/10.5739/jfps.33.91

[4]   Cai, M.L. and Kagawa, T. (2007) Energy Consumption Assessment and Energy Loss Analysis in Pneumatic System. *Chinese Journal of Mechanical Engineering*, **43**, 69-74. http://dx.doi.org/10.3901/JME.2007.09.069

[5]   Li, K.W. (1995) Applied Thermodynamics: Availability Method and Energy Conversion. Taylor & Francis, London.

[6]   Cai, M.L., Kawashima, K. and Kagawa, T. (2006) Power Assessment of Flowing Compressed Air. *Journal of Fluids Engineering, Transactions of the ASME*, **128**, 402-405.

12

A Theoretical Approach to Estimating Bird Risk of Collision with Wind Turbines Where Empirical Flight Activity Data Are Lacking

Robert W. Furness[1], Mark Trinder[1], David MacArthur[1], Andrew Douse[2]

[1]MacArthur Green, Glasgow, UK
[2]Scottish Natural Heritage, Great Glen House, Inverness, UK
Email: bob.furness@macarthurgreen.com

Abstract

There are standard procedures for collecting data on numbers of birds at sites being proposed for wind farm development and evaluating collision risk for each key species. However, methods do not work well for all species. Where a local bird population is depleted, empirical data cannot provide estimates of likely collision mortality numbers if that population returns to satisfactory conservation status. Field survey methods are also inadequate for cryptic bird species. Both these problems can be important for evaluation of impacts of proposed wind farms on bird populations protected by the EU Birds Directive. We present an alternative method, based on energy constrained activity budgets and natural history, which permits assessment of likely collision numbers where empirical data are inadequate. Two case studies are presented where this approach has been successfully used to resolve disputed planning applications, one for a hen harrier population where numbers present are much below the population size at designation, and one for a cryptic species (greenshank). Our novel method helps reduce uncertainty in assessments constrained by difficulties in collecting suitable empirical data.

Keywords

Wind Farm, Collision Risk Modelling, Activity Budget, Flight, Birds Directive

1. Introduction

The European Union (EU) requires at least 20% of its total energy needs to be produced by renewable sources by 2020 [1], and the UK government is legally committed to meeting 15% of the UK's energy needs from renewables by 2020 [1]. Member states may aim higher than the threshold set by the EU. Germany aims to produce 45% of its energy needs from renewable sources by 2030 [2]. Wind farms provide a major part of the strategy to increase energy production from renewable sources. Germany installed 5.2 GW of wind power capacity in 2014, and the UK installed 1.7 GW, contributing to an increase of 11.8 GW across the EU [3].

Although making a major contribution to reducing emissions of carbon dioxide, wind farms represent a hazard for some bird populations of conservation importance, through risk of collision with turbine blades, barrier effects, and habitat loss through displacement [4] [5]. Some proposed wind farms may have predicted impacts on bird populations protected under EU law by Special Protection Areas (SPAs) designated by Member States under the Birds Directive (2009/147/EC). Where a proposed wind farm may impact upon an SPA population, a Habitats Regulations Assessment (HRA) is required under planning law. Similarly, many bird populations in the wider countryside (with no association with SPAs) receive greater protection due to their status as Annex 1 birds under the Birds Directive. Where a proposed wind farm may impact on an Annex 1 species in the wider countryside, an Ecological Impact Assessment (EcIA) is often required; an obligation which originates from the Environmental Impact Assessment Directive (2011/92/EU).

Guidance from UK Statutory Nature Conservation Bodies (SNCBs) is for studies of bird flight activity at proposed wind farms to be carried out by Vantage Point (VP) surveys of bird activity in order to assess numbers of birds at risk of collision mortality [6]. Where sensitive bird species exist, such surveys should be carried out throughout the year, over a period of at least one and often two years, with observations made for a minimum of 36 hours per season (e.g. breeding, non-breeding, migratory) per VP, and using multiple VPs to provide adequate visual coverage of the entire proposed wind farm area [6].

HRA and EcIA can be complicated, and potentially compromised, for several reasons. HRAs can be particularly affected by the SPA bird population being in unfavourable condition (e.g. the population size being lower than the reference level for which the SPA was designated). In such circumstances, it is necessary to establish if the development may affect the population's ability to recover. If the birds are absent, then VP watches cannot establish risks of collision mortality, compromising such assessment. A good example of this problem is provided by hen harrier *Circus cyaneus*, a species on Annex 1 of the Birds Directive that breeds in upland habitat where many wind farms have been (and are proposed to be) situated in the UK. Fielding *et al.* [7] identified that conservation status of hen harrier was "Unfavourable" throughout England and Wales, and in 15 out of 20 Natural Heritage Zones in Scotland. The hen harrier population of Bowland Fells SPA was "at least 12 pairs" in the years 1986-90 immediately before SPA designation [8], but no hen harriers bred there in 2012 [9]. Similarly, the species has been lost from North Pennine Moors SPA [9], and the conservation status of hen harrier is now classified as Unfavourable at Forest of Clunie SPA, Langholm-Newcastleton Hills SPA, and Muirkirk and North Lowther Uplands SPA due to large declines in breeding numbers [10].

Proposed wind farms close to these SPA populations (defined as <2 km by Scottish Natural Heritage, SNH) require HRA, but VP watches are unlikely to record levels of hen harrier flight activity representative of the designated populations because numbers are severely depleted. SNH initially objected to a wind farm proposal at Stranoch because there was insufficient information to properly assess the impact of collision mortality on the hen harrier populations of Glen App and Galloway Moors SPA. That SPA was designated on the basis of 10 breeding female hen harriers on average between 1994 and 1998, but survey work for the wind farm developers was carried out in 2009-2011 when only a single female hen harrier bred on the SPA. Survey data suggested that only 0.015 hen harriers would collide with turbines per year and therefore that the wind farm would have no impact on the integrity of the SPA. SNH rejected that argument on the grounds that the impact when the hen harrier population was at the designated population size was the relevant impact to address and not the impact when the population was close to local extinction.

That led the consultants for that project, MacArthur Green, to develop the modelling work described in this paper. The same approach was also used successfully for another wind farm (Kennoxhead) near the Muirkirk and North Lowther Uplands SPA where hen harrier breeding numbers had fallen from 29.2 females at designation to 11.7 in 2009-11 when VP surveys were carried out.

Some bird species are highly cryptic and therefore liable to be under-recorded in VP surveys of flight activity. For example, many flights by merlins *Falco columbarius* and by breeding waders are not detected during VP watches [6]. A recent wind farm proposal at Strathy South, Scotland, adjacent to an SPA with greenshank *Tringa nebularia* as a feature, was supported by VP watches that recorded very few greenshank flights despite large numbers of greenshanks known to be nesting in the area. This anomaly led SNH to question whether collision risk may have been underestimated. SNH used the approach developed for hen harrier to provide a theoretical estimate of the collision risk for greenshanks at Strathy South in order to inform their position on that planning application which had gone to public enquiry.

Case law (Case C-127/02 "the Waddenzee Case") has concluded that all aspects of plans or projects which can affect an SPA's conservation objectives must be identified, using the "best scientific knowledge in the field" and that "...the competent national authorities ... are to authorise such activity only if they have made certain that it will not adversely affect the integrity of that site. That is the case where no reasonable scientific doubt remains as to the absence of such effects".

Reaching a conclusion that a development will not affect site integrity may be impossible where that decision has to be based on compromised VP data. In this paper we outline a theoretical approach to estimating the amount of flight activity likely to occur at a proposed wind farm. We do not suggest that this novel approach should replace VP studies that provide empirical data, but rather that our theoretical approach may help to provide an alternative where empirical data appear to be compromised either by problems of low detectability, or by depleted numbers of birds in the study population. Our theoretical approach is founded on the metabolic constraint to flight activity in birds which sets a ceiling on flight activity levels, coupled with information on seasonality of activity budgets, and scenario modelling of population densities and home range use.

This approach led to SNH withdrawing its objections to the Stranochand Kennoxhead wind farms by providing confidence that it would not affect the integrity of the respective SPAs and was presented by SNH to the public enquiry for Strathy South as a key part of their case regarding the adverse impact of that proposed development on the greenshank population of Caithness and Sutherland Peatlands SPA. It has therefore played a key role in decision making for three recent planning applications for wind farms in Scotland where empirical data were judged inadequate.

Scenario modelling may have a number of useful applications for developers and SNCBs: it could be used as an early screening tool in the assessment of sensitivity of a site to collisions affecting bird populations, it could be used to increase the robustness of EcIAs and HRAs, and an agreed standardised approach (such as this) could help to reduce uncertainty in impact assessments, and therefore help to reduce delays in planning.

2. Modelling Approach

2.1. The Concept of a Metabolic Ceiling to Time Spent Flying

A constraint on flight activity that can be used to set an upper limit on activity levels is the energy cost of flight, which is far higher than can be sustained for prolonged periods except for a few bird species with special adaptations to gliding flight such as albatrosses and frigatebirds [11] [12]. Empirical evidence from many sources shows that all birds can only increase sustainable energy expenditure up to a limit of four times Basal Metabolic Rate (BMR) [13]. The development of a metabolic approach to setting an upper limit to flight activity is explored here.

The energy expenditure of birds during breeding has been measured for a wide range of species by use of the doubly labelled water method [14]. The daily energy expenditure (kj/day) during breeding (known as Field Metabolic Rate, FMR) is expressed by the equation:

$$FMR = 10.5 \, M^{0.681} \tag{1}$$

This has been tested for a large sample of bird species and with a high statistical confidence ($n = 95$, $r^2 = 0.938$, $p < 0.0001$) [14], so can be estimated for any bird species for which it has not been measured directly, from a knowledge of the mean weight of birds. Although birds can expend energy at higher rates, for example by using fat stores to fuel migratory flight, such rates of energy expenditure require metabolism of stored reserves and so are not sustainable.

The maximum sustainable work rate of birds (and of mammals) has been shown to be four times BMR, a ceiling that is set by physiological constraints on energy processing [13]-[17]. Flight represents a critical part of the daily energy budget of flying birds because flight costs are especially high and particularly so in birds that are larger than typical small passerines, so that the amount of time birds can spend in flight within their maximum sustainable work rate is constrained [12] [17]-[22].

2.2. Setting the Flight Time Ceiling

Where flight is by flapping flight at relatively high speed and without major use of economies of soaring, the mean energy cost of sustained flight is around 10 times BMR [12] [17]-[19] [21]-[24]. Numerous short flights, or flight with frequent changes of direction and altitude, as in display flights or territorial or courtship chases, will have higher energy costs [19] [23] [24]. Birds with high wing loading also have higher flight energy costs than birds with low wing loading [12] [18]. The mean cost of flapping flight for birds with high wing loading is therefore around 12 times BMR whereas for birds with low wing loading it is around 8 times BMR [12] [18] [24]. There is, therefore, a limit to the amount of time that a bird can spend in flight, especially display flights, and remain within the ceiling of four times BMR. Since most other routine activities such as foraging, walking, preening, resting, incubating eggs, digesting food, and general maintenance behaviour cost 2.6 times BMR [13] [20] [25], this sets a maximum of no more than three hours per day in flight in order to keep total daily (24 hours) metabolism below the ceiling of four times BMR for species with high wing loading [12] [18]. An example of this calculation for a high wing-loading species, assuming 3 hours of flapping flight (at 12 BMR) and 21 hours of other activity (at 2.6 BMR) is shown in Equation (2):

$$24 \times 4 \, \text{BMR} \approx (21 \times 2.6) + (3 \times 12) \tag{2}$$

This theoretical conclusion was demonstrated empirically by experimental work with starlings *Sturnus vulgaris*, a species with relatively high wing loading. Breeding adult starlings declined in body mass as they were forced to spend more than two hours per day in flight, and this decline increased dramatically when birds had to spend three hours per day in flight, indicating that this work rate had exceeded a sustainable level [13]. This interpretation has been supported by subsequent studies, for example demonstrating that kestrels *Falco tinnunculus* forced experimentally to increase work rate above four hours flight per day incurred a large survival penalty [15].

For bird species with a low wing loading, and slow flapping flight, where the metabolic cost of flight is around 8 times BMR, the same consideration indicates a ceiling of no more than six hours of flight per day as shown in Equation (3):

$$24 \times 4 \, \text{BMR} \approx (18 \times 2.6) + (6 \times 8) \tag{3}$$

While birds may be able to exceed the mean four times BMR ceiling for a short period, this will be at the cost of depleting body stores such as fat and protein, and so would be likely to compromise survival prospects. It is, therefore, reasonable to assume that birds will generally avoid exceeding the four times BMR ceiling except in the short term and for brief periods. Some species, such as pelagic seabirds, are adapted to cope with fluctuations in foraging conditions by carrying large fat reserves even when breeding [26], but this does not apply to most bird species. Although migratory birds may store large quantities of fat to fuel long-distance migrations [27], they do not carry large reserves of fat when breeding [28], and indeed the evidence is that most birds regulate body stores of energy at levels that reduce risk of starvation but minimize increase in body mass because higher body mass increases risk of capture by a predator [29].

Birds that use soaring or gliding flight will have a lower flight energy cost, so will be able to sustain higher periods of time in flight. In the extreme case of birds highly adapted to dynamic soaring, such as albatrosses, the energy cost of flight is below four times BMR [11], so that continuous flight is possible without reaching an energy-based constraint. The approach we propose would therefore not be applicable to such species because no ceiling to flight activity could be set. However, many bird species of conservation concern with regard to wind farm collision risk use flapping flight, and so there is an upper limit to the amount of time spent in flight, and therefore an assessment can be made of the collision risk associated with this flight activity.

The simplest form of ceiling set by energy costs would be to assume that birds work to their maximum capacity at all times. This would set a limit of three hours of flight per day by birds with a high wing loading that use flapping flight, or six hours of flight per day by birds with a low wing loading that use flapping flight. That approach would be precautionary since birds may not reach their metabolic ceiling all the time, and so if an assessment of collision risk based on such a model indicated no adverse effect on populations then there would be no need to carry out a more complex assessment based on knowledge of the natural history of the species. However, this metabolic ceiling approach can be combined with knowledge of the natural history of bird species in order to provide more realistic estimates, as described in the following sections.

2.3. Time Budgets Based on Natural History Knowledge

The energy costs of birds are generally highest when rearing chicks [14], and for this reason, birds time their breeding so that chick-rearing coincides with the seasonal peak of food resources [30]. It is therefore during chick-rearing that birds are most likely to be working at sustained levels close to their physiological limits. Setting aside the seasonal flights of long-distance migrant species (for which specific stores of energy need to be accumulated), flight activity of birds is therefore likely to peak during chick-rearing. The amounts of flight activity can therefore be scaled to a peak during chick-rearing, taking account of natural history knowledge of the dates of arrival, laying, hatching, fledging and departure of birds, and seasonal changes in amounts of display flight and foraging activity during early and late breeding season relative to the peak of activity during chick-rearing.

3. A Case Study: Hen Harriers

3.1. The Requirement for Scenario Modelling

It proved difficult to assess likely impacts of a proposed wind farm development at Stranoch adjacent to Glen App and Galloway Moors SPA in southern Scotland because the breeding population of hen harriers on this SPA was considerably depleted (in "unfavourable condition") and breeding success was low during the three years in which surveys of bird activity around the development site were carried out. Empirical evidence from Vantage Point surveys suggested little hen harrier activity in the proposed wind farm area, but provided no indication of how this might differ if the population was in "favourable condition" (defined as conditions similar to when the population was originally designated as an SPA).

Where an SPA population is below the levels for which it has been designated, the assessment needs to consider how a development may affect the population's ability to recover to its size at designation. SNH therefore advised that it would be necessary to demonstrate that the development will not be detrimental to the full recovery of the SPA, and that could not be achieved using empirical data on hen harrier flight activity at the site.

3.2. Defining the Breeding Cycle of Breeding Hen Harriers

Hen harriers start to visit breeding sites in southern Scotland during March, but pairing displays generally occur at the end of March, so establishment or re-establishment of nesting sitesoccurs in early April to late April, with egg laying starting from late April [31]. There is a 30 day incubation period which starts around laying of the third egg, followed by a 30 - 35 day nestling period, after which nestlings remain dependent for a further 2 - 4 weeks [31]. Thus a typical breeding attempt would involve 30 days of pre-laying activity centred around the nest site in April, followed by 31 days of egg laying and incubation during May, followed by 30 days of chick rearing in June, followed by 31 days of fledgling rearing in July when the young are still mostly fed at the nest site [32], with dispersal of the young in August.

3.3. Defining the Foraging Range of Breeding Hen Harriers

Radio tracking of breeding male hen harriers shows that their foraging activity is highest close to the nest site, and decreases with distance, with very few foraging trips taking birds more than 2 km from the nest [33] [34]. Radio tracking studies at Langholm, Orkney and Galloway [33] [34] estimated the ranging area used 90% of the time by male hen harriers as between 650 and 1180 ha (equivalent to foraging distances of 1.4 km and 1.9 km, assuming circular foraging areas). Equivalent figures for females ranged up to a maximum of 350 ha (equivalent to a maximum foraging distance of about 1.1 km).

Visual observations of breeding male hen harriers showed that they spend the majority of time foraging within 2 km of the nest site [33] [35] and only habitat within 2 km of the nest site influences the type of prey brought to the nest [36]. On the basis of the proportional effort reported from the tagging studies, the amount of time spent by foraging birds within bands 500 m wide radiating up to 2 km from the nest location was estimated.

3.4. Defining the Flight Seconds of Breeding Male and Female Hen Harriers

The main prey of breeding hen harriers is voles and small passerines such as meadow pipits *Anthus pratensis*. A vole weighing around 40 g provides about 200 kJ, while a meadow pipit weighing around 19 g provides about

100 kJ (allowing for some indigestible content and the energy costs of assimilating food). Applying the Field Metabolic Rate Equation (1) to the hen harrier, a male weighs around 350 g [37], so has an estimated average field metabolic rate of 567 kJ/day. So a male hen harrier would need to eat about 3 voles per day, or about 6 meadow pipits a day to obtain its energy requirements. Brown [38] used empirical data on food consumption by hen harriers to conclude that their daily food requirement can be provided by about "two large voles or four small ones". So Brown's estimate based on observed consumption of food by hen harriers agrees closely.

A nonbreeding hen harrier can easily obtain its daily food requirements in about an hour of hunting [38]. The energy budget of a breeding male hen harrier is greater as it includes food required for self-maintenance plus food provided to the female pre-laying in April to assist in egg formation, provided to the incubating female in May, and provided to the young in June and July. Peak flight activity of breeding males is therefore likely to occur in June-July.

Because hen harriers have a relatively low wing loading and fly slowly when hunting, their flight may be less energetically expensive than that of some birds which fly faster and have higher wing loadings. So a ceiling of between four and six hours of flight activity per day may be consistent with keeping the daily metabolic rate below four times BMR. This theoretical estimate is consistent with a small data set for tracked male hen harriers breeding in Ireland indicating that chick-rearing males flew for about four to six hours per day [39].

Females mainly guard the nest site, but spend smaller amounts of time foraging close to the nest. Males also spend small amounts of time in display flights early in the season, although these mostly involve birds that are establishing territories rather than established males.

Estimation of the amount of foraging activity (flight seconds) was therefore based on the following: Male foraging in April is one hour of flight activity per day, centred at the nest site and decreasing with distance from the nest; Male foraging in May is two hours of flight activity per day, centred at the nest site and decreasing with distance from the nest; Male foraging in June and July is six hours of flight activity per day, centred at the nest site and decreasing with distance from the nest.

3.5. Using the Flight-Seconds Estimate to Generate a Theoretical Collision Risk Estimate

Flight-seconds estimates were used in conjunction with a 17-year data set on breeding hen harriers within the Glen App and Galloway Moors, published data on hen harrier flight heights [40], and collision risk guidance from the regulator to generate a scenario based collision risk estimate. Using the data on breeding hen harriers for Glen App and Galloway Moors SPA it was possible to determine the frequency of nesting attempts within 2 km of the turbine closest to the SPA and an average nest location (average of eastings and northings) from which to estimate flight activity in the scenario model.

When considering how activity levels may relate to collision risk, about 90% of hen harrier flight activity occurs below potential collision risk height; the vast majority of flight activity taking place close to the ground [40]. Combining the estimated amount of time spent by hen harriers flying at rotor height in each 500 m band around the nest location with the proportion of each band which was swept by wind turbine rotors enabled estimation of the amount of time birds would be considered at risk of collision. Using the average flight speed this was then converted into an estimate of the number of passes made through the rotor swept area and multiplied by the probability of collision on any given transit (from the Band model [41]) to give an estimate of the number of collisions. These data permitted a collision risk analysis to be carried out for the hypothetical scenario where the hen harrier population had recovered to achieve favourable conservation status.

Empirical data collected from vantage point watches suggested 0.015 collisions per year but based on a depleted population of hen harriers during the survey period and breeding failure. By comparison, the energetics-based modelling suggested 0.27 collisions per year for a population of the size present at SPA designation and with successful breeding. This estimate was below levels that would affect the integrity of the SPA and so allowed SNH to remove their holding objection to the development.

4. A Case Study: Greenshanks

4.1. The Requirement for Scenario Modelling

Vantage point surveys at a proposed wind farm site at Strathy South in Caithness, surrounded by the most important SPA for breeding greenshanks in the UK, recorded few flights by greenshanks despite the fact that large numbers were known to be breeding in the area. This was probably due in part to the cryptic nature of breeding greenshanks [42] [43].

Several aspects of greenshank biology make their flight activity budgets particularly difficult to observe. Greenshanks often nest several kilometres away from their feeding territory. They also defend a display area, which can be some distance away from their nest site and their preferred feeding area [43]. This behaviour appears to be an adaptation to minimize risk of nest predation, which is consistent with the cryptic activity of the species when breeding. Change-overs at the nest are very infrequent, with a tendency for females to incubate from very early morning to late evening, and males to incubate during the night [43]. As a result, flights between feeding areas and nests may occur predominantly very early in the morning and late at night [44]. During most of the day, when many vantage point surveys are carried out, there may be very little flight activity taking place. After hatching, adults may escort chicks long distances away from the nest site to suitable feeding areas, so that activity may occur in very different areas at different stages in the breeding season [43].

4.2. Defining the Activity Budget of Greenshanks

Although greenshank breeding behaviour has been studied in great detail [43] [44], there are no quantitative data on flight activity budgets, so the detailed qualitative information needed to be converted into quantitative scenarios to estimate collision risk. Greenshanks use direct and fast flapping flight, and have a high wing loading, so the energy cost of flight is likely to be high [12] [18]. Assuming that most other routine activities such as foraging, walking, preening, resting, incubating eggs, digesting food, and general maintenance behaviour costs about 2.6 times BMR [13] [20] [25] [45], this is likely to set a maximum of no more than three hours per day in flight for a bird the size of the greenshank in order to keep total metabolism below the ceiling of four times BMR [13].

This constraint of a maximum of three hours of flight per day allows construction of alternative activity budget scenarios based on the maximum sustainable amount of flight activity, a more realistic scenario with a lower level of flight activity, and a likely minimum scenario with a level of flight activity set at the minimum plausible. Comparison among scenarios then allows assessment of likely collision risks and the range of variation due to uncertainty in the approach.

The timing of the greenshank breeding season is well documented. Males generally arrive back on the breeding grounds in Scotland in late March or early April, with females following soon after [43] [46]. Males and females take part in display flights, though females less than males. Display flights may occur for several hours per day and may involve long chases over several km; birds breeding in treeless areas and at high population density seem to spend more time in display flight [44]. Most display activity occurs between 0300 and 0700 hrs [47]. Breeding birds usually fly at least 500 m from nest to feeding site, but often fly five km, and may travel up to 14 km [44]. Display activity is much reduced in May while birds are incubating [44], although birds that lose the clutch may resume displaying and lay a replacement clutch.

Females normally lay four eggs, with intervals of about 44 hours between eggs [43]. Clutch completion varied from year to year and among studies but was mainly around 5th to 10th May [43]-[46], indicating that egg laying starts around 1st May. Incubation continues for 24 - 26 days [43] [44], and may be by both sexes although in many pairs most incubation is by the female. During incubation, which is predominantly by females during the day, females will make two (sometimes one to four) feeding trips per day of ca. 30 minutes, flying typically 2 km to feeding grounds [44]. Males will explore different possible feeding sites and will song-dance overhead to establish territorial ownership of suitable feeding areas [43]. Broods may move several kilometres from nest to feeding grounds, depending on local geography. Chicks fledge when about 29 days old [44], so start to fly around the beginning of July. Females tend to migrate south first (early to mid-July). Shortly after chicks can fly strongly, males and fledglings migrate south (mid to end of July or early August) [43] [46].

4.3. Defining Flight-Seconds of Breeding Greenshanks

Flight activity can therefore be estimated in relation to the following phenology: males present for 30 days in April (display and foraging), 31 in May (incubation period), 30 in June (chick-rearing) and 20 in July (tending fledglings). Females present for 30 days in April (display and foraging), 31 in May (incubation period), 30 in June (chick-rearing) and 10 in July (tending fledglings). Fledglings present for 20 days in July.

Females spend less time in display flight activity than males, but probably spend more time in commuting flights than males as they need to feed more intensely to obtain resources for egg formation and are likely, on average, to do this in a larger number of foraging bouts than required by males. In May, incubation requires females to commute between the nest and feeding areas, and the constraint of having to return to the nest site after each foraging bout will increase their commuting costs relative to the pre-laying period. The need to visit the nest site as well as feeding areas and to display will probably increase commuting flight activity of males in May compared to pre-laying levels in April, but display activity is known to decrease from peak pre-breeding levels so is assumed to be much less in May than in April.

When eggs hatch, commuting flight costs for males and females will reduce because chicks are led towards feeding grounds within the first few days after hatching. The modelling uses an estimated 50% reduction in commuting flight activity between incubation (May) and chick-rearing (June), but assumes a small increase in the amount of display/territorial defence flight activity because movement of chicks to feeding grounds results in adult territorial behaviour (*i.e.* display flights) over the area being used by chicks. In July, when chicks are larger, adult flight activity associated with commuting between sites is assumed to reduce further, whereas display/territorial activity is assumed to continue but at a lower level, at half the level in June.

Fledglings begin flying in July. The literature provides very little guidance as to how much time fledglings spend in flight in the breeding areas in July before dispersing away from the area in late July. For the modelling it is assumed that fledglings spend between 0 and 60 minutes in flight per day during July, but the most likely amount seems to be towards the lower end of this range. In 60 minutes of flight a fledgling flying at an average of 10 m/sec could travel 36 km. While it is known that fledglings will explore their natal area before dispersing, it seems very unlikely that fledglings would explore over such large distances on a daily basis.

Greenshank sustained level flight speed is around 12.3 m/sec [48]. However, birds need to take off and land, and may need to gain height during flights between nest sites and feeding sites, so commuting flights of only 500 m to a few km will average a slower speed than this, perhaps around 10 m/sec for a typical commuting flight. The activity budget assumes that, during April, males commute between display sites, foraging sites and potential nest sites, typically making three return trips of ca 3 km (1.5 km each way based on the typical distance between nest sites and foraging sites [43] [44]), at an average flight speed of 10 m/sec, so spend about 15 minutes flying. Around this estimate, are possible minimum estimates of only six minutes commuting, and maximum of 30 minutes commuting, based on descriptions of the minimum and maximum distances between nest sites and feeding sites [43] [44]. The large difference between the lower and upper estimates is due to high uncertainty in the evidence, and also to the indication in the literature that these values vary considerably among pairs and among locations, at least in part likely to be a result of local ecological features and the spatial scale of habitat mosaics.

Males show considerable song-flight activity in April and may chase other males and females in display flights. Such flights involve high aerial agility, but are likely to average around 10 m/sec. The amount of time spent in display flight is uncertain, but may be around 90 minutes per day. A minimum of 60 minutes per day is assumed. A ceiling on all flight activity is set by the metabolic cost which will limit this to no more than three hours total in flight per day for both commuting and display, so, allowing for up to 30 minutes of commuting, a maximum of about 150 minutes in display flight is assumed for days in April.

Females spend less time in display flight activity than males, but probably spend more time in commuting flights than males as they need to feed more intensely to obtain resources for egg formation and are likely, on average, to do this in a larger number of foraging bouts than required by males. In May, incubation requires females to commute between the nest and feeding areas, and the constraint of having to return to the nest site after each foraging bout will increase their commuting costs relative to the pre-laying period. The need to visit the nest site as well as feeding areas and to display will probably increase commuting flight activity of males in May compared to pre-laying levels in April, but display activity is known to decrease from peak pre-breeding levels so is assumed to be much less in May than in April. When eggs hatch, commuting flight costs for males and females will reduce because chicks are led towards feeding grounds within the first few days after hatching. The modelling uses an estimated 50% reduction in commuting flight activity between incubation (May) and chick-rearing (June), but assumes a small increase in the amount of display/territorial defence flight activity because movement of chicks to feeding grounds results in adult territorial behaviour (*i.e.* display flights) over the area being used by chicks. In July, when chicks are larger, adult flight activity associated with commuting between sites is assumed to reduce further, whereas display/territorial activity is assumed to continue but at a lower level, at half the level in June.

Fledglings begin flying in July. The literature provides very little guidance as to how much time fledglings spend in flight in the breeding areas in July before dispersing away from the area in late July. For the modelling it is assumed that fledglings spend between 0 and 60 minutes in flight per day during July, but the most likely amount seems to be towards the lower end of this range. In 60 minutes of flight a fledgling flying at an average of 10 m/sec could travel 36 km. While it is known that fledglings will explore their natal area before dispersing, it seems very unlikely that fledglings would explore over such large distances on a daily basis.

4.4. Using the Flight-Seconds Estimate to Generate a Theoretical Collision Risk Estimate

Estimation of the amount of flight activity was therefore based on three scenarios, summarised in **Table 1**. Those amounts then sum to minutes of flight activity, summarised in **Table 2**. From these data, standard methods can be used to assess collision risk for the three scenarios which provide a high, low and mid-range estimate

Table 1. Assumed flight activity budgets of greenshanks in terms of time spent commuting between nest sites and foraging areas, and in terms of time spent in territorial or sexual display flights, in minutes flight per day, of adult males, adult females, and fledglings, based on three scenarios; high level set by maximum sustainable metabolic level, low based on the minimum flight activity consistent with knowledge of breeding behaviour, and average set at a level intermediate between these extremes.

Class	Activity budget levels	Mean minutes in flight per day (commuting to forage)				Mean minutes in flight per day (display)			
		April	May	June	July	April	May	June	July
Males	Low	6	12	6	3	60	10	30	15
	Average	15	30	15	10	90	20	60	30
	High	30	60	30	20	150	40	120	60
Females	Low	6	12	6	3	0	0	0	0
	Average	30	45	20	10	10	10	10	5
	High	60	90	45	20	30	30	30	15
Fledgling	Low	0	0	0	0	0	0	0	0
	Average	0	0	0	10	0	0	0	0
	High	0	0	0	60	0	0	0	0

Table 2. Assumed flight activity budgets of greenshanks in terms of total minutes flight per day, of adult males, adult females, and fledglings (based on components in **Table 1**).

Class	Activity budget levels	Mean minutes in flight per day (commuting to forage)			
		April	May	June	July
Males	Low	66	22	36	18
	Average	105	50	75	40
	High	180	100	150	80
Females	Low	6	12	6	3
	Average	40	55	30	15
	High	90	120	75	35
Fledgling	Low	0	0	0	0
	Average	0	0	0	10
	High	0	0	0	60

[49]. This modelling allowed SNH to take a clear position regarding the impact of the proposed Strathy South wind farm on the Caithness and Sutherland Peatlands SPA at the public enquiry, giving much greater confidence than had been provided by vantage point survey data.

5. Discussion

In the case studies outlined above, an energetics approach combined with knowledge of the natural history of the species has allowed theoretical scenarios to be tested to evaluate likely collision risk. This approach was helpful to the developers and statutory nature conservation advisors in presenting a range of plausible outcomes despite the short-comings of empirical vantage point data on flight activity budgets. This approach has enabled uncertainty within the assessment to be adequately addressed by using the best available scientific knowledge and in turn this has allowed informed decisions to be made by competent authorities.

We suggest that this approach may be broadly applicable to certain birds of particular conservation concern at many onshore and offshore wind farm developments, and therefore represents a useful tool contributing to HRA, EcIA and also potentially Strategic Environmental Assessments, SEA (2001/42/EC).

In particular, this approach could be of use in a number of situations including: where SPA populations are in unfavourable condition; where flight activity is difficult to observe due to behavioural characteristics; for modelling the impact of land-use change (forest felling/restructuring); where historic regularly used nest sites are not used during vantage point surveys; where breeding fails and vantage point surveys are therefore not representative of a season where breeding is successful; or as an early risk assessment tool by developers. There is also scope to use this type of scenario modelling in Strategic Environmental Assessments (SEAs). For example, SEAs for offshore wind consenting rounds could consider impacts on seabird SPA populations using scenario modelling to ensure the best available approaches are used to inform the appropriate assessment for the proposed Plan.

In the absence of knowledge about the natural history of the species, a precautionary approach would be to assume that birds work at their energetic ceiling all the time. This would define a maximum plausible level of flight activity, and therefore would indicate a maximum plausible level of collision mortality. However, more refined estimates can be derived where there is information on the natural history of the species, allowing more realistic assessment of seasonal variation in the activity budget of birds with the metabolic ceiling used to define the seasonal peak of flight activity.

The development of new techniques is allowing the time budgets of birds to be quantified by deployment of data loggers on a sample of individuals. This will make it increasingly possible to obtain species-specific empirical evidence of the amount of time spent in flight, and therefore at potential risk of collision with wind turbines. Such evidence will not only provide greater confidence in the use of an energetics constraint modelling approach to assessment of risk, but would also permit assessment of collision risk for bird species that use soaring flight (such as golden eagles *Aquila chrysaetos*). In such species, the amount of time birds can spend flying may not be so clearly constrained by energetic considerations as seen in species that use flapping flight, but could be measured empirically. Furthermore, it may be possible to refine models further through the addition of quantitative spatial and temporal data on prey availability and habitat suitability. We therefore strongly advocate this general approach as a useful new tool in the assessment of collision risk and population-level impacts of wind farms on birds, and as an approach that is complementary to traditional vantage point survey and collision assessments.

Acknowledgements

We thank SNH colleagues for expert advice and discussion on the approach developed in this paper, especially Andrew Stevenson and Paul Taylor for discussions on the hen harrier modelling, and Des Thompson for advice on greenshank ecology. Paul Haworth kindly made available a copy of his unpublished report on impacts of wind farms on hen harriers. Mark Wilson and John O'Halloran kindly provided unpublished information on their tracking studies of hen harriers in Ireland.

References

[1]   European Commission (2011) Energy Roadmap 2050.COM885/2.

[2] Federal Ministry of Economics and Technology (2010) Energy Concept for an Environmentally Sound, Reliable and Affordable Energy Supply. Federal Ministry of Economics and Technology, Berlin.

[3] European Wind Energy Association (2015) Wind in power 2014 European Statistics. European Wind Energy Association, Brussels.

[4] Drewitt, A.L. and Langston, R.H.W. (2006) Assessing the Impacts of Wind Farms on Birds. *Ibis*, **148**, 29-42. http://dx.doi.org/10.1111/j.1474-919X.2006.00516.x

[5] Dahl, E.L., Bevanger, K., Nygard, T., Roskaft, E. and Stokke, B.G. (2012) Reduced Breeding Success in White-Tailed Eagles at Smøla Windfarm, Western Norway, Is Caused by Mortality and Displacement. *Biological Conservation*, **135**, 79-85. http://dx.doi.org/10.1016/j.biocon.2011.10.012

[6] Scottish Natural Heritage (2013) Recommended Bird Survey Methods to Inform Impact Assessment of Onshore Wind Farms. SNH Guidance. Scottish Natural Heritage, Battleby.

[7] Fielding, A., Haworth, P., Whitfield, P., McLeod, D. and Riley, H. (2011) A Conservation Framework for Hen Harriers in the United Kingdom. JNCC Report No. 441. Joint Nature Conservation Committee, Peterborough.

[8] Natural England (1992) EC Directive 79/409 on the Conservation of Wild Birds: Special Protection Area Bowland Fells (Lancashire). Secretary of State for the Environment, London.

[9] Hayhow, D.B., Eaton, M.A., Bladwell, S., Etheridge, B., Ewing, S.R., Ruddock, M., *et al.* (2013) The Status of the Hen Harrier, Circus Cyaneus, in the UK and Isle of Man in 2010. *Bird Study*, **60**, 446-458. http://dx.doi.org/10.1080/00063657.2013.839621

[10] SNH Sitelink. https://gateway.snh.gov.uk/sitelink/

[11] Shaffer, S.A., Costa, D.P. and Weimerskirch, H. (2003) Foraging Effort in Relation to the Constraints of Reproduction in Free-Ranging Albatrosses. *Functional Ecology*, **17**, 66-74. http://dx.doi.org/10.1046/j.1365-2435.2003.00705.x

[12] Pennycuick, C.J. (2008) Modelling the Flying Bird. Elsevier Academic Press, Amsterdam and London.

[13] Drent, R.H. and Daan, S. (1980) The Prudent Parent: Energetic Adjustments in Avian Breeding. *Ardea*, **68**, 225-252.

[14] Nagy, K.A. (2005) Field Metabolic Rate and Body Size. *Journal of Experimental Biology*, **208**, 1621-1625. http://dx.doi.org/10.1242/jeb.01553

[15] Moore, S.J. (Ed.) (2005) Seeking Nature's Limits: Ecologists in the Field. KNNV Publishing, Utrecht.

[16] Green, J.A., Aitken-Simpson, E.J., White, C.R., Bunce, A., Butler, P.J. and Frappell, P.B. (2013) An Increase in Minimum Metabolic Rate and Not Activity Explains Field Metabolic Rate Changes in a Breeding Seabird. *Journal of Experimental Biology*, **216**, 1726-1735. http://dx.doi.org/10.1242/jeb.085092

[17] Johnston, D.W. and McFarlane, R.W. (1967) Migration and Bioenergetics of Flight in Pacific Golden Plover. *Condor*, **69**, 156-168. http://dx.doi.org/10.2307/1366605

[18] Pennycuick, C.J. (1989) Bird Flight Performance—A Practical Calculation Manual. Oxford University Press, Oxford.

[19] Nudds, R.L. and Bryant, D.M. (2000) The Energetic Cost of Short Flights in Birds. *Journal of Experimental Biology*, **203**, 1561-1572.

[20] Rogers, D.I., Piersma, T. and Hassell, C.J. (2006) Roost Availability May Constrain Shorebird Distribution: Exploring the Energetic Costs of Roosting and Disturbance around a Tropical Bay. *Biological Conservation*, **133**, 225-235. http://dx.doi.org/10.1016/j.biocon.2006.06.007

[21] Schmidt-Wellenburg, C.A., Biebach, H., Daan, S. and Visser, G.H. (2007) Energy Expenditure and Wing Beat Frequency in Relation to Body Mass in Free Flying Barn Swallows (Hirundorustica). *Journal of Comparative Physiology B*, **177**, 327-337. http://dx.doi.org/10.1007/s00360-006-0132-5

[22] Weber, J.M. (2009) The Physiology of Long-Distance Migration: Extending the Limits of Endurance Metabolism. *Journal of Experimental Biology*, **212**, 593-597. http://dx.doi.org/10.1242/jeb.015024

[23] Butler, P.J. and Bishop, C.M. (2000) Flight. In: Whittow, G.C., Ed., *Sturkie's Avian Physiology*, Chap. 15, Academic Press, London, 391-435. http://dx.doi.org/10.1016/B978-012747605-6/50016-X

[24] Gavrilov, V.M. (2011) Energy Expenditures for Flight, Aerodynamic Quality, and Colonization of Forest Habitats by Birds. *The Biological Bulletin*, **38**, 779-788. http://dx.doi.org/10.1134/S1062359011080024

[25] Piersma, T., Dekinga, A., Van Gils, J.A., Achterkamp, B. and Visser, G.H. (2003) Cost-Benefit Analysis of Mollusc Eating in a Shorebird. 1. Foraging and Processing Costs Estimated by the Doubly Labelled Water Method. *Journal of Experimental Biology*, **206**, 3361-3368. http://dx.doi.org/10.1242/jeb.00545

[26] Jacobs, S.R., Edwards, D.B., Ringrose, J., Elliott, K.H., Weber, J.M. and Gaston, A.J. (2011) Changes in Body Composition during Breeding: Reproductive Strategies of Three Species of Seabirds under Poor Environmental Conditions. *Comparative Biochemistry and Physiology Part B*, **158**, 77-82. http://dx.doi.org/10.1016/j.cbpb.2010.09.011

[27] Newton, I. (2010) Bird Migration. HarperCollins, London.

[28] Furness, R.W. and Greenwood, J.J.D. (1993) Birds as Monitors of Environmental Change. Chapman & Hall, London.
       http://dx.doi.org/10.1007/978-94-015-1322-7

[29] Metcalfe, N.B. and Ure, S.E. (1995) Diurnal Variation in Flight Performance and Hence Potential Predation Risk in
       Small Birds. *Proceedings of the Royal Society of London B*, **261**, 395-400. http://dx.doi.org/10.1098/rspb.1995.0165

[30] Gienapp, P., Lof, M., Reed, T.E., McNamara, J., Verhulst, S. and Visser, M.E. (2013) Predicting Demographically
       Sustainable Rates of Adaptation: Can Great Tit Breeding Time Keep Pace with Climate Change? *Philosophical
       Transactions of the Royal Society B*, **368**, 1-10.

[31] Forrester, R.W., Andrews, I.J., McInerny, C.J., Murray, R.D., McGowan, R.Y., Zonfrillo, B., *et al.* (2007) The Birds of
       Scotland. Scottish Ornithologists Club, Aberlady.

[32] Newton, I. (1979) Population Ecology of Raptors. T & AD Poyser, Berkhamsted.

[33] Arroyo, B., Leckie, F., Amar, A., Aspinall, D., McCluskie, A. and Redpath, S. (2004) Habitat Use and Range Man-
       agement on Priority Areas for Hen Harriers: 2003 Report. CEH Project Number: C02018, NERC/Centre for Ecology &
       Hydrology, 31 p.

[34] Arroyo, B.E., Leckie, F. and Redpath, S.M. (2006) Habitat Use and Range Management on Priority Areas for Hen
       Harriers: Final Report. CEH Report to SNH.

[35] Arroyo, B., Amar, A., Leckie, F., Buchanan, G.M., Wilson, J.D. and Redpath, S. (2009) Hunting Habitat Selection by
       Hen Harriers on Moorland: Implications for Conservation Management. *Biological Conservation*, **142**, 586-596.
       http://dx.doi.org/10.1016/j.biocon.2008.11.013

[36] Amar, A., Arroyo, B.E., Redpath, S.M. and Thirgood, S. (2004) Habitat Predicts Losses of Red Grouse to Individual
       Hen Harriers. *Journal of Applied Ecology*, **41**, 305-314. http://dx.doi.org/10.1111/j.0021-8901.2004.00890.x

[37] BTO BirdFacts. http://www.bto.org/about-birds/birdfacts

[38] Brown, L. (1976) British Birds of Prey. Collins, London.

[39] Professor John O'Halloran, Personal Communication.

[40] Haworth, P. and Fielding, A. (2014) A Review of the Impacts of Terrestrial Wind Farms on Breeding and Wintering
       Hen Harriers. Draft Report to Scottish Natural Heritage. Haworth Conservation, 34 p.

[41] Band, W., Madders, M. and Whitfield, D.P. (2007) Developing Field and Analytical Methods to Assess Avian Colli-
       sion Risk at Wind Farms. In: De Lucas, M., Janss, G. and Ferrer, M., Eds., *Birds and Wind Power*, Quercus Editions,
       Madrid, 259-275. www.quercus.pt

[42] Hancock, M.H., Gibbons, D.W., Thompson, P.S. (1997) The Status of Breeding Greenshank (Tringanebularia) in the
       United Kingdom in 1995. *Bird Study*, **44**, 290-302. http://dx.doi.org/10.1080/00063659709461064

[43] Nethersole-Thompson, D. and Nethersole-Thompson, M. (1979) Greenshanks. T & AD Poyser, Berkhamsted.

[44] Nethersole-Thompson, D. (1951) The Greenshank. Collins New Naturalist, London.

[45] Hale, W.G. (1980) Waders. Collins New Naturalist, London.

[46] Thompson, D.B.A., Thompson, P.S. and Nethersole-Thompson, D. (1986) Timing of Breeding and Breeding Perfor-
       mance in a Population of Greenshanks (Tringanebularia). *Journal of Animal Ecology*, **55**, 181-199.
       http://dx.doi.org/10.2307/4701

[47] Pendlebury, C., Zisman, S., Walls, R., Sweeney, J., McLoughlin, E., Robinson, C., *et al.* (2011) Literature Review to
       Assess Bird Species Connectivity to Special Protection Areas. Scottish Natural Heritage Commissioned Report No.
       390.

[48] Alerstam, T., Rosen, M., Bäckman, J., Ericson, P.G.P. and Hellgren, O. (2007) Flight Speeds among Bird Species: Al-
       lometric and Phylogenetic Effects. *PLoS Biology*, **5**, e197. http://dx.doi.org/10.1371/journal.pbio.0050197

[49] MacArthur Green (2016) Greenshank Collision Mortality Estimated Based on Ecological and Behavioural Studies.
       SNH Commissioned Research Report. (in press)
       http://www.snh.gov.uk/publications-data-and-research/publications/search-the-catalogue/

13

Integration of Renewable Energy Resources in Microgrid

Manzar Ahmed, Uzma Amin, Suhail Aftab, Zaki Ahmed

Electrical Engineering Department, Faculty of Engineering, University of South Asia, Lahore, Pakistan
Email: azaki786@usa.edu.pk

Abstract

Microgrid is a new concept in power generation. The Microgrid concept assumes a cluster of loads and micro sources operating as a single controllable system that provides both power and heat to its local area. Not much is known about Microgrid behavior as a whole system. Some models exist which describe the components of a Microgrid. In this paper, model of Microgrids with steady state and their transient responses to changing inputs are presented. Current models of a fuel cell, microturbines, wind turbine and solar cell have been discussed. Finally a complete model built of Microgrid including the power sources, their power electronics, and a load and mains model in MATLAB/Simulink is presented.

Keywords

Microgrid, Diesel Engine, Fuel Cell, Microturbines, Wind Turbine, Photovoltaic, Genetic Algorithms

1. Introduction

The main components of Microgrid are mini-hydro, solar cell, wind energy, fuel cell and energy storage system. These are integrated for electricity generation, energy storage, and a load that normally operates connected to a main grid (macro grid). Microgrid can operate in two modes: one is grid-connected and the other is stand-alone mode. The main benefit of Microgrid is that it can operate in standalone mode or main grid disconnection mode. The Microgrid can then function autonomously. Generation and loads in a Microgrid are usually interconnected at low voltage. But one issue related to Microgrid is that operator should be very vigilant because numbers of power system are connected to Microgrid. In the past, there was single entity to control.

In Microgrid generation resources can include such as fuel cells, wind, solar, or other energy sources as shown in **Figure 1**. These multiple different electric power supply generation resources have ability to isolate the Microgrid from a large network and will provide highly reliable electric power. Produced heat from gen-

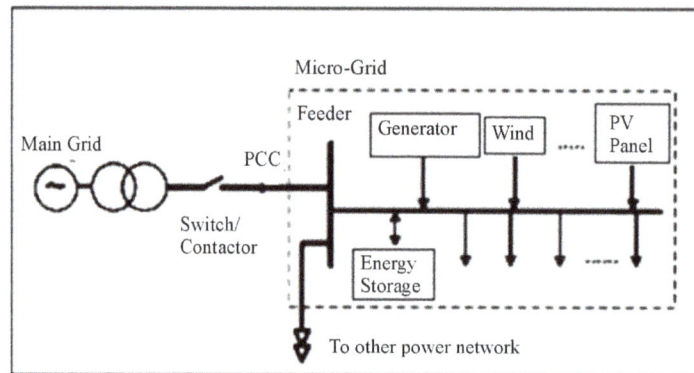

Figure 1. Microgrid systems [1].

eration sources such as microturbines could be used for local process heating or space heating, allowing flexible tradeoff between the needs for heat and electric power.

The followings are parameters of Microgrid:

- Small Microgrid covers 30 - 50 km radius;
- The small Microgrid can produce power of 5 - 10 MW to serve the customers;
- It is free from huge transmission losses and also free from dependencies on long-distance transmission lines.

2. DC Microgrid

The concept of the DC Microgrid closely parallels Thomas Edison's original concept of local DC power generation [1]. This concept could be implemented in 21st century power generation and utilization system. Although the distance between electricity generation sources and loads must be at a minimum, cost-effective solar and wind farms at a particular site also meet the requirements of the DC Microgrid. Minimum conversion from DC to AC and or AC to DC must take place. This increase is partly due to the compatibility of local DC electricity infrastructures, which co-exist with existing electrical infrastructures that are based upon alternating current (AC).

Regarding power storage, DC storage devices such as batteries, capacitors, and fuel cells also meet the requirements of local DC electricity. In essence, the self-sufficient power network of energy generation and energy storage sources, known as the Microgrid is basically a smaller version of the larger power grid. In the absence of no external connectivity of the Microgrid with the main grid is required. This self-sufficient PV-based "Nanogrid" can generate, store and distribute its own power, which is ideally suited for rural electrification.

3. The Need of DC Local Grid

Followings are the reasons Microgrid system used in generation of local DC power:

- The traditional model of large base-load AC centralized electrical power generation long haul distribution via high voltage transmission and low-voltage lines causes huge losses of energy and costs required to operate such systems. Direct Current (DC) electricity locally generated by renewable energy sources such as solar panels, windmills used with a minimum conversion (DC to AC or AC to DC) and minimum transmission can reduce energy losses by as much as 30% or more energy. That is typically lost in AC generation, transmission, and distribution infrastructures.
- Unlike 20th century technologies, the cost of generating local power generated from solar PV and wind systems is decreasing daily, with the substitution of DC for AC power further reducing that cost. Since 2008, solar PV panel prices have fallen well over 70 percent cost of wind turbines decreasing by 40 percent during that same percent. The cost of centralized AC power generation has increased. Wind and solar generated power is cheaper than coal-fired power plants when considering the social costs of carbon foot prints. Some utilities are now using more PV as it has become more cost effective with the natural gas.
- DC-based PV and wind power systems are more reliable than AC based systems. While the inverter cost is less than about 20% of PV system cost, any system malfunction can shut the system down, with a total loss in energy production [2]. The Wind energy can be more reliable in DC form because it greatly reduced complexity of the mechanical transmissions system, which is required for wind turbine which generates AC [3].

- Batteries, capacitors and fuel cells can be used to store DC electricity. The use of AC in place of DC increases the cost of storage device, as with batteries in which AC based storage systems increase their cost to as much as 50% [4].
- Integrated circuits and other solid-state devices revolutionized virtually every facet of human life. But a very few of these cases, (e.g. certain motor-based systems), all other loads require DC power, with more of these loads increasingly being DC. For example cathode-ray tube televisions, solid-state TVs do not use AC current. Similarly, though lighting consumes about 20% of the electricity produced worldwide, it too uses DC power. Also, unlike DC current, typical AC based cell phone chargers waste approximately 20% - 35% energy used [5]. Electrical vehicles do not require AC power for charging batteries. With revolution in the IT industry, more semiconductor based electronics are being used, with a concurrent increase in DC loads and a decrease in AC loads.
- The battery-based hybrid and electrical vehicles and solid-state based LED lighting are transforming the transportation and lighting industries, both of which are powered by direct current.
- Energy-efficient appliances use adjustable speed motor drives in which a rectifier converters the AC from the grid into an internal DC bus voltage.
- A DC Microgrid is the key enabler of the "zero energy building model". With minimum wastage in transmission and conversion, the use of locally generated DC electricity can provide 100% energy needs of a building.
- The worldwide adoption of DC power can prevent such a redundancy of efforts by providing uniform voltage standards worldwide, thus reducing the cost of related power electronics to yield an overall lower manufacturing cost of every DC-based electrical system. Local DC electricity as affordable electricity to underprivileged people in worldwide 2.6 billion people in the developing countries depend on bio fuel energy (fuel from wood, charcoal, and animal dung) to meet their energy needs for cooking and other daily necessities.

The worldwide adoption of DC power can prevent such a redundancy of effort by provide uniform voltage standards worldwide, thus reducing the cost of related power electronics to yield an overall lower manufacturing cost of every DC-based electrical system. The Microgrid architecture is shown in **Figure 2**.

4. AC and DC Transmission System

In past we are using AC transmission system and it become now mature system. The output of an AC transmission line is directly proportional to the square of the voltage and inversely proportional to the impedance of the transmission line. But it depends on the distance between load and line. Losses increases with distance. To attain high level transmission capability in long distance transmission line AC lines, a simple way is increase the voltage level. For small-to-medium scale RE power supply transmission lines below 330 kV are usually used. For large scale, long-distance RE power, transmission lines above 500 kV are usually needed. An example of AC transmission above 500 kV for RE integration is given below. For transmission & distribution, major AC/DC grids will be interconnected and the consumer will combine consumption and production of energy. Grid design will evolve to a network of interconnected small and large grids shown in **Table 1**.

5. Power Generation & Storage Issues

Followings are the related power generation system and energy storage issues with related dc system:
- Maintaining **grid reliability requires precise synchronization of voltage and current**. The ratio of actual power to theoretical power is the "power factor", and typically runs between 85% - 95%. As power factor falls, generators still make the same amount of power and burn the same amount of fuel, but less gets to the load, so the effect is to lower system-wide fuel efficiency.
- Motors, capacitors, and other electrical devices cause current to shift out of phase with voltage, so power factor degradation is unavoidable and grid operator must take actions to correct.
- Location **is always play important role while calculating losses**. The resistance of a wire is directly related to the length of the wire. Voltage is directly proportional to the resistance and current. When more wire used to separate a generator and the load, it greater current and therefore greater energy losses through that wire for any given voltage. These line losses typically run 3% - 6% on average but increase rapidly during peak hours when wires are congested and often exceed up to 25%.

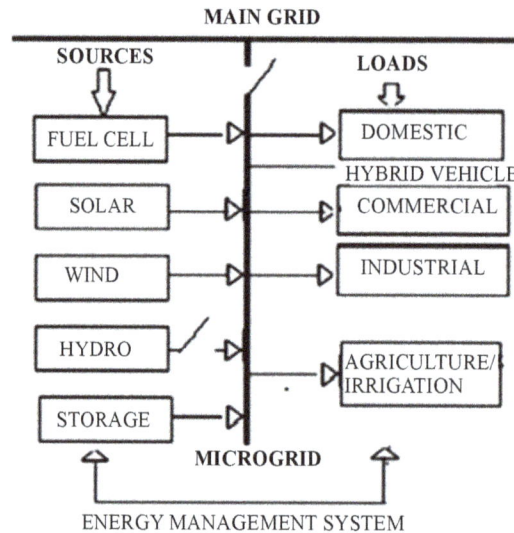

Figure 2. Architecture of Microgrid.

Table 1. Microgrid components.

Power Generation System	Storage issues
Resources, devices, machine, electrical control	• Medium size and micro-turbines • Permanent magnet and doubly fed induction generators • Efficient inverters • Self-excited induction generators • Asynchronous or synchronous generators • Induction machine and DC machine based storage system such as flywheels
Advanced electrical design of devices and components	• Dynamic design of wind turbines • Bio-fuel based turbines • PV Solar cell system with high insulation system and sun tracking system to get maximum power • Minimum looses smooth output of wind farms
Types of loads	• Dynamic modeling of loads • Adoptive and frequency/voltage dependent characteristics of loads • Real time load shedding planning • Adoptive load management • SCADA based control
Energy management methods	• Both systems adoptable either centralized vs. decentralized control system. • Impedance matching to control reactive power and minimizing losses • Efficient protection methods • Implementation of hybrid sources to control fault current levels in the micro-grid • Smooth coordination with other micro-grid with the conventional grid • Intelligent Interface system • Operate in both modes grid connected or Islanding and detection methodologies • Low-voltage controlling system

6. Methodology

Following steps are proposed for implementation of Microgrid system:

- First we need to select electric power supply system such as renewable energy resources according to requirement and availability of input source;
- Then we need to integrate all these resources in Microgrid such as renewable energy resources;
- Energy storage and Management system;
- Integration of renewable energy in Microgrid;
- Energy control and management in Microgrid.

7. The Proposed Model

Figure 3 shows the proposed model for Microgrid system. The inputs of the system are renewable energies such as solar energy, wind energy, hydro and fuel cell energy. After minimizing the losses and increase the efficiencies of the systems the energies can be combined standard are applied on the energy input and finally stored energy can be integrated into grid.

7.1. Selection of Renewable Energy Resources

The most important step is to select the types of renewable energy resources because it depends on the location and environment. In this research work three renewable energy systems are considered such as solar, wind and fuel cell. The advantages of selected system are given in **Table 2**. For all these systems we need efficient energy storage and monitoring system. With revolution in the IT industry, more semiconductor based electronics are being used, with a concurrent increase in DC loads and a decrease in AC loads.

7.2. Energy Storage and Management in Microgrid

There are following reasons for energy storage:
- Smart grid;
- Increasing use of demand response;
- Commonly available electricity price signals;
- Regulatory incentives;
- Transmission capacity constraints;
- Increasing usage of electric vehicles;
- Increasing usage of renewable energy sources;
- Distributed energy sources;
- Environmental concerns due to fossil-based fuel use;
- Advancements in storage technology.

7.3. Integration of Renewable System in Microgrid

Distributed generation can support weak grids, adding grid voltage and improving power quality. In certain circumstances, distributed generation can be used in conjunction with capacitor banks for management of power flows or to manage active and reactive power balance. In **Figure 4** the proposed method of integration of energy in girds is shown. In the block diagram all the resources such as wind turbine energy, solar energy (PV), Fuel cell and hydro is integrated in the main grid (IG). The output of fuel cell hydrogen (H_2) is provided to the vehicles for transportation as a fuel. If harvested and taken care of control system, it can reduce environmental impacts such as:
- Can reduced emission of green house gases hazard for environment;
- Can reduced dependencies on local or imported fuels and increased energy security due to distribution generation and share of all energy sources.

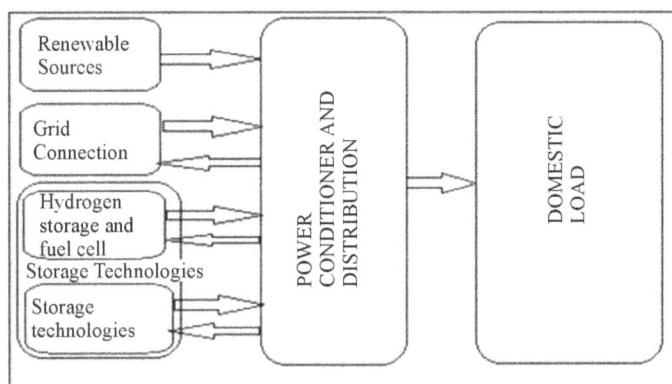

Figure 3. Proposed Microgrid distribution system.

Table 2. System components.

Main Sources	Macro Grid Generators
Renewable Energy Resources	• Small hydro system • Wind energy • Bio fuel energy • PV solar cell • Ocean energy and geothermal energy
Energy Storage Systems	• Fuel cells (PEM, SOFC and alkaline) • Batteries (lithium) • Super capacitor • Pump storage • Fly wheel
Types of Loads	• Small (domestic) • Medium (commercial) • Large (agriculture/irrigation three phase) and industrial-three phase
Technical Parameters	• Preferably linear • Balanced line not unbalanced • System should be dynamic

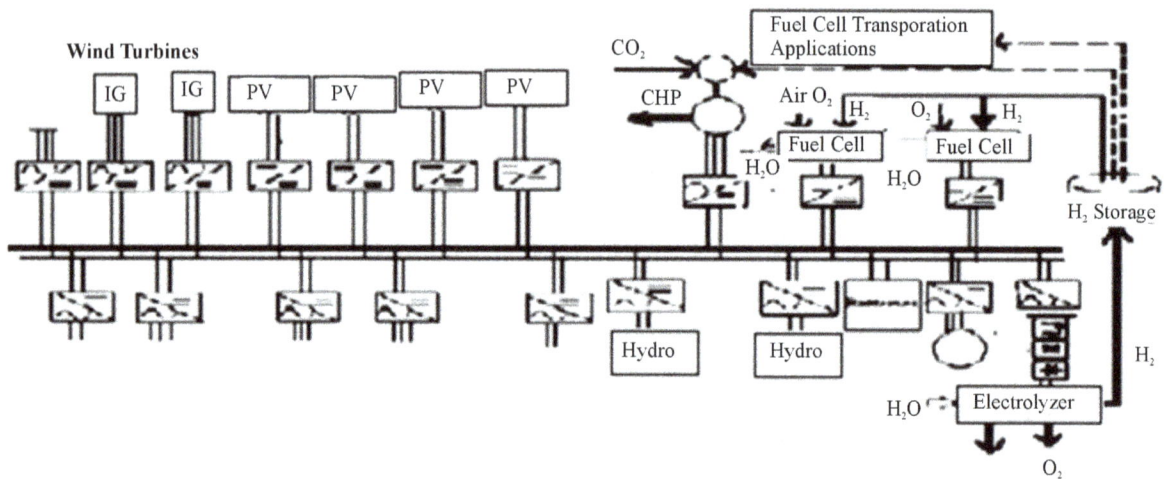

Figure 4. Integration of energy resources.

7.4. Proposed Modes of Operation

In future each home would have its own uninterruptible renewable energy system and has the capability to work in two main modes of operation such grid-connected mode and stand-alone mode. This system will be capable of producing a smooth, uninterrupted transition between these modes by using an advanced islanding detection and resynchronization algorithm. The both mode of transition power cycle is shown in the **Figure 5**. There are two modes of operations as shown in figure below, stand alone and grid connected. When the power is shut off from the utility grid the system goes to the islanding mode and when the power is available from the utility grid the system will synchronize and connected to the utility grid. Output power from the fuel cell is available for Plug in hybrid electric vehicles (PHEV) system. Bidirectional Power Converter (BPC) controls two parameters active current and active/reactive power. In standalone mode BPC control two parameters AC frequency and voltage.

The following paragraph will provide working principle of these modes. There are always changes in the condition of weather and speed of wind. Wind and solar system are condition dependents. These systems should be adoptable. This means that these systems should change operating parameters according to the changes in weather conditions. These parameters are very unpredictable. These parameters cannot be predicted even the operators in the grid cannot predict the speed of wind and sun light conditions for solar cell farms accurately. The output of these systems still always varies and operator needs to be very careful in the grid. Still there are lots of issues not to solve.

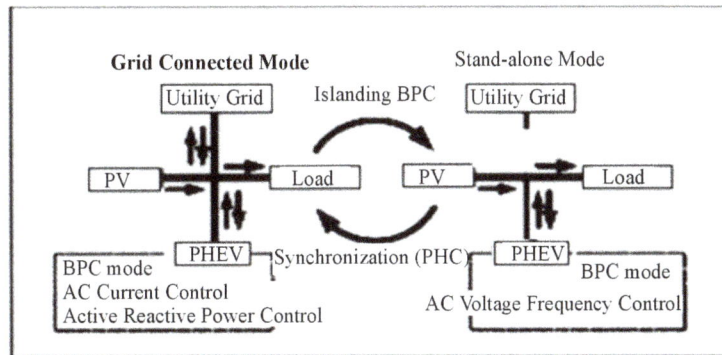

Figure 5. Mode of operation.

In grid control of voltage and frequency are very necessary because in seconds to minutes time scale the grid operators must know how to deal with fluctuations in frequency and voltage on the transmission in transmission line system. From these parameters if anyone left unchecked it could damage the system as well as equipment used in the grid. In that case operator may need to start generators to insert power (may be active or reactive) into the Microgrid but this power is free of cost not for sale to consumers but it used in order to balance the actual and predicted forecasted generation of electrical power which is necessary to minimize the variation in frequency and voltage on the grid. These necessary services must be provided by the grid when required.

Typical services are:
- Variation in the frequency occurs rapidly and must be control. These variations can control by using device Automatic generation control (AGC) signals to grid generators;
- Some generator must be in the reserve when fault occurs in the main generators reserve generators should provide power within 10 minutes. These reserves generators are used when main generator on the system become faulty or deactivates from the system unexpectedly;
- Some generators run in the same way as reserve generators but they have same function but may have a slower response time;
- When voltage goes down in main grid the generators used for reactive power to boost up the voltage to desire level when required in grid;
- In Microgrid generators should be available to restart the power system if the load shading persists for long time.

8. Power Management System

In Microgrid system all energy recourses can be control such as renewable resources and also control changes in the grid's operating conditions. It can provide additional benefits because it is distributed generation system or when installed at the transmission level. The power management system is shown **Figure 6**. In below diagram there are three main parts of the system, distributed generation (DG), Microgrid and Loads and storage. Main components of loads and storage are variable speed drives (HVAC, heat pumps), lighting control and battery storage for electrical vehicles (EV).

Key components in the Microgrid are:
- Smart metering, which enables two-way communication between utilities and customers (including electrical energy storage facilities such as rechargeable batteries and electric vehicles) or dispersed generation (DG) [6];
- Information technologies, which enable optimal control of the total grid even where very many DG units are integrated;
- Energy management systems, which implement the most efficient possible use of electrical energy for customers;
- Advanced control and protection systems, which improve the security and reliability of both small- and large-scale power networks [7];
- The technologies for Microgrid have mainly been debated in the context of optimization of small power networks.

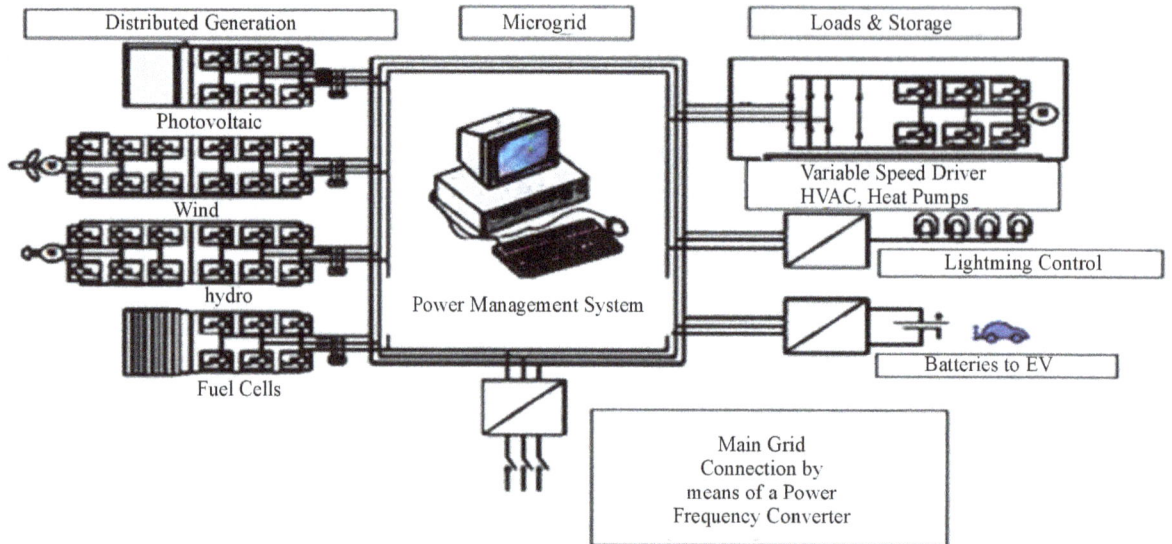

Figure 6. Complete circuit of green energy integration.

9. Implementation of Distribution System

- In future to provide clean environment we must reduce CO_2. One way is to reduce it by using it in other system and convert it to another form. In micro grid the preferred system is decentralizing system. By decentralizing the micro grid the system the system efficiency would be increased and hence reduce the amount of electrical energy lost in transmission system. The transmission lines may be reduced by increasing the efficiency of the power system and this will be more economically beneficiary to reduce number of power lines that will need to be installed in the future to fulfill the demand. CHP system will play very important role in micro grid system. CHP Power and heat in micro grid system could provide additional benefits and will increase efficiency by up to 85% - 95%. This would be significant increase in Grid system the current fossil fuel system only have an efficiency of 34% [8].
- Green energy system must be implemented to satisfy the customers those customers who are not satisfied with the local electric supply system they can approach to green energy via the electrical power grid or they can install their own energy system such as renewable energy system. The consumer who are not satisfy can install their own electric power supply system such as mini-hydro, fuel cell, PV solar cell and wind turbines and many other types of renewable energy systems which are available in countries. In future Pakistan must install Microgrid system for green, clean and pollution free environment.

10. Mathematical Modeling

The following equation used for calculation of active reactive power and harmonics.

$$YU = P + JQ \tag{1}$$

$$U_K \in \left\{ U_{k,1}, \cdots\cdots, U_{k,m} \right\}, \left| I_k \right| \leq I^{\max} \tag{2}$$

For $k \in$ Step-down

$$U_i^{\min} \leq \left| U_i \right| \geq U_i^{\max}, \left| I_{i,k} \right| \geq I_i^{\max} \tag{3}$$

$$P_i^{\min} \leq P_i \left(U; \pi_i \right) \leq P_i^{\max} \tag{4}$$

$$Q_i^{\min} \leq Q_i \leq \left(U_i; \mu_i \right) \leq Q_i^{\max} \tag{5}$$

For the selected area value of ΔU selected as
$\Delta U \geq 4\%$

$\Delta U \geq 3\%$

$\Delta U \geq 2\%$

$\Delta U < 2\%$ (at main point)

11. Simulink Model for Microgrid

In Microgrids all renewable energies are integrated in DC system and stored in grid. Both AC and DC bus can be used. Proposed system block diagram is shown in **Figure** 7. For the load profile new elements like PV, E-car, batteries, fuel cells, wind etc can be used.

For simulation methods, load profile simulation used for domestic load. Unbalanced dynamics protection/ restoration strategies used for in proposed model. The Simulink model used in this research work is shown in **Figure 8**. In Simulink model as shown in figure the number of load points are 16 and power sources used in this model are fuel cell (FC), wind turbines and photovoltaic cells in Microgrid. For the simulation purpose the solar cells power is used with main hydro power.

The active power supply remains constant, independent of the voltage. If this is above or below predefined voltage limits, the DC-In feeder is removed from the network. The reactive power demand for the inverter is simulated by a factor for the required reactive power. Photovoltaic in feeders behave the same way in any procedure based on load flow.

Short Circuit—Maximum: the current supply is constant and calculated from the installed peak power.

Short Circuit—Minimum: this method ignores DC-In feeders.

Short Circuit—Standard: the current supply is constant and calculated from the installed peak power and the current factor for active power. In all the procedures based on short circuit (multiple fault, protection, etc.), photovoltaic supply behaves in the same way.

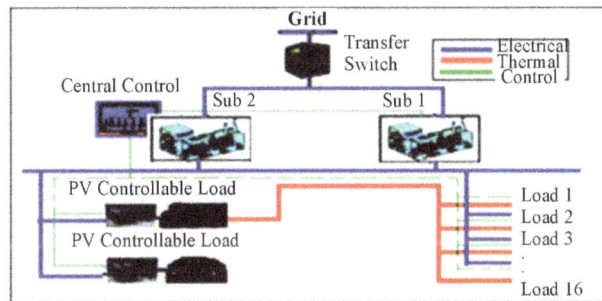

Figure 7. Blocked diagram of proposed system.

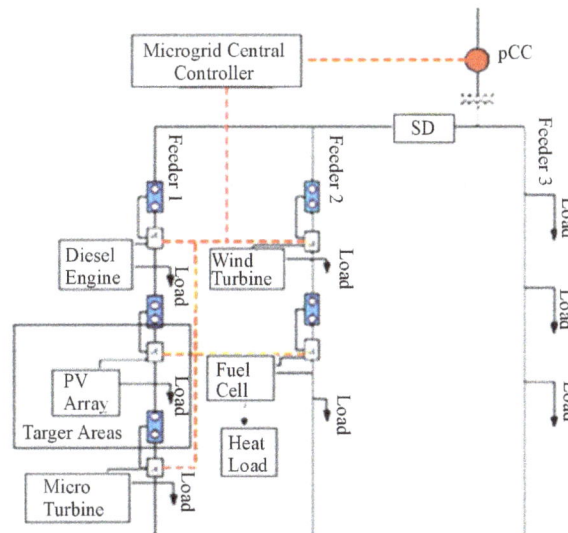

Figure 8. Microgrid Simulink model.

11.1. Smart Generation

- Distributed generation like PV cell, E-cars, batteries, fuel cells, wind, etc.;
- Virtual power plants;
- Decentralized energy management.

11.2. Load Profile Simulation

- Smart metering – Meter reading + Active meter controlling;
- Load management;
- Cost models.

For simulation the total power is 89.76 KW to 255 KW for PV Load efficiency 30% to 100% selected for area. Single line diagram with 30% to 100% load efficiency is shown in **Figure 9**, **Figure 11**, **Figure 13**, **Figure 15**, **Figure 17**, **Figure 19**, **Figure 21** and **Figure 23** respectively. The total load is distributed among the eight domestic loads. The curve shows the covered area and total distribution in smart grid. Load efficiency graphs with 30% to 100% Photovoltaic efficiency shown in **Figure 10**, **Figure 12**, **Figure 14**, **Figure 16**, **Figure 18**, **Figure 20**, **Figure 22** and **Figure 24** respectively. It can be concluded from graphs that power decreases increasing covered area.

It can be concluded that as the load efficiency of the PV cells increase the distribution curve limited to support loads. It can be seen at 30% load efficiency it covered maximum area as the efficiency increased to 112% in Microgrid the curve limited to wide area. Eight points of loads were selected. In future all alternative energy resources can be applied in Microgrid using same method.

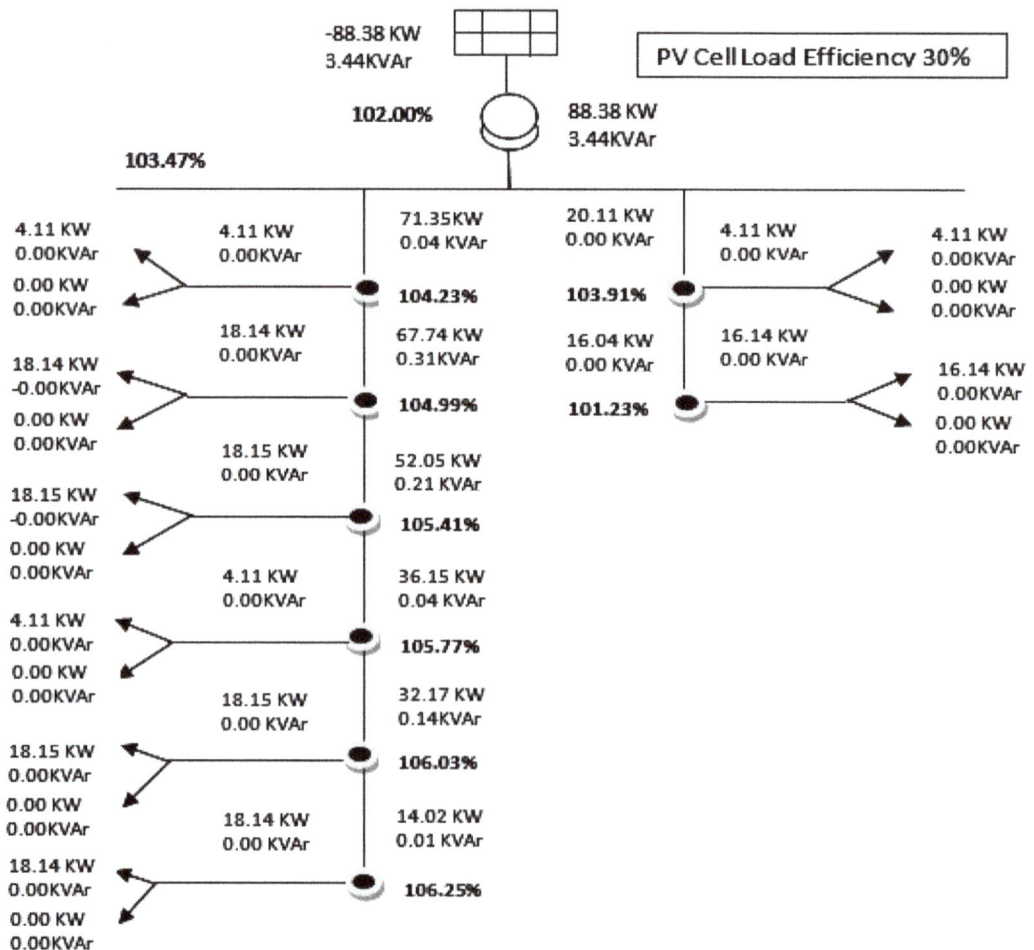

Figure 9. Single line diagram with 30% load efficiency.

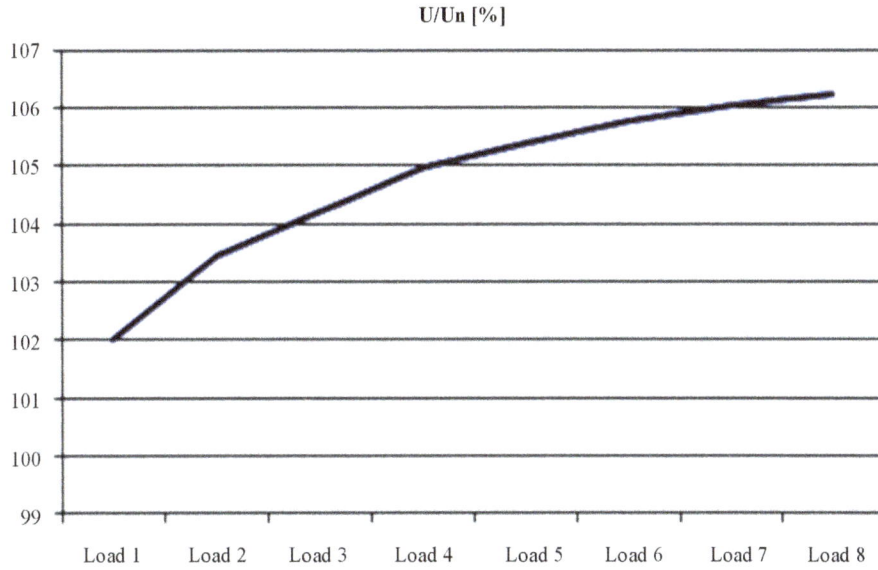

Figure 10. Loads efficiency with 30% PV efficiency.

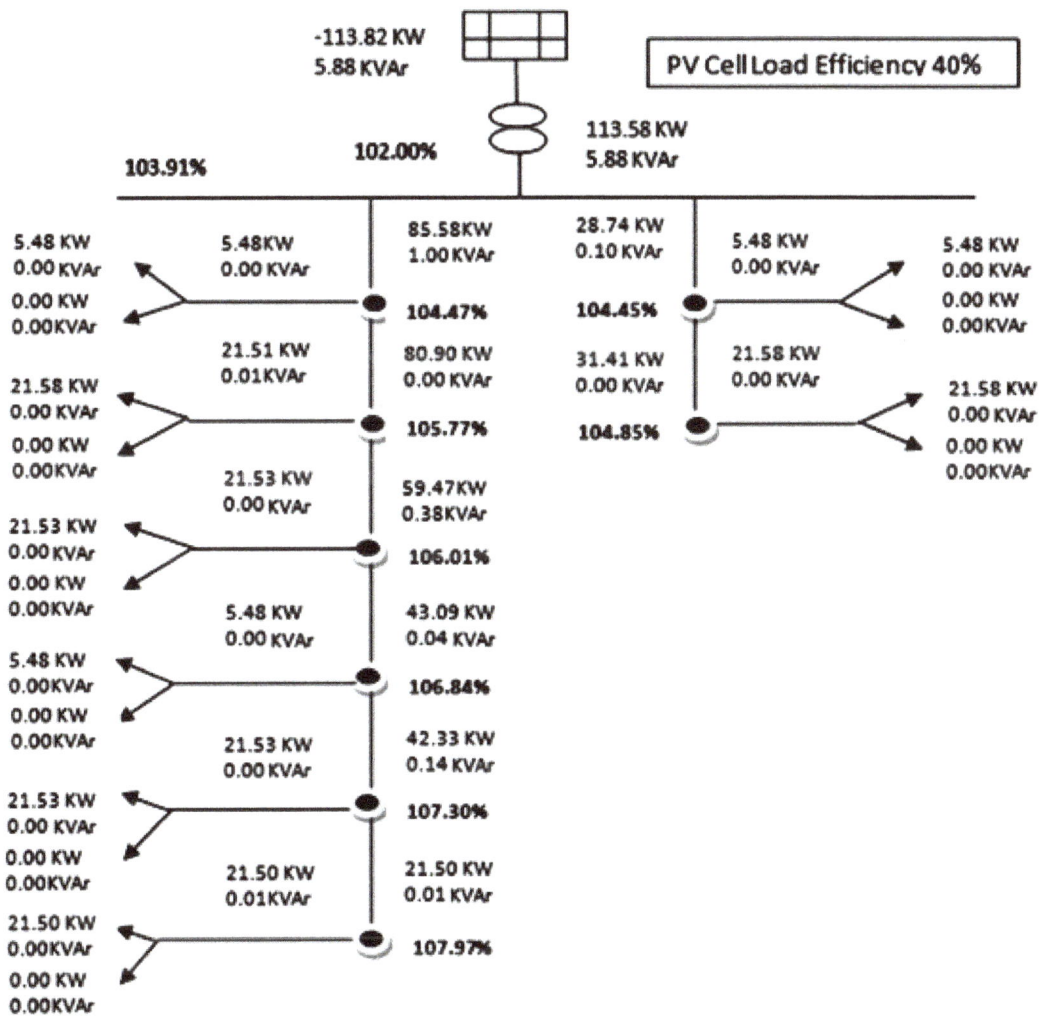

Figure 11. Single line diagram with 40% load efficiency.

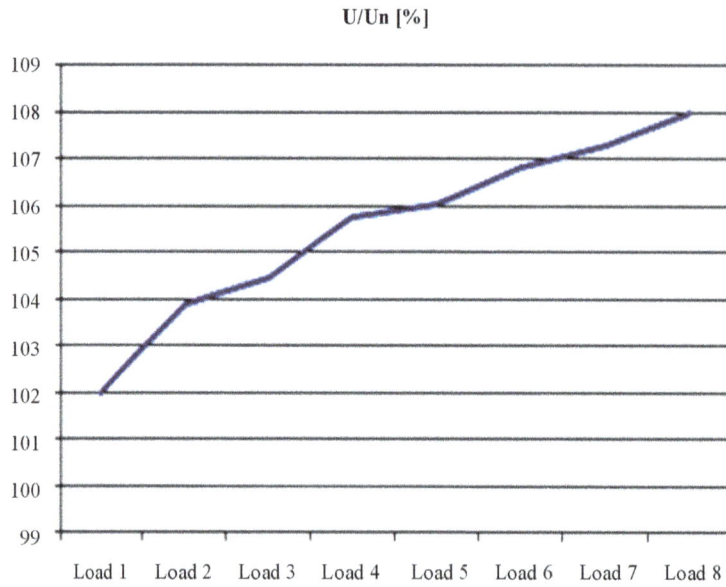

Figure 12. Loads efficiency with 40% PV efficiency.

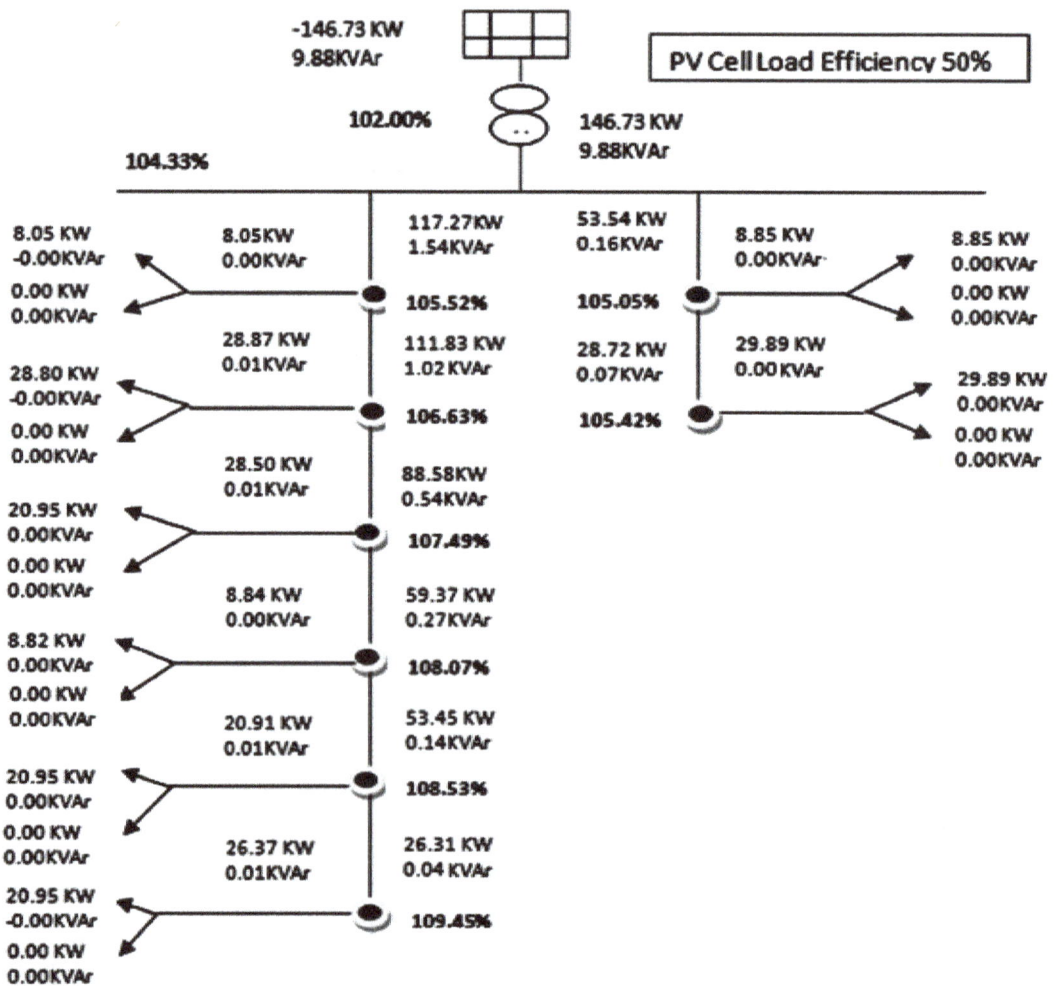

Figure 13. Single line diagram with 50% load efficiency.

U/Un [%]

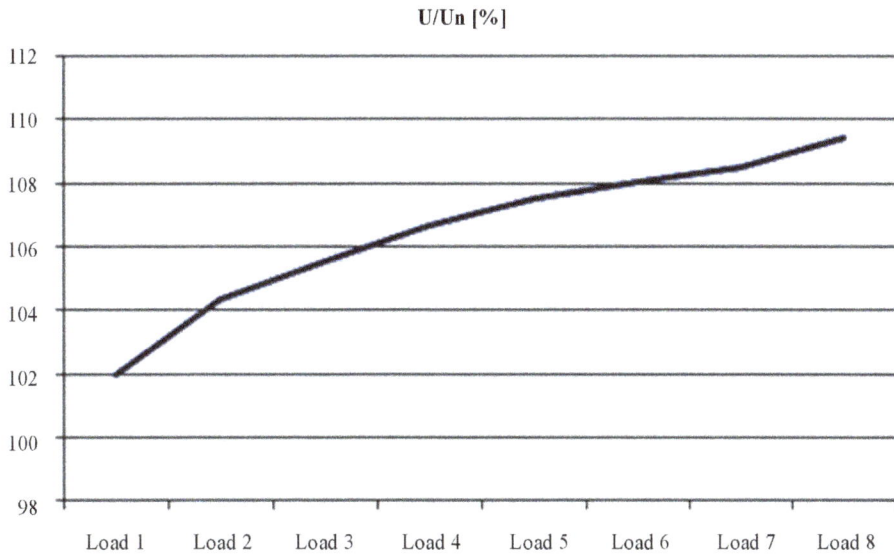

Figure 14. Loads efficiency with 50% PV efficiency.

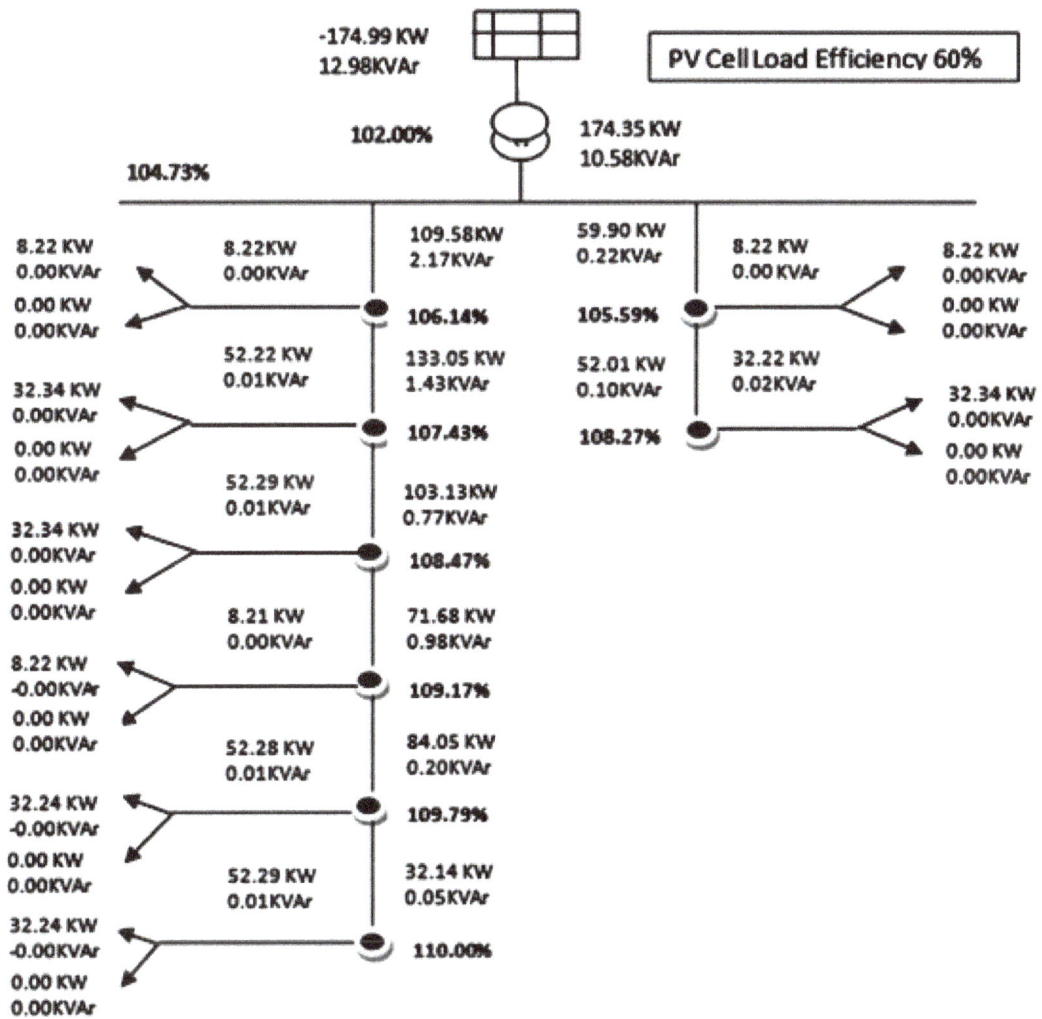

Figure 15. Single line diagram with 60% load efficiency.

U/Un [%]

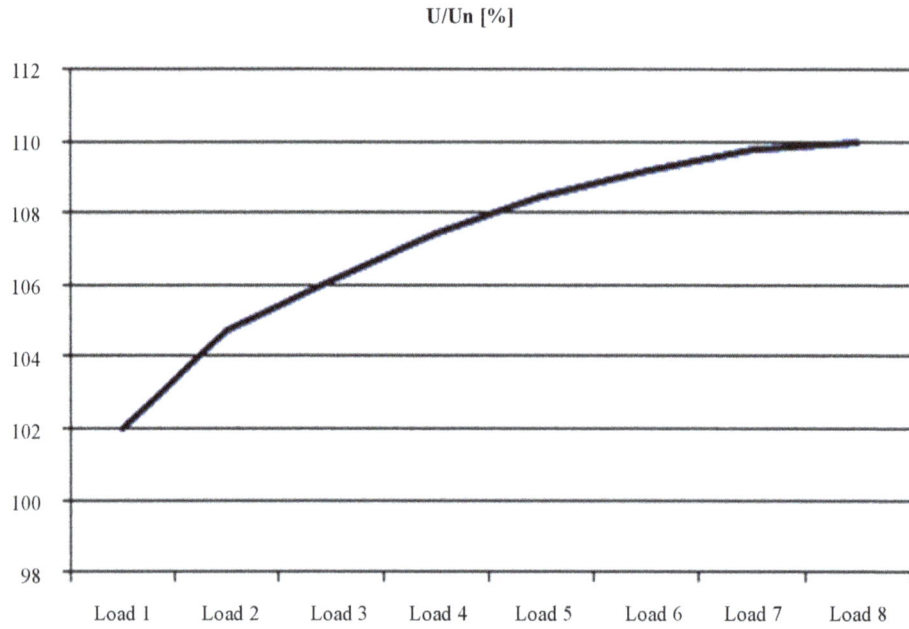

Figure 16. Loads efficiency with 60% PV efficiency.

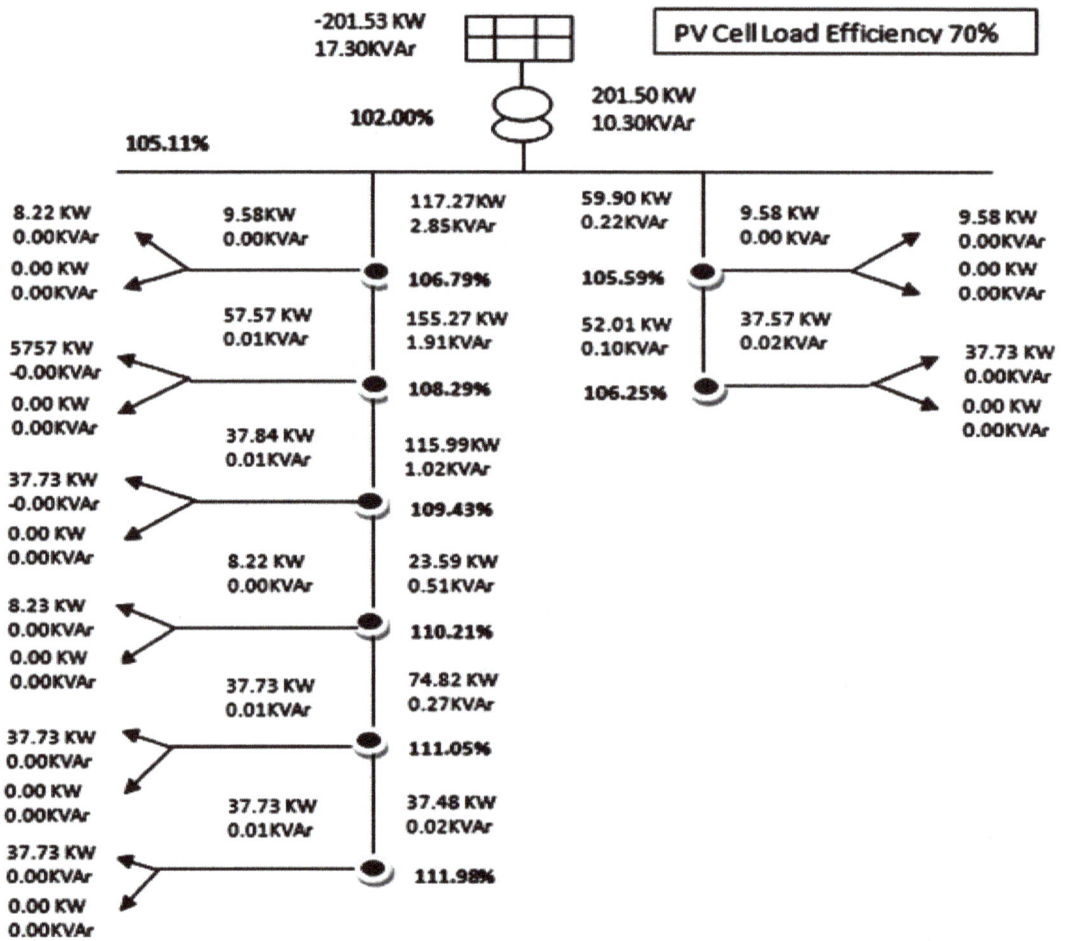

Figure 17. Single line diagram with 70% load efficiency.

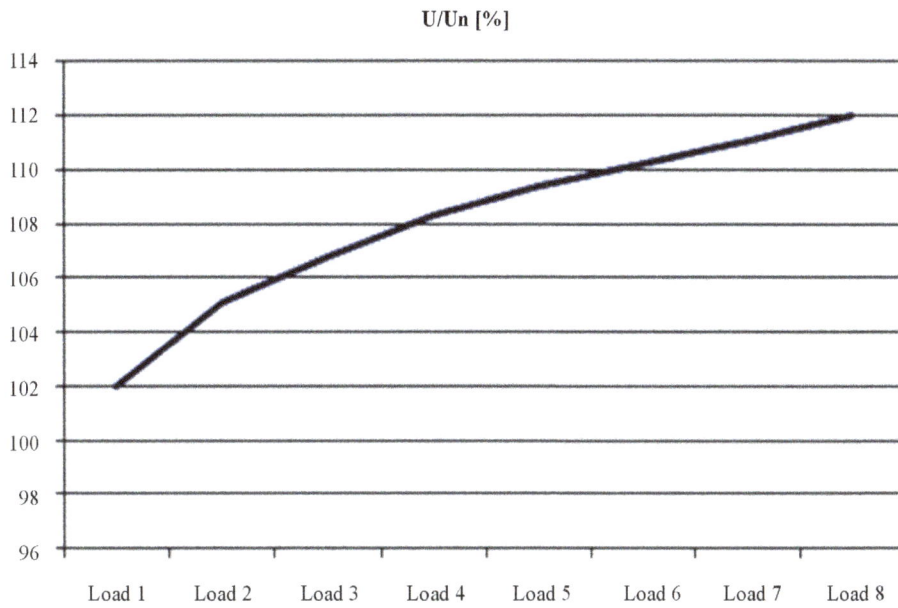

Figure 18. Loads efficiency with 70% PV efficiency.

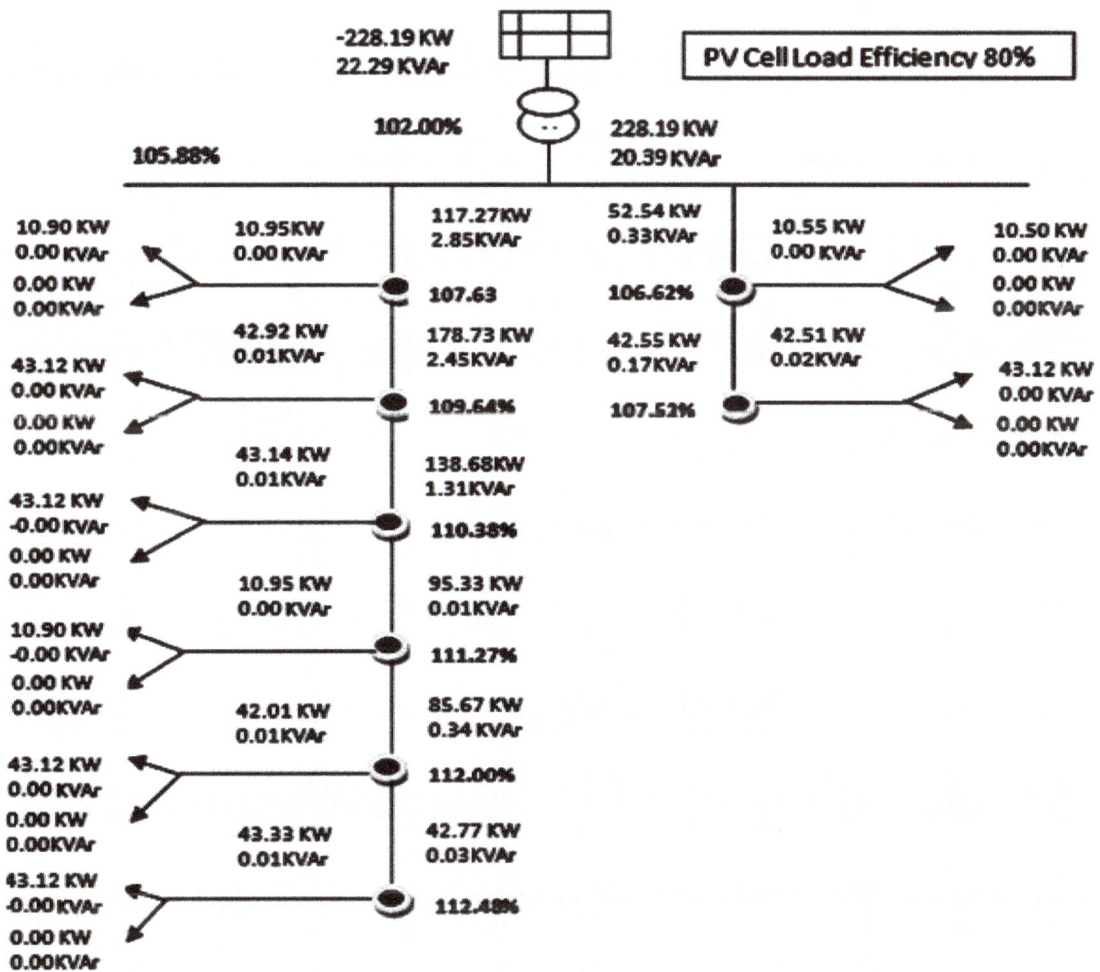

Figure 19. Single line diagram with 80% load efficiency.

U/Un [%]

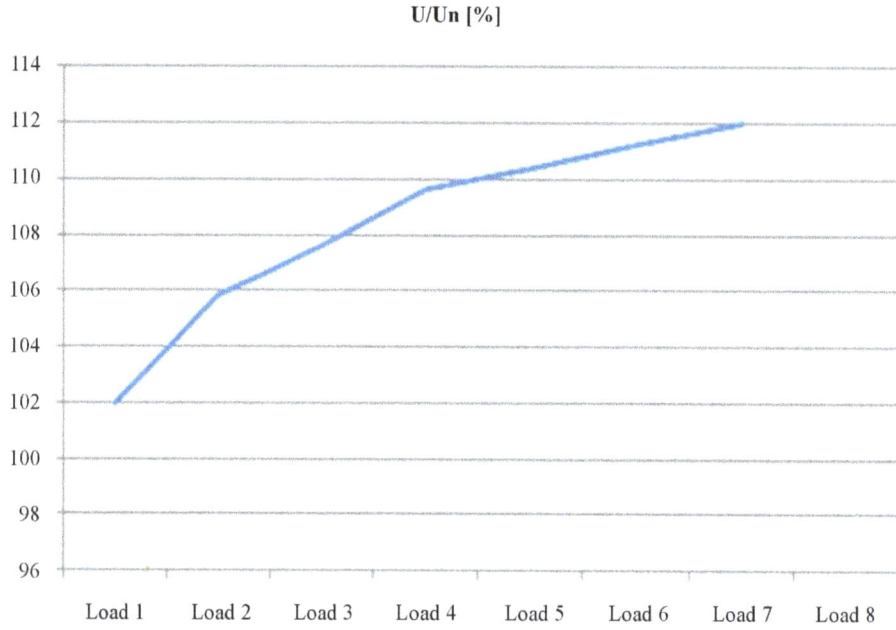

Figure 20. Loads efficiency with 80% PV efficiency.

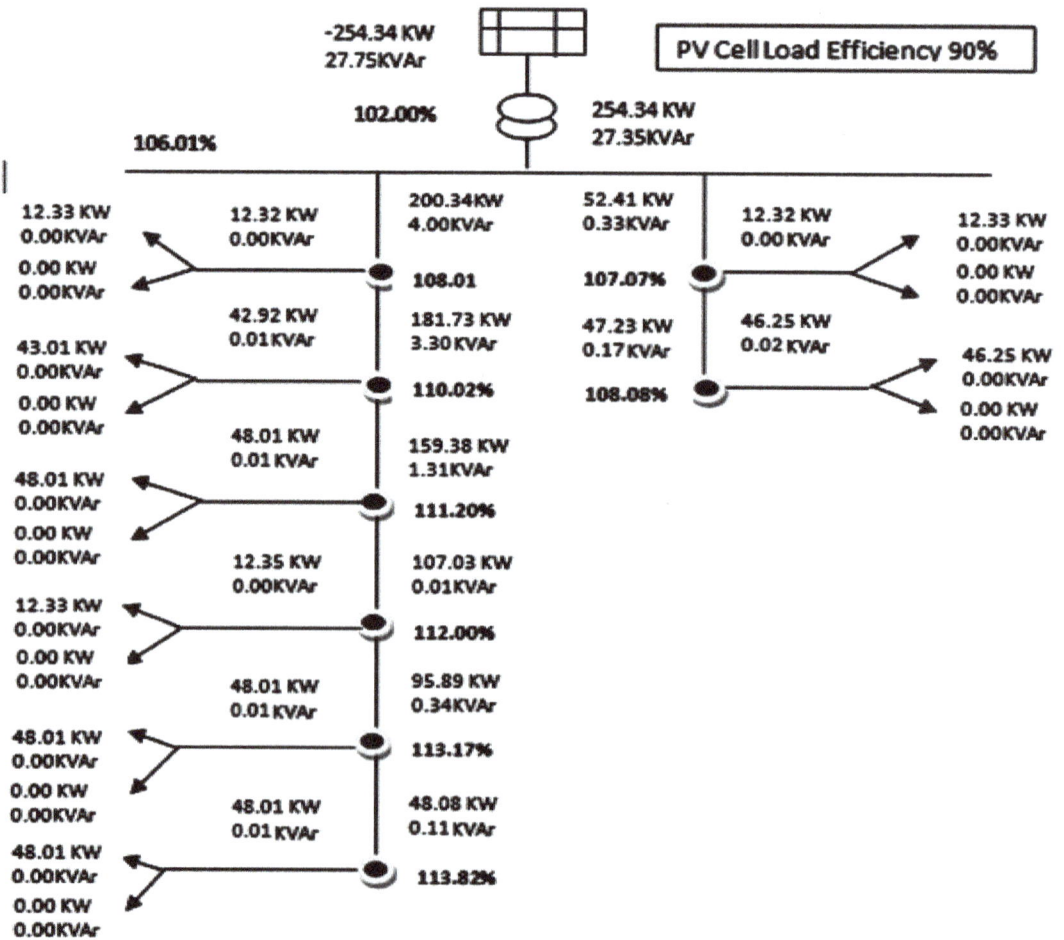

Figure 21. Single line diagram with 90% load efficiency.

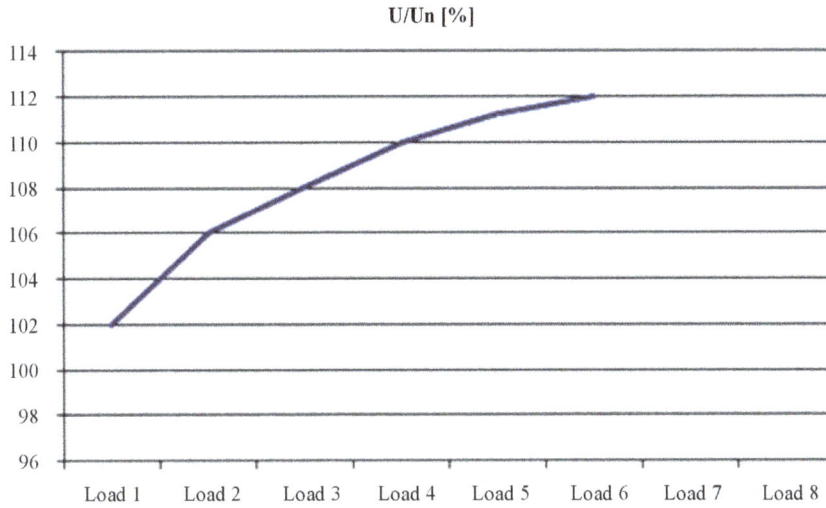

Figure 22. Loads efficiency with 90% PV efficiency.

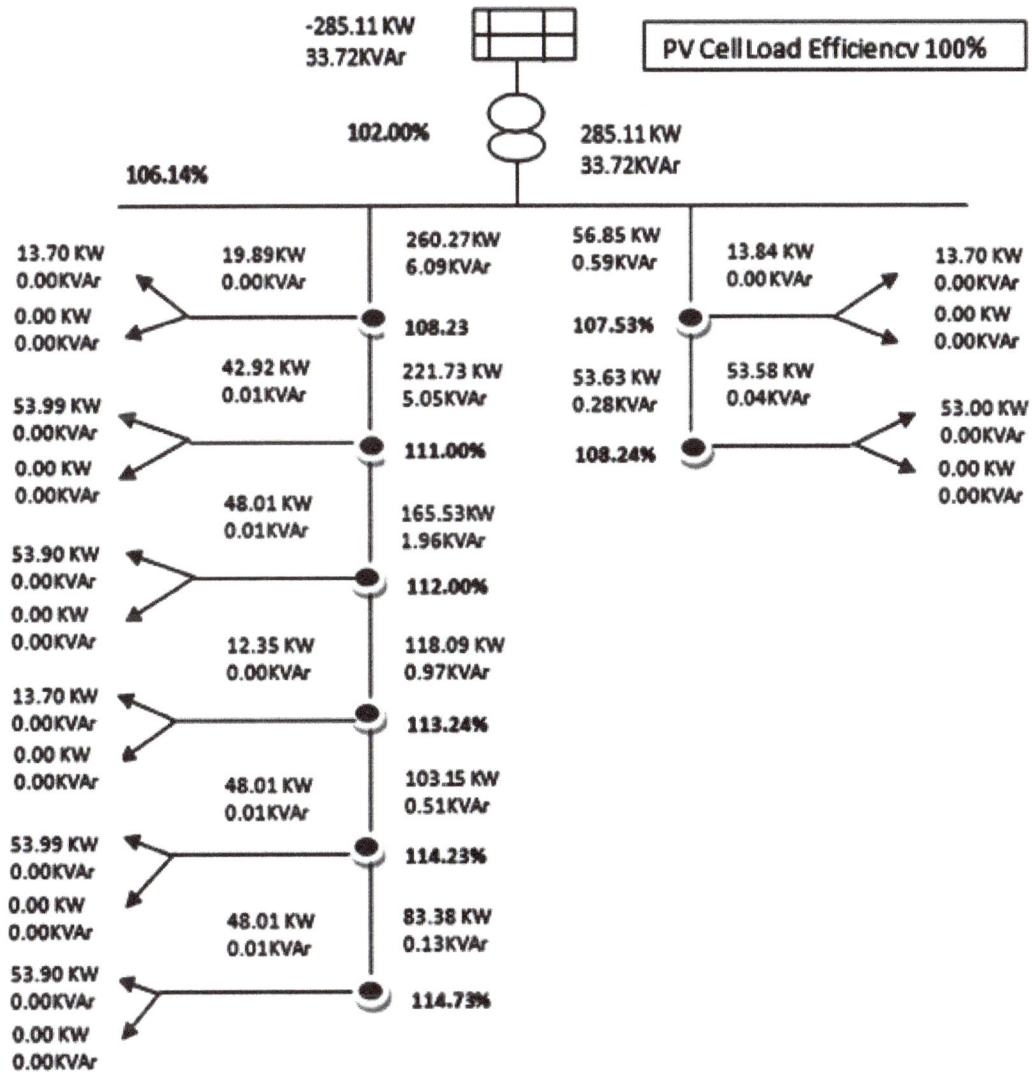

Figure 23. Single line diagram with 100% load efficiency.

U/Un [%]

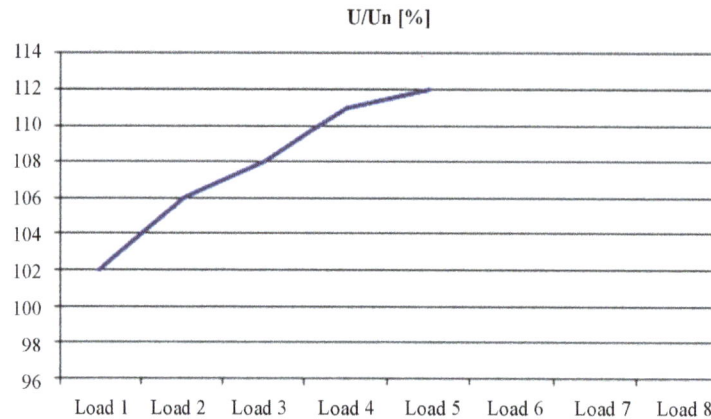

Figure 24. Loads efficiency with 100% PV efficiency.

12. Conclusions

In this research paper, sixteen load points have been selected and these load points were supported by Microgrid. Microgrid uses alternative power systems to support these load points such as solar power, wind power etc. In this research work solar power was selected as an alternative energy source. After calculating harmonics and load efficiency, it can be concluded that when PV load efficiency increases gradually, the load supporting points decrease and loss increases and reaches out of limit. In the future we can use wind power and also fuel cell power to support Microgrid system. Microgrid technology can control renewable resources to affect changes in the grid's operating conditions and can provide additional benefits as distributed generation assets or when installed at the transmission level. Distributed generation can support weak grids, add grid voltage and improve power quality.

The demand of energy generation worldwide grows rapidly, because energy generation is low but energy consumption is on a high rate. Electricity companies cannot satisfy consumers with this electricity generation rate for population and must use renewable energy system by using Microgrid technology.

Microgrid system should be deployed in Pakistan as well as in other developing countries. Pakistan has taken no steps to adopt Microgrid system with its prime role in energy system. Integration of renewable energy system in Microgrid has been proposed in this paper. If this system properly developed and implemented, it would make it much easier to adopt emerging technologies and improve economies conditions. If Microgrid system implemented in Pakistan, it would support the green energy and sustainable energy program in Pakistan. Microgrid system would reduce stress on the transmission and the distribution systems which is currently overloaded with a decaying infrastructure. This system would overcome the current energy crisis.

References

[1] Piwko, R. (2012) Grid Integration of Large-Capacity Renewable Energy Sources and Use of Large-Capacity Electrical Energy Storage. White Paper, IEC.

[2] Farret, F.A. and Simoes, M.G. (2006) Integration of Alternative Sources of Energy. John Wiley & Sons, Hoboken.

[3] Mehrotra, P. (2011) Nanotechnology Applications in Energy Sector. Reinste Nano Ventures Nano Science and Technology.

[4] Vader, N.V. and Bhadang, M.V. (2010) Smart Grid with Renewable Energy. *Renewable Research Journal.* http://rexjournal.org.managewebsiteportal.com/files/documents/System-Integration-Smart-Grid-with-Renewable-Energy---Mrs.-N.-V.-Vader.pdf

[5] Keyhani, A., Marwal, M.N. and Dai, M. (2010) Integration of Green and Renewable Energy in Electric Power Systems. John Wiley and Sons, Hoboken.

[6] Siril, P.F. (2003) Nanotechnology and Its Application in Renewable Energy. *Science*, **300**, 1127.

[7] Farret, F.A. and Simoes, M.G. (2006) Integration of Alternative Energy Sources of Energy. John Wiley & Sons, Hoboken.

[8] DOE's Office of Energy Efficiency and Renewable Energy (2004) Solar Energy Technologies Program Multi-Year Technical Plan 2003-2007.

14

Environmental Integration of Wind Farms: The Territorial Governance

Francesco Ruggiero, Graziarosa Scaletta

Dipartimento di Scienze dell'Ingegneria Civile e dell'Architettura, Politecnico di Bari, Bari, Italy
Email: ruggiero@poliba.it, arch.g.s.scaletta@gmail.com

Abstract

This research arises from the need to investigate the phenomenon of the development of wind farms in Puglia and the aspects related to the environmental impact that these systems generate on the territory. This represents a sign of change and adaptation on landscapes for people and local governments. The demand and the need to install renewable energy systems must be mediated by the preservation of the landscape and governed by planning instruments, which in this case should be expanded with a strategic energy planning in the anthropized environment that is being examined. With a careful analysis of the current situation, this paper suggests, a model of integrated development in which technology, landscape and bureaucracy reach an almost perfect balance between the protection of the territory and the incessant vicious speculative and criminal process.

Keywords

Wind Farm, Environmental Planning, Renewable Energy and Sustainability

1. Introduction

This work extends a recent research on the development of energy production from wind power in the Mediterranean area and, in particular, in Puglia [1] [2]. Here the focus is on the landscape evaluation for wind farms. This is a subject of great relevance due to the increasing development of wind energy in recent years in some European countries (e.g., France, Spain, England), and especially in Italy. The need for this study is the result not only of the growing commitment to sustainable development, but also of more general policies to ensure a widespread landscape quality for which the principles of the European Landscape Convention (Florence, 2000) represent a fundamental reference.

Several researchers have addressed the issues of territorial integration [3]-[10] often focusing on specific aspects such as environmental impact, visual impact, noise impact, and social impact. Here we propose a different methodology of systematic approach to plan the placement of wind farms on the landscape.

In recent years, an exponential growth of wind power plants in Puglia has been observed. In fact, in December 2011 the installed capacity was 1393.5 MW, an increase of 220% over the previous year [11], and these data are still growing.

Looking at the data, the landscape sustainability issue is becoming a necessity, an urgency, but also an opportunity. It is combined with ethical choices related to the most important issues of our time: peace, security and social environmental, sustainable and equitable development, a better distribution of resources and social opportunities.

This proposal aim not simply to insert the wind generator in the area as a foreign object, but as a project able to rethink the area, actualizing the meanings and uses, and ensuring that the transformations become an integral part of the existing area, that is to say the new landscapes: the energy landscapes.

The technological innovation of a wind farm and the benefits of clean energy need to coexist with a strongly characterized and anthropized area, without generating damage to the regional landscape.

For these reasons, the knowledge of the physical characteristics of the current regional landscape contexts, their historical formation, historical meanings and those attributed to them by their communities is fundamental. Each site must be read and interpreted so that the wind farm project will itself become a characteristic of the landscape and its forms contribute to the recognition of its own peculiarity, building a coherent relationship within the existing context. The wind project must become a new landscape plan to better integrate the context.

Wind plants are linked to a form of energy that depends on the availability of wind resources that requires the location of the plants in specific areas. Unsuitable locations, which most often pertain to beautiful countryside, has originated a heated debate on the benefits produced that are insufficient compared to the territory impact and, above all, on the perception of the landscape.

The energy planning must take into account the great and diffused historical, architectural, morphological and natural quality of Puglia landscapes. The evident impacts of wind turbines on landscape (especially in cases of wind generators over 50 m height) have slowed projects which, although their impacts are not comparable with thermo-electric systems, they represent a necessary way to achieve the objectives to protect the future of our planet agreed by the international community [12].

The real impacts on the environment and landscape are shown in **Table 1**.

The main problem is a proper contextualization of the plants and we should not underestimate the effects generated by the presence of several plants (cumulative impact). To consider the cumulative effects on the landscape means to estimate the distance between the plants, the relationships between the respective visual areas of influence as well as the general characters of the landscape [13]. Hence it is required to rationalize the number of wind turbines in the area in a single intervention in a way which avoids "the forest effect", in other words a "crowding" of turbines on the affected area.

Another important aspect concerns the bureaucratic issues, the problems related to the choice of a good design

Table 1. Wind farm impacts on the environment and landscape (*).

Causes	Effects
Visual impact	changes to the landscape and visual scenario in the surrounding context of the wind farm
Territorial impact	wind farms modify the ground but not the visual impact so much (so that it is always possible to restore the original sites), allowing crops and usual agricultural practices
Noise impact	it is mainly due to the movement of the rotors of the wind turbines which generates noise especially at the ends of the blades
Electromagnetic disturbances	it is due to the presence of big rotors, but they are restricted only to surrounding areas of the wind farm and they mainly concern interference with radio waves
Interaction with the birds and migratory	integration of wind farms in areas with high wildlife interest (especially wetlands) in which there are birds considered protected or endangered species and migratory routes

(*) Source: Guidelines for the planning and siting of renewable energy plant, Puglia Region.

and, above all, how to move into a jungle of proponents and inside different phases of the proposal-review and approval of a wind farm.

Authorization procedures for the construction of renewable energy plants in the Puglia region are provided by a national Law n. 387/2003) [14] and a regional law [15].

The incentive system used until a few months ago has resulted in a high financial speculation in the renewables sector with ethical problems such as the sale of agricultural land, including high-qualified and productivity in order to achieve a faster and easier gain thanks to the sale of energy rather than the sale of agricultural products.

At present the result is that only for the Puglia region, the authorization procedures in progress and completed cover plants with an energy production higher than the national electricity demand, demonstrating the futility of the continuous destruction of the territory. Therefore it is necessary to regulate the relations between plants and tools in territorial governance and the transformation projects that follow.

This task is entrusted to the Town Planning, which should lay down rules, conditions and opportunities for the use of local energy capacity in relation to specific tissues and forms of settlement of the territory as well as the good sense of designers and production companies.

2. Criteria for the Correct Design of a Wind Farm and Its Inclusion on the Territory

The features that contribute to a potential impact on the landscape (positive or otherwise) for wind power plants are due mainly to their physical characteristics: the towers and the height of turbines, the number of towers, movement, colours and materials, etc.

The state of the impact depends on its size, thought as a rated power of energy produced, the plant itself, as well as by the distribution choices on the territory.

Tables 2-4 describe the characteristics and parameters of a wind power plant insertion on the territory.

The considerable size of a system is not often accompanied by a lay-out consistent with the structural elements of the landscape in which they occur, causing confusion and perceptual disturbance (forest effect). It is therefore necessary to control certain parameters linked to a location such as density, land use, and land form.

In addition to the critical nature of the perception, the construction of a system involves changes and transformations and if they are not controlled by a project, that is respectful of its natural setting or hydrogeological problems, or historical features of the site, could damage irreversibly the landscape.

For example, the opening of new roads has interrupted in some cases the continuity of natural areas for grazing; in other cases the wind turbines and the service roads have been located in areas classified as strongly dangerous from the geomorphological point of view, and this contribute to weaken the hillside.

Moreover, in respect to the settlement characteristic, there are also examples of proximity or overlap of windfarms to sites of archaeological interest. Finally, there are other examples on the proximity of the plants to urban centres in a position that does not take care of the structural elements and the identity elements of the site. This generates high levels of criticality and visual disturbance (**Table 5**).

3. The Research Steps

The aim of this research is the development of a model through the tool of GIS that is a source of information, technical support and critical guidance for policy-makers, administrators, operators and technicians involved in

Table 2. Different wind farm for size (*).

Types of Wind Farm	Generators (n°)	Rated Power P (MW)
Large Size	1 or more	>1
	1 or more	$0.50 < P < 1$
Medium Size	1	$0.50 < P < 1$
Miniwind	1 or more	<0.50
Microeolic	1	<0.01

(*) Source: Guidelines for the planning and siting of renewable energy plant, Puglia Region.

Table 3. Characteristics of the elements of a wind turbine.

Reference	Characteristic	Description
Turbine	Typology	- Horizontal-axis machines - Vertical-axis machines single-blade, bi-blade, tri-blade (the choice is irrelevant, it depends on the geographical context; it is important that the same wind area will adopt the same machine type to avoid visual disturbances)
	Size	Diameter: max 90 m
	Power	(*cf.* **Table 1**)
	Colour	Better neutral colours and opaque: light grey, beige, cream, but it is important that they are integrated in the environment and that they comply aeronautical provisions.
Tower	Typology	- wind pole - pylon
	Height	max 80 - 100 m
	Width	
	Colour	Better neutral colours and opaque: light grey, beige, cream, but it is important that they are integrated in the context.
	Density (distance between the blades and between plants)	- Concentration rather than dispersion: homogeneous groups of plants are preferable to individual machines scattered throughout the territory. (It is less the visual impact of a smaller number of larger turbines that a greater number of smaller turbines). The minimum distance between wind turbines is suggested 3 - 5 times the diameter of the rotor in the same row and 5 - 7 times the diameter of parallel rows (Guidelines for the planning and siting of renewable energy plant, Puglia Region) - To place the machines in groups of no more than eight turbines with a relatively large distance between them (Danish Guidelines for the design of windfarms) - German guidelines for the design of the windfarms provide more than 5 km between wind plants
	Distance from the town	buffer of 1000 m for large wind farm and 500 m for small wind farm, both for reasons of perception than urban planning (production areas are not considered because suitable for the location of wind generators)
	Distance from vegetation	buffer of 200 m
	Distance from coast	- buffer of 300 m from wind farm of any size and number of wind turbines (except industrial and port areas, to be regulated in an appropriate way) - buffer of 2 km from power wind farm with more than 1 MW
	Distance from restricted areas	buffer of 500 m
Infrastructure and Services	Electrical substation	It is preferable to use the existing substations, to reduce their number in the area It is clear that this often leads to a lengthening of the electric transmission lines
	Access roads	It is preferable to use the existing road network and if necessary to adapt it to the requirements of transport, construction and maintenance. It is important, finally, paved roads with permeable surfaces (macadam or similar)
	Power transmission lines and substations	The distance from the network of high and medium voltage is of fundamental importance to avoid problems of over-infrastructure of the territory. The distance between layout and connection point must not be greater than 8 km. Furthermore, to ensure a low impact underground electrical power grid lines is preferred, alongside the existing road network, in accordance with local regulations
	Service/Disposal	- it must be guaranteed a constant access to the site - in the case of decommissioning of the wind farm (the operating life is approximatively thirty years), it is necessary to restore the area, ensuring the original use
Noise		the noise coming from a generator must be inferior to 45 dB near houses (value equivalent to a quietly conversation)

Table 4. Relationship between height of the wind turbine and visibility in the territory.

Height including the Wind Turbine Rotor [m]	Sight Distance [km]*
Up to 50	15
51 - 70	20
71 - 85	25
86 - 100	30
101 - 130	35

*The sight distance is the maximum distance in km from where you can see a wind turbine of given height (the height of the radius of the rotor together with those of the structure up to the hub).

the design and in the evaluation of projects, so that the landscape quality of sites constitutes an important reflection on the choice for territorial transformations.

This procedure will help:

- to propose rational development for the installation of windfarms in the region, through a computerized framework of knowledge, integrated to regional and local planning established by DRAG (Regional Document of General Layout) [16];

- to create an application protocol, as a system oriented both to the government control for the state of the works constantly in changing and updating, and to a framework for the development of the wind energy system, useful for private companies, to intervene according to the criteria established by regulation planning;

- to create a replicable, repeatable and adaptable model for different municipal needs with a degree of autonomy and above all a model accessible to any type of user, through computer channels (official website of the region), the same that will be constantly updated in a single database. This will help to ensure that data are accessible and that procedures on the information produced will be based on transparent criteria.

The instrument used to achieve the research objectives is the PUG (General Urban Plan) (Regional Law n. 20/2001) by including the requirements and guidelines for the proper and adequate planning of the municipality using the Cognitive Frameworks, *i.e.*, Invariants Structural and Regional Contexts, such as those areas excluded from the construction of large and small size systems.

This research proposal uses the planning tool in a simple way together with the regional regulatory for a proper insertion of wind power plants in the area, and seems to be a good basis for building a tool for the identification of suitable areas, not suitable or partially suitable areas for the construction of windfarms, providing the results of the research.

The survey phase has started from the construction of the Regulatory Framework of reference, an international, national and local level, very useful to understand the management rules of the matter. Then it has been proposed an update of the cognitive frameworks of Puglia Landscape about environmental resources, landscapes, settlements and infrastructures (hydrography, morphology, vegetation, land use, urbanization, natural reserves, historic and scenic landscapes of international, national and local interest, scenic routes, areas of strong tranquillity or naturalness or full of symbolic meanings).

Finally, a census of the wind turbines already placed on the territory was created, their related features paying greater attention on any incompatibilities with landscape.

The final result is a database containing all the information relating to wind farms already present in the region above and below to 1 MW, including decommissioned plants too, as a result of a malfunction or a pure speculative action, specifying forms and procedures of their disposal.

4. Suggested Model for the Environmental Integration of Wind Farms in the Countryside of Puglia

The final aim of this research is the proposal of a model for the rational development of the installation of wind farms in the region, through a computerized framework of knowledge created after the survey phase, described in the previous paragraph, and integrated with the regional and local planning established by DRAG.

The model is developed on Arcgis, with the official map in Gauss-Boaga UTM 33, based on cadastral references

Table 5. Areas suitable and unsuitable and their associated requirements for the integration of wind power plants (*).

Wind Farm Type	Areas Suitable	Areas Unsuitable
Large size	- *planned production areas*: without prejudice to safety distances set by legislation and to acoustic compatibility: along the main entrance roads, the internal distribution roads, in areas with urban standards; - *farmland near industrial areas:* while maintaining the agricultural use of the land. The plant design in this case must relate to the signs of the agricultural landscape (roads, walls, divisions between farms); - *areas close to mining basins*: according to the standards.	
Medium size	- *planned production areas*: without prejudice to safety distances set by legislation and to acoustic compatibility: along the main entrance roads, the internal distribution roads, in areas with urnan standards; - *farmland near industrial areas:* while maintaining the agricultural use of the land. The plant design in this case must relate to the signs of the agricultural landscape (roads, walls, divisions between farms); - *areas close to mining basins*: according to the standards; - *agricultural areas*; - *areas classified as rural populated and urbanized.*	- SIC areas (Site of Community Importance), SIN areas (Site of National Importance) and SIR areas (Site of Regional Importance), National and Regional nature reserves, National and Regional nature parks, Important Bird Areas, Ramsar Wetlands, Public waterways land their buffer zones (standards of AdB, the River Basin Authority), Italian Law 1497/39; - Architectural constraints (Italian Law 1089), Archaeological areas, other standards for landscape, provision from the Chart of monumental heritage: distance buffer more than 500 m; - Coastal zone and lakes (with the exception of industrial and port areas): distance buffer more than 2 km; - Areas with a slope greater than 20%; - Areas with Geomorphic vulnerability (as defined by AdB); - Hydrographic areas, erosive furrows, sinkholes and ravines (AdB); - Natural areas (forests and maquis shrubland, wetlands, and grazing lands): distance buffer more than 500 m; - Roads with landscape relevance: distance buffer more than 200 m; - Urban centres (with the exception of production areas): distance buffer more than 1 km.
Mini-wind	- on the ground; - on the roof of buildings.	- Special Areas of Conservation (ZSC) and Special Protection Areas (ZPS); - Architectural constraints (Italian Law 1089), Archaeological areas, other standards for landscape, provision from the Chart of monumental heritage: distance buffer more than 500 m; - Coastal zone and lakes (with the exception of industrial and port areas): distance buffer more than 300 m; - Areas with a slope greater than 20%; - Natural areas (forests, wetlands).
Micro-wind	- on the ground; - on the roof of buildings.	- Special Areas of Conservation (ZSC) and Special Protection Areas (ZPS); - Architectural constraints (Italian Law 1089), Archaeological areas, other standards for landscape, provision from the Chart of monumental heritage: distance buffer more than 500 m; - Coastal zone and lakes (with the exception of industrial and port areas): distance buffer more than 300 m; - Natural areas (forests, wetlands).

(*) Source: Linee guida sulla progettazione e localizzazione di impianti di energia rinnovabile-Regione Puglia.

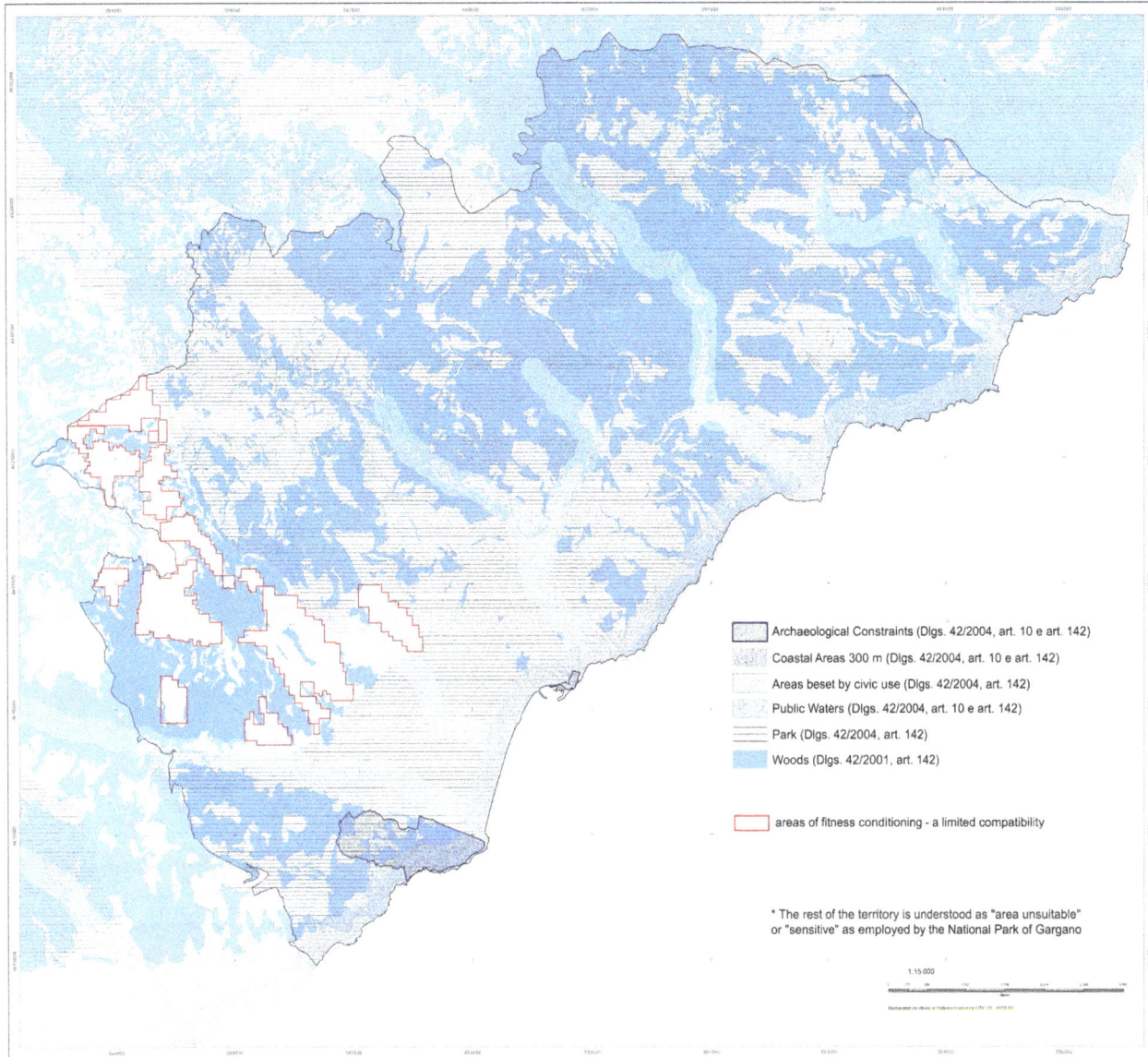

Figure 1. Structural Invariants represents the superordinate to PUG, according to National Law 42/2004.

updated to 2012 and the Regional Technical Map of 2008.

The goal of the system is to proceed overlaying invariant characteristics of a municipality reproduced in the cartography representative of the interested area (habitats and vegetation system, hydro-geomorphological system, historical-architectural system, settlement and infrastructure, topography, wind conditions, birds system, territorial and urban contexts defined by the PUG).

The result is summed in a series of inscribable maps gradually achieved, discarding areas in dangerous for the landscape and by the general characteristics, from which the result "in negative" allows the identification of areas:

a) not suitable-"sensitive": those areas characterized by elements of the natural system, settlement or town; in this case is prohibited to include any element to alter the status quo;

b) conditioned suitability-limited compatibility: for small size of wind turbines and directed to self-consumption, in those characterized areas that do not have ministerial and regional restrictions or safeguards. The new Regional Plan for Landscape (PPTR) addresses actions and projects towards self-consumption policies, addressed to municipalities and individual users;

c) appropriate-compatible: mainly those production areas and near the mining basins, where there are no ele-

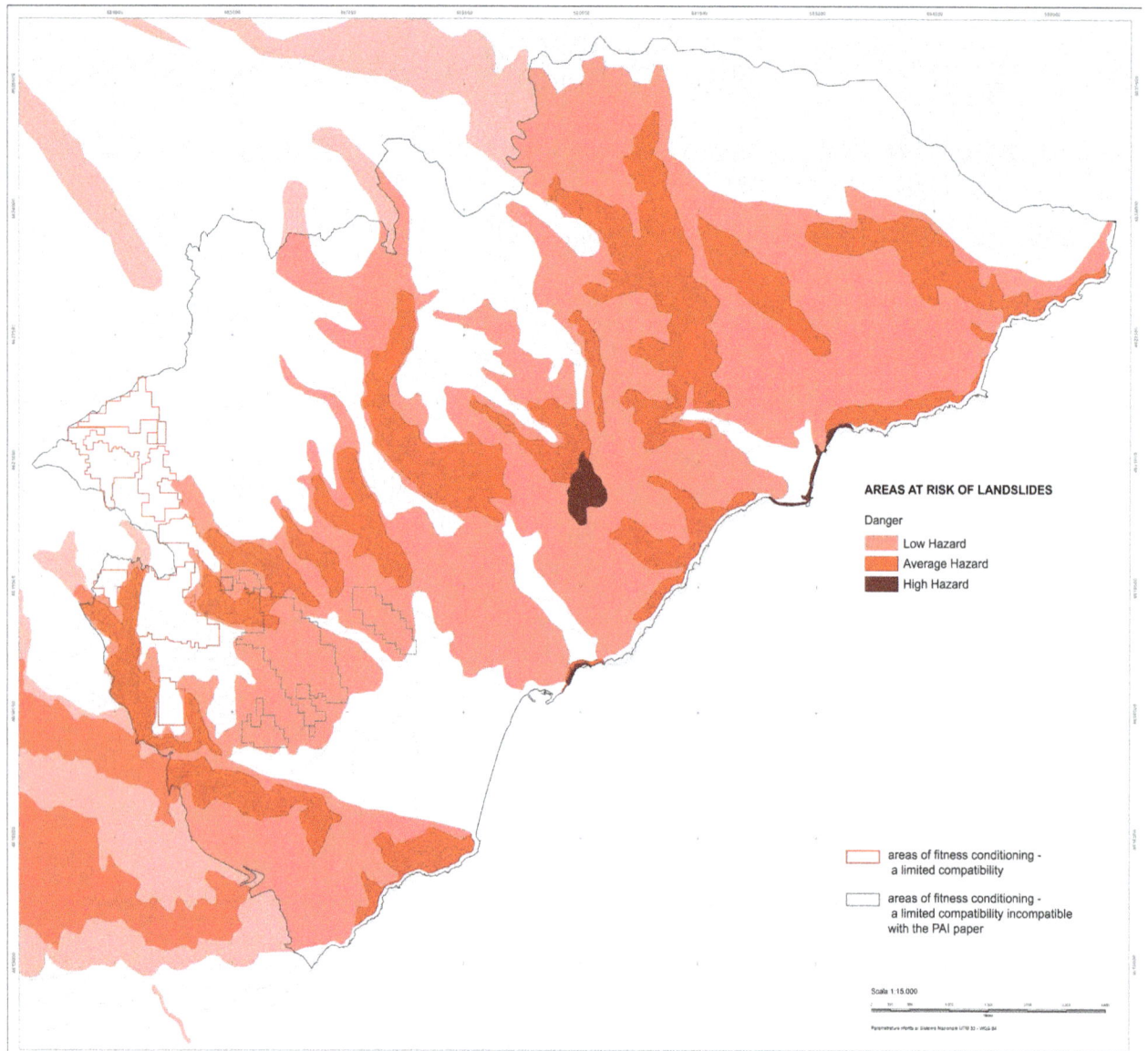

AREAS AT RISK OF LANDSLIDES

Danger
- Low Hazard
- Average Hazard
- High Hazard

areas of fitness conditioning - a limited compatibility

areas of fitness conditioning - a limited compatibility incompatible with the PAI paper

Scala 1:15.000

Figure 2. Structural Invariants represents the superordinate to PUG, according to the Hydrogeological Plan (PAI).

ments to protect. The PPTR aims to encourage the concentration of wind farms, photovoltaic and biomass plants in the industrial areas. In this direction it is important to reconsider the production areas as real power energy production areas where it is possible to design an integration of different technologies in symbiosis among them, for the benefit of the companies that use the electric and thermal energy produced. All this takes part of a wider scenario of design for Productive Areas for landscape and Ecologically Outfitted (APPEA).

To illustrate the application of the proposed model, it was applied to a case study, to the City of Mattinata, district of Foggia in the North of Puglia, characterized by good values of wind. The application of the protocol to install wind turbines in small and large sizes is divided into the following phases:

Phase 1-identification of structural invariants and spatial features (habitats and vegetation system, hydro-geomorphological system, historical-architectural system, settlement and infrastructure, topography, wind conditions, birds system, Italian Law n. 42/2004, Implementing Technical Standards of the Thematic Territorial Urban Plan (PUTT/P), Wind Atlas, regional laws and municipal ordinances);

Phase 2-identification of local urban and rural contexts areas (with the exception of the built-up area in rural zones and primarily to industrial and handcraft areas for large size plants);

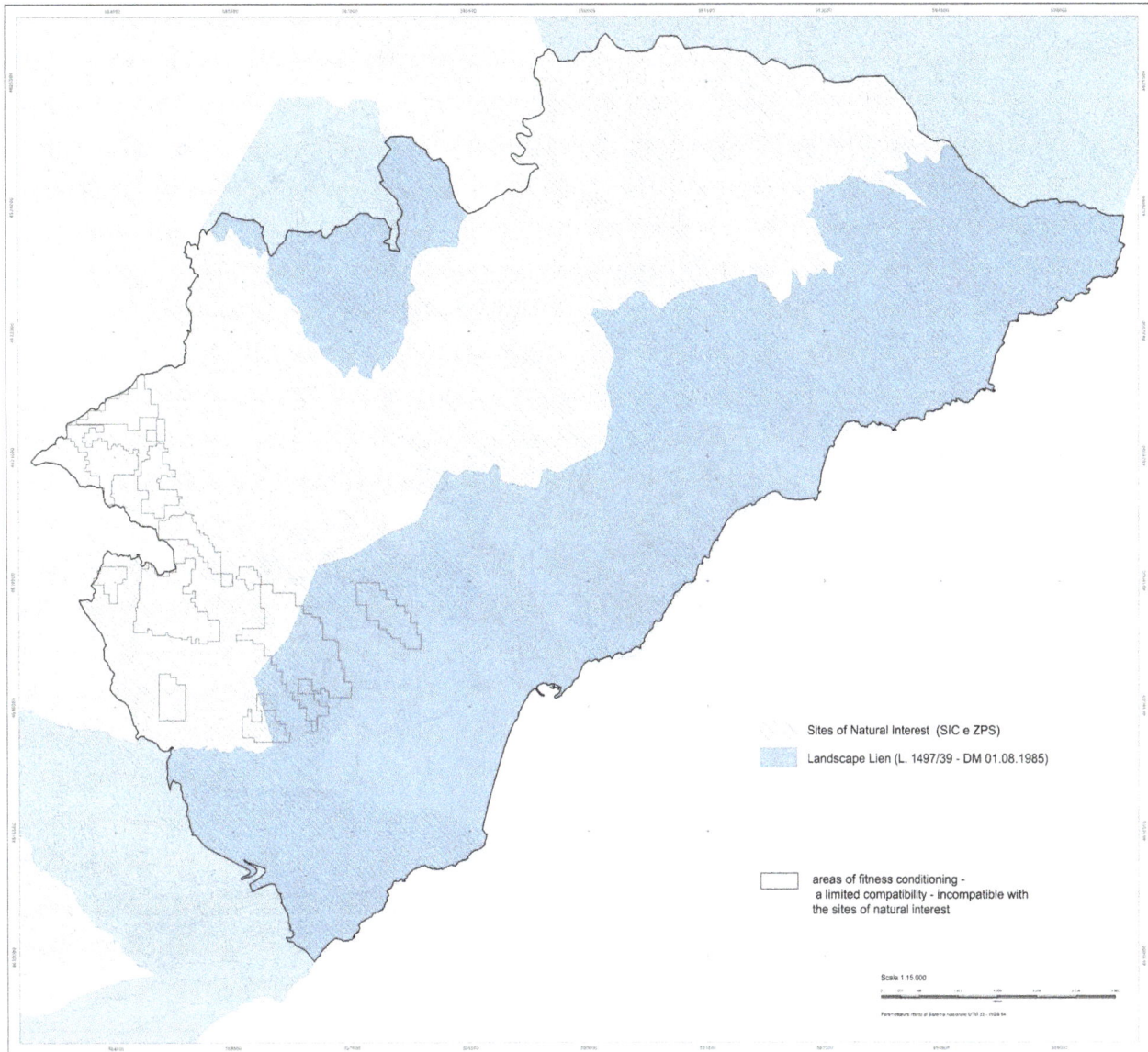

Figure 3. Sites of Nature represents a national scale (SIC and ZPS).

Phase 3-classification of macro-areas: not suitable-"sensitive"; conditioned suitability-limited compatibility and appropriate-compatible as described above. In such areas, it will be possible to predict or not the localization of the plants according to type and size permitted in the area and in the context;

Phase 4-in the previous phase we have identified macro-areas, that will be superimposed on the Cadastral Map, to point out in which particles it is possible to install a wind turbine and make the case clear and known to people involved.

The figures below indicate how the proposed model works.

Figure 1 shows the restrictions imposed by National Law 42/2004 [17] (archaeological restrictions, coastal areas with a buffer of 300 m, public areas, public water, parks, woods) from which, in red, are indicated those areas in "conditioned suitability". The rest of the territory is "not suitable", a risk area, because it covers the 85% of the Gargano National Park.

Figure 2, instead, shows the Hydrogeological Plan (PAI) constraints. In this overlap is clear that the areas of "conditioned suitability" (red) are reduced by eliminating those one drawn in grey because with a low and high hydro-geological risk.

Figure 3 shows that Natural Sites of national interest (SIC and ZPS) and the Landscape constraints

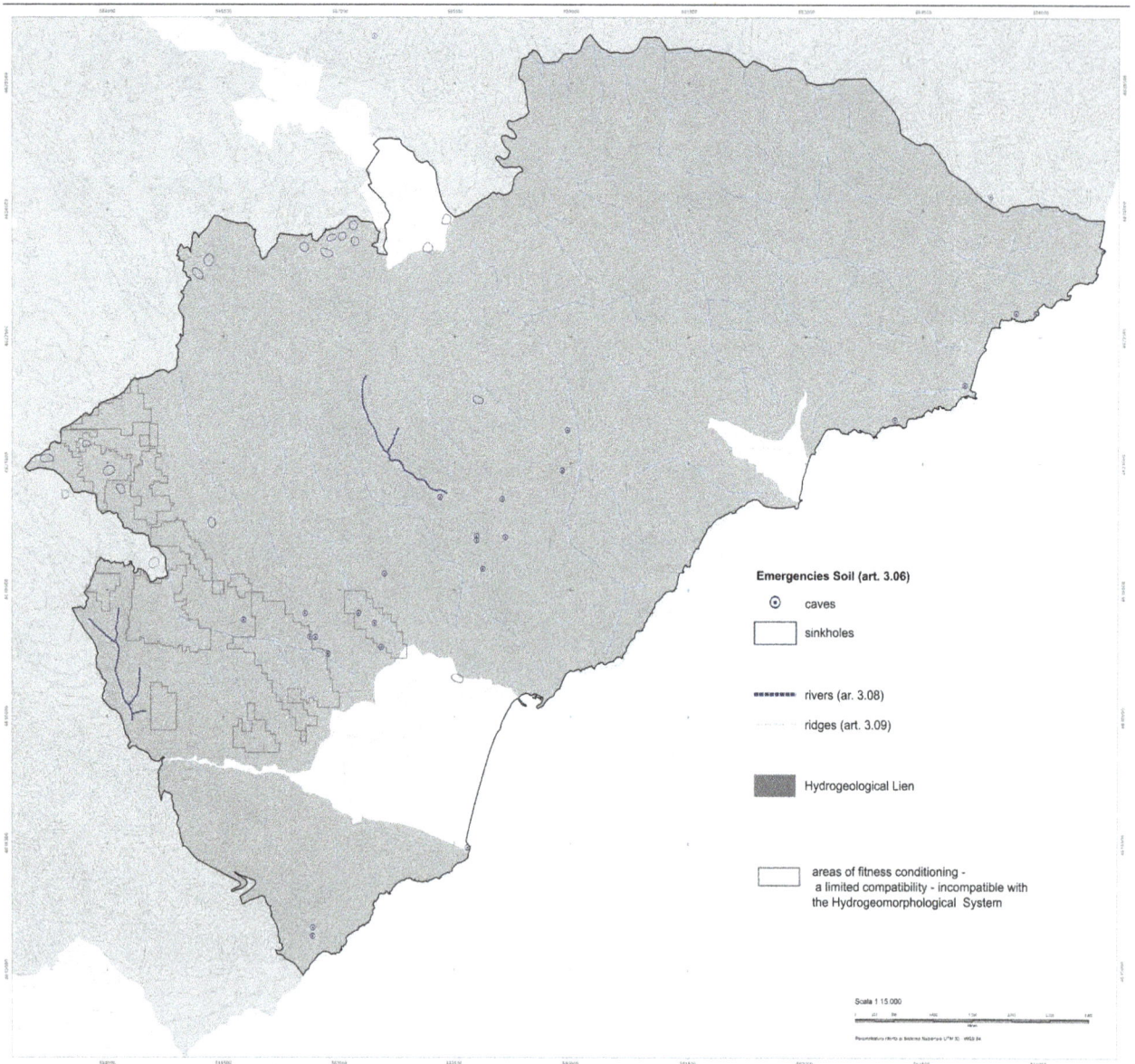

Figure 4. National constraints Invariants Structural System hydro-geomorphological (caves, sinkholes, rivers, ridges, hydrogeological restrictions) at the regional scale of the PUTT/P.

according to the Law 1497/1938. In this overlap all areas (gray) are "not suitable" for any insertion of plants.

Other overlap maps show on the **Figure 4** an overwrite of hydro-geomorphological constraints (caves, sinkholes, rivers, ridges, hydrogeological restrictions) at the regional scale of the PUTT/P, according to the DGR 1748/2000, while the **Figure 5** shows overlaying areas considered as "conditioned suitability" according to the constraints of national and regional habitats and vegetation system (woods, soil, natural assets and protected areas, sites of natural interest). Even in this case, the areas initially considered as "conditioned suitability" become "not suitable" areas (grey).

Figure 6 shows the additional overlaying areas considered as "conditioned suitability" according to the national and regional historic-cultural system (constraints, archaeological and architectural settlement, agricultural landscape-DGR 1748/2000).

Those areas initially considered as "conditioned suitability" become "not suitable" areas (grey).

Finally, **Figure 7** shows an overlap of the areas considered as "conditioned suitability" according to national constraints listed in the draft of the PUG of Mattinata City (FG). According to this plan, all the areas should be

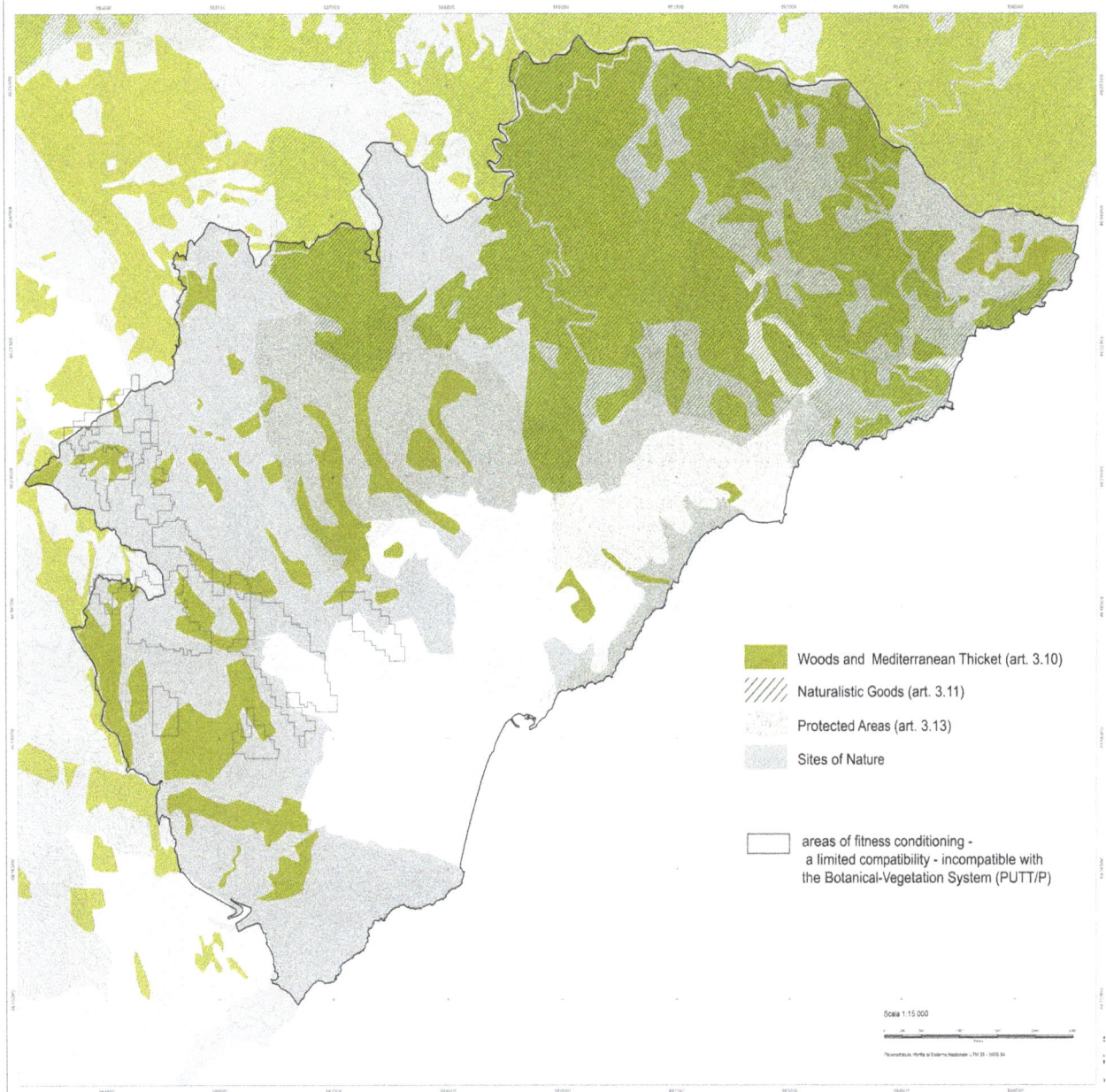

Figure 5. National constraints Invariants Structural System botanical-vegetation (woods, soil, natural assets and protected areas, sites of natural interest) at the regional scale of the PUTT/P.

part of agricultural contexts.

As additional information, **Figure 8** shows the map of the wind according to the National Wind Atlas [18] with an indication of the wind speed in the examined areas.

The final results of the superposition of these maps shows that the application of the model did not produce any "suitable area" for the installation of wind farms; almost the entire territories is "not suitable" and under protection, since the territory include the 85% of the Gargano National Park.

In this borderline case, priority will be given to mini-wind finalized to self-consumption, applied to residential and industrial areas, which in this case are very reduced.

Therefore, even if the territory for geological and geographical location is "suitable" for the installation of the plants, the landscape value and the various constraints forbid their insertion.

The thesis supported by this research aimed to create a tool for the identification and location of those areas

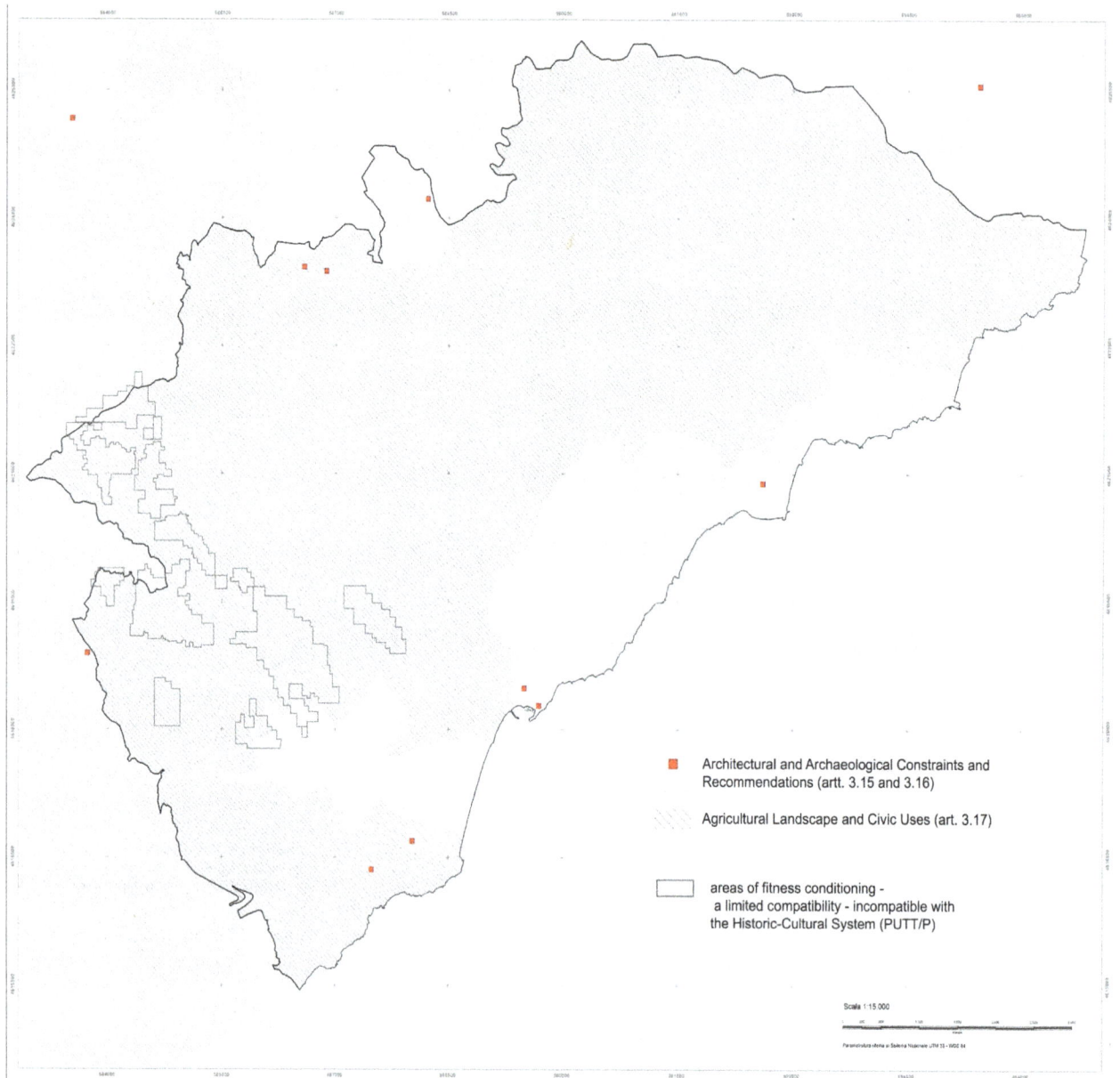

Figure 6. National constraints Invariants Structural System historic-cultural (constraints, archaeological and architectural settlement, agricultural landscape. DGR 1748/2000) at the regional scale of the PUTT/P.

with suitable or partially suitable characteristics for the insertion of a wind farm, thanks to this method and tools, it is considered useful and applicable to other cases to plan a PUG.

In this way it is possible to create an interactive and available method for companies and individuals interested in the installation of wind turbines, available on the municipal authorities' official websites to check the suitability of areas instead of a failed authoritative process by local government.

This would also facilitate the drafting of a Landscape Report which is attached to the authorizations and to the Environmental Impact Assessment; it could be simpler to have for all different cases the data available for a proper construction and permit's procedures.

Anyway, it is possible to check all the cases because the identification of the suitable areas does not deplete all areas right for wind plants, rather it suggests potential areas that can be examined in depth in local planning.

Other areas may be identified also in regional planning processes after the check of the energy producibility based on data collected by more accurate anemometers and of the environmental compatibility of selected areas.

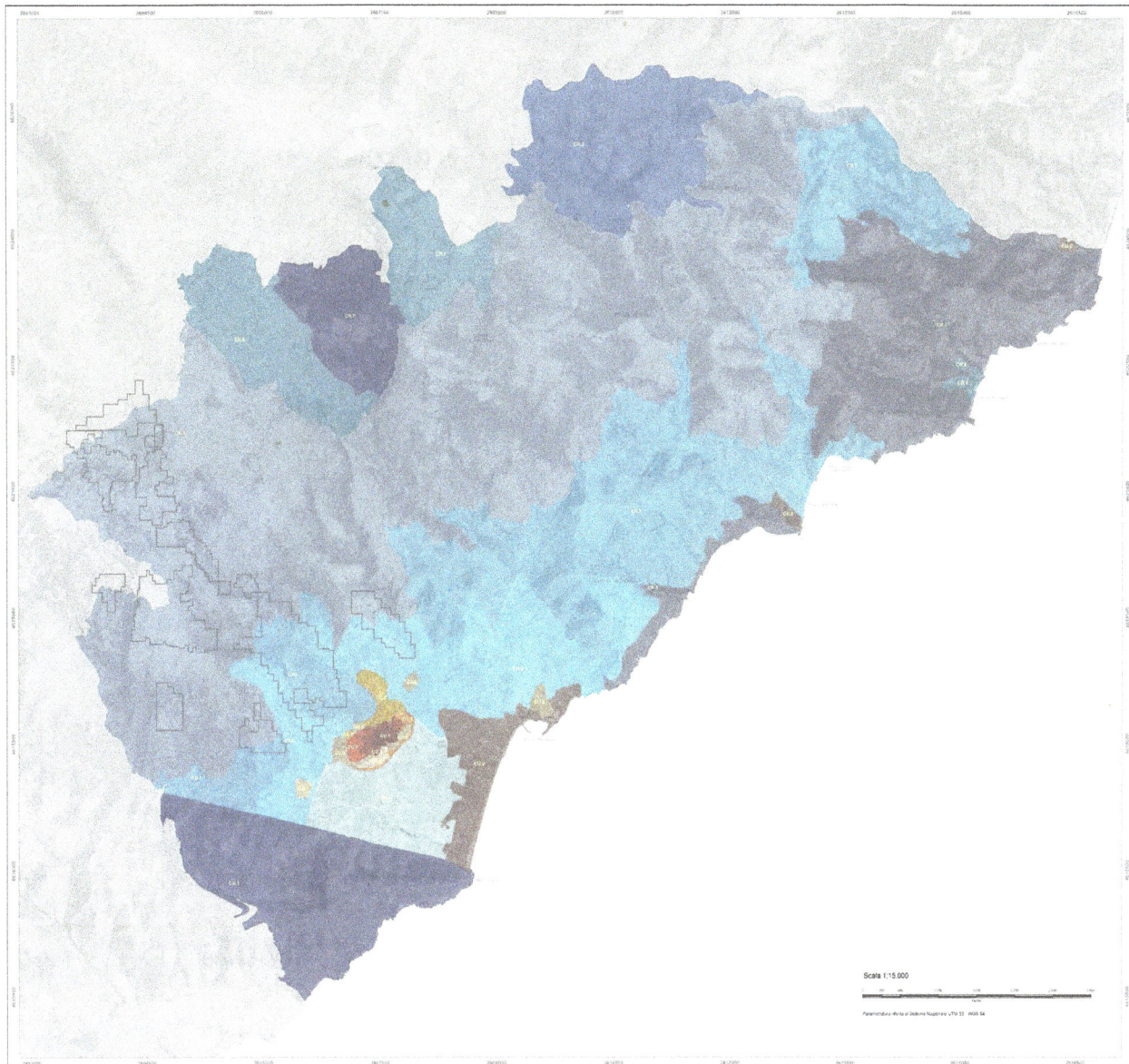

Figure 7. Overlap of conditioned areas PUG of Mattinata City (FG).

6. Conclusions

In this paper, it has been tested an application protocol to find the suitable area for the installation of wind farms, according to criteria defined in areas established by planning and standards. This procedural model is directed both to a government to control the state of the constantly changing and updating works, and to private companies as a framework for the check of the feasibility of the project to avoid the additional costs of design.

The model is replicable, repeatable and adaptable to different municipal needs with a degree of autonomy and, above all, is a model accessible to any type of user, through the official website of the Region continuously updated in a single database.

This ensures the accessibility of the data with a transparency on the procedural criteria and on the accuracy of the information.

This procedural model facilitates the understanding of the procedures for the evaluation of a wind farm project and those relating to the landscape authorizations necessary to overcome some delays and bureaucratic-

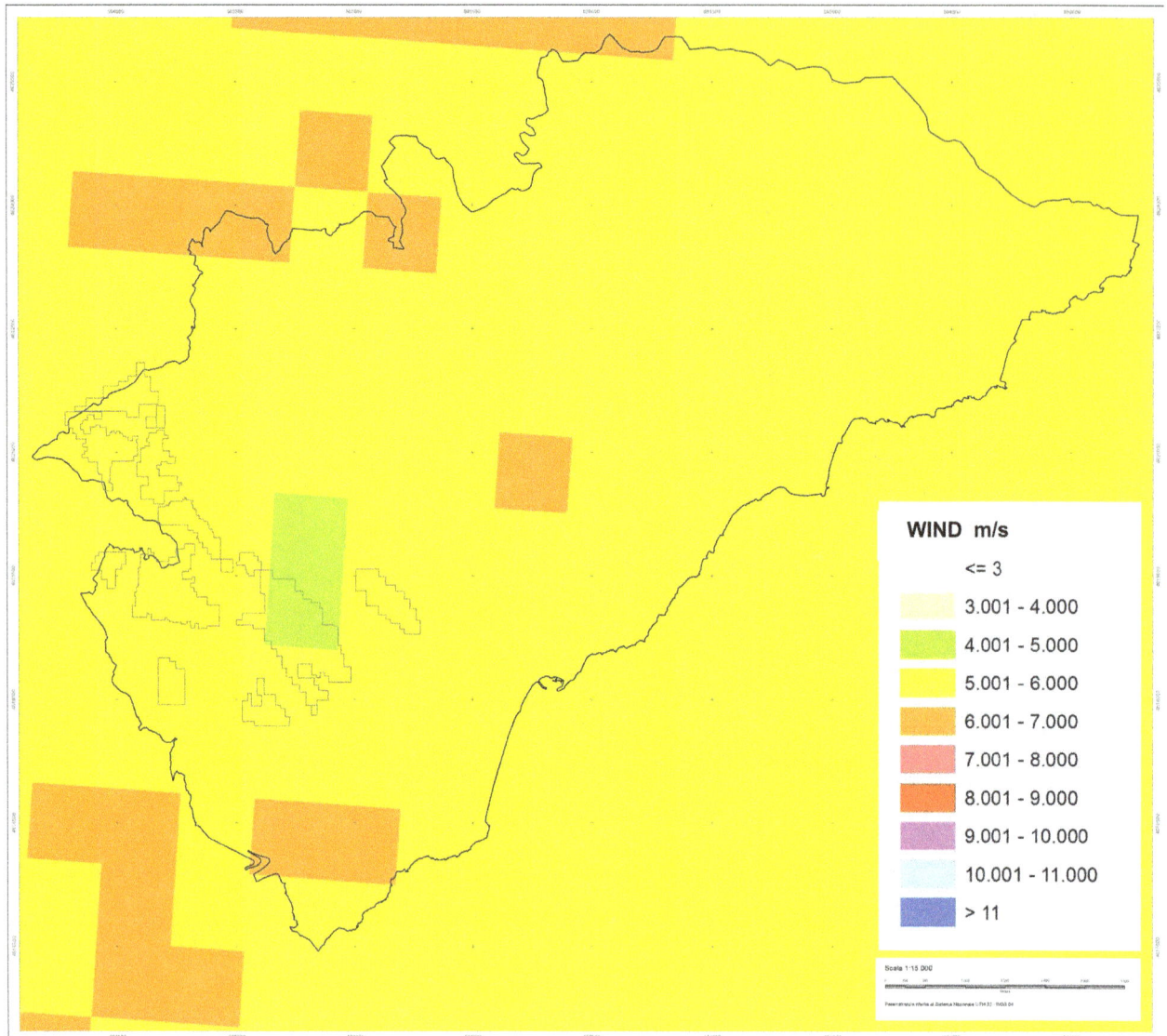

Figure 8. National wind atlas CESI.

complexities that often degrade the quality of the project.

References

[1]   Pantaleo, A., Pellerano, A., Ruggiero, F. and Trovato, M. (2005) Technical and Economical Feasibility of Off-Shore Wind Farms. *Solar Energy*, **79**, 321-331. http://dx.doi.org/10.1016/j.solener.2004.08.030

[2]   Ruggiero, F., Forte, G., Dicorato, M. and Trovato, M. (2009) Energy Potential from Off-Shore Wind Farms in the Mediterranean Sea. Proceedings of SEEP 2009 Dublino (Irlanda), August 2009.

[3]   Kondili, E., Kaldellis, J.K. (2012) 2.16-Environmental-Social Benefits/Impacts of Wind Power. Comprehensive Renewable Energy, Volume 2: Wind Energy, 503-539

[4]   Mays, I.D. (1994) The Environmental Benefits and Implications of Wind Energy. *Renewable Energy*, **5**, 537-541.

[5]   Evans, A., Strezov, V. and Evans, T.J. (2009) Assessment of Sustainability Indicators for Renewable Energy Technologies. *Renewable and Sustainable Energy Reviews*, **13**, 1082-1088. http://dx.doi.org/10.1016/j.rser.2008.03.008

[6]   Saidur, R., Rahim, N.A., Islam, M.R. and Solangi, K.H. (2011) Environmental Impact of Wind Energy. *Renewable and Sustainable Energy Reviews*, **15**, 2423-2430. http://dx.doi.org/10.1016/j.rser.2011.02.024

[7]   Gross, C. (2007) Community Perspectives of Wind Energy in Australia: The Application of a Justice and Community

Fairness Framework to Increase Social Acceptance. *Energy Policy*, **35**, 2727-2736. http://dx.doi.org/10.1016/j.enpol.2006.12.013

[8]   Clarkea, A. (1989) Wind Farm Location and Environmental Impact. *International Journal of Ambient Energy*, **10**, 129-144. http://dx.doi.org/10.1080/01430750.1989.9675132

[9]   Bell, D., Gray, T., Haggett, C. (2005) The "Social Gap" in Wind Farm Siting Decisions: Explanations and Policy Responses. *Environmental Politics*, **14**, 460-477. http://dx.doi.org/10.1080/09644010500175833

[10]   Szarka, J., Cowell, R., Ellis, G., Strachan, P.A. and Warren, C. (2012) Learning from Wind Power Governance, Societal and Policy Perspectives on Sustainable Energy. Series: Energy, Climate and the Environment, Palgrave Macmillan Editor.

[11]   GSE (Gestore Servizi Energetici) (2011) Statistical Report 2011 Plants to Renewable Energy Sources. GSE, Italy.

[12]   Directive 2009/28/EC of the European Parliament on the Promotion of the Use of Energy from Renewable Sources and Amending and Subsequently Repealing Directives 2001/77/EC and 2003/30/EC.

[13]   Scottish Guidelines (2006) Visual Analysis of Windfarms. Good Practice Guidance. Scottish Natural Heritage.

[14]   Italian Law n. 387/2003 (2003) Implementation of the Directive 2001/77/EC on the Promotion of Electricity Produced from Renewable Energy Sources in the Internal Electricity Mark.

[15]   Regional Law n. 25/2012 (2012) Regulation of the Renewable Energy Sources.

[16]   Regional Law n. 20/2001 (2001) General Rules of Governance and Land Use.

[17]   Italian Law n. 42/2004 (2004) Code of Cultural Heritage and Landscape, in Accordance with Article 10 of the Law No. 137 July 6, 2002.

[18]   CESI (2009) Wind Atlas of Italy.

Upcoming Transitions in the Energy Sector and Their Impact on Corporations Strategies

Jose M. "Chema" Martínez-Val Piera

ETSI Minas, Universidad Politécnica de Madrid, Madrid, Spain
Email: chemaval@gmail.com

Abstract

An analysis is presented on a set of enabling technologies which are opening new routes for energy conversion and consumption. This portfolio of innovations is complemented by a new framework in hydrocarbon production. This integration yields an optimization of energy uses that can result in lower greenhouse gases emissions and expand the lifecycle of current available resources. These options are confronted with the need for higher quantities of energy, at affordable costs in order to maintain the economic development. The conclusion is that there are no contradictions among the general objectives in global energy policy and the goals of corporations. Companies can take advantage of their previous expertise to remain competitive, but have to further develop new skills to operate in a new energy sector that is likely to be highly interlinked; evolving for the previous model that had markets segmented by specialty. New goods, such as the electric vehicles or the advanced high temperature high power fuel cells for generating electricity, should pave the way for a more synergetic and efficient energy sector.

Keywords

Energy Enabling Technologies, Integral Energy Efficiency, Hydrocarbon New Uses, Electrochemical, Thermal Hierarchies

1. Introduction

A minimum of $20 trillion investments is estimated for the next 25 years to meet increasing energy demand and to offset the declining reserves hydrocarbons [1] [2]. This is an indicator of the challenges that the sector has to address. Further, local and global contamination is posing a severe problem, which will be intensified as the number of cars increases from a circa 1 billion to 1.7 billion.

Such a global market expansion will concur with a wide portfolio of new technologies that can produce a solution to the otherwise inevitable energy crisis [3]-[5]. This solution will have to go beyond alternative and renewable energy and embrace the economic harvesting of remaining oil and gas resources, including the so called unconventional.

New trend in energy and currency accounting is likely to appear. This will take into account the efficiency of different technologies and the fit among sources, technologies and end-uses. Such scenarios could be characterized by the Integral Energy Efficiency (IEE), a concept presented in this paper. The IEE can be used to find the optimum cost-benefit ratio at a global scale, at a given, worldwide.

If such a deep transition materializes, industrial corporations will have to undergo changes in structure, scope and methodologies. This will be both a threat and an opportunity [6] [7].

Energy policies are primary established at national level. Basic principles apply everywhere: security of supply, environmental quality, and minimum cost. The latter is, however, an incomplete concept if a time frame is not defined during cost minimization. Most of the policies are established only considering a very short term (*i.e.* four years or less).

Spain can be taken as an example, notably for the electricity industry [8]-[13]. The boom of renewable energy sources was the consequence of a generous framework of feed-in tariffs established in the Royal Decree RD 661/2007. Collateral effects of this subside based policy [14] have become a notorious problem to maintain a proper balance between the economy and the profits of the system.

When the energy problem is addressed with a longer run perspective, underlying uncertainties are too broad as to enable the reach of the optimum solution in terms of energy policy. Moreover, collateral effects can appear in policies implementation, and correction measures must be enforced. A similar problem regarding uncertainties appears in the elaboration of a paper on these subjects. Some of the relevant variables are not of physical nature, but rather evanescent and quickly fade or disappear. These variables are related with geopolitics, financial pressures on a currency, environmental trend, ideology and other social developments.

A technical analysis can be aimed at optimizing a given energy problem with defined boundary conditions [15]-[17]. However, the result can become useless because of the interference of the evanescent variables, which usually appear with enormous strength over short periods of time. They are crucial in actual life, but they are almost not admitted in a technical paper. If a paper dares commenting this type of evanescent variables, the paper risks to be deemed unsuitable for standard scientific publications, where the formal procedure has been established according to the scholar tradition.

Let us consider the problem of Global Warming and the risk of increasing the atmospheric greenhouse effect by methane emissions. Methane has a Global Warming Potential that is 70 higher than CO_2. However, accounting of methane emission is much less accurate than that of CO_2. The latter comes from chemical combustion, and CO_2 emission rate is directly calculated from fuel composition and stoichiometric balance. On the contrary, methane can come from natural reservoirs, leaks from gas-pipes, or leaks from extraction and production process. These mechanisms do not have proper instruments for measuring the potential flow of methane.

In 2007, Global Warming [18] was so high in the cultural agenda that the Nobel Prize for Peace was awarded to the former US Vice President, Al Gore, and the Intergovernmental Panel on Climatic Change (IPCC). Seven years later, the IPCC is less emphatic in the declaration that "unambiguously, there is relation between human activity and climatic change". Moreover, President Obama's Administration is favoring fracking [19]-[21], as a way to reach energy independence which represents a milestone in the US international policy. A reliable estimate of methane leaks from fracking-based gas production is not available, which means that the famous "IPCC scenarios" for reaching a certain level of temperature rise must be revisited.

Such abrupt and evanescent changes in energy policy from the biggest economic power in Planet Earth is rather difficult (*i.e.* almost impossible) to model in any scientific analysis [22] and it is, however, second to none in terms of ranking of critical variable.

From the point of view of methodology, the difference between physical variables and evanescent variables is similar to the difference between "optimization process" and "decision-making process". To optimize a system, this must be well known and comprehensively defined so that a minimum and/or a maximum can be found (including local min/max, or a saddle point).

When uncertainties ranges and evanescent variables dominate the description of the system, optimization techniques cannot be applied. In such circumstances different methods must be used, from fuzzy logic to purely stochastic.

In scientific attempt to express all internal relations of a system in terms of integral-differential equations, evanescent variables are usually omitted because they disturb the ideal picture we can get from the energy world, and compromise our goal.

In the presentation of the work pursuing the goal of dissecting the energy world, I focus on the possibilities and potential of the physical mechanisms available to extract, convert or apply energy by the end user. It goes without saying that the path to the goal and the goal itself can be disturbed by an evanescent variable, but an analysis must be carried out on the basis of physical facts and laws [23] [24].

In this context, we propose a new concept to complement the guidance of energy policy making, which is the Integral Energy Efficiency. IEE should lead to maximize the total amount of End Uses of Energy, compatible with a minimum cost at a global scale (relative to the produced benefit).

In this paper, this concept is analyzed considering that some of the enabling technologies currently under research will actually achieve industrial maturity. Hence, opening the energy sector to more degrees of freedom to optimize global efficiency. Those technologies would include deep changes in the transport sector. For instance, the current dominance of petroleum products could be challenged by other hydrocarbons and biological products [25]-[28] or other sources of energy (*i.e.* the electric vehicles).Energy storage would represent a fundamental element in such long term energy scenario. It is therefore mandatory to analyze the role and features of corporations in a much more integrated Energy sector. The classical objective for a corporation is to achieve a niche in a given market. Very likely, a further objective of the corporation will be to consolidate and/or enhance that market quota. However, both the framework and the boundary conditions for energy corporations will not be the same as those experienced in the past. This is why the analysis pays attention to the problem of fitting corporation skills to future energy sector requirements, and accounts for alternative tools to deliver energy policies in the near future.

It seems we are in a crossroad similar to that the energy sector was in mid-seventies of the 20th Century. At that time, new agencies and developers were created (*i.e.* the Energy Research and Development Administration (ERDA) later substituted by the Department of Energy). Technology was one of the key elements for these types of agencies. However, one should underline that some of the most relevant enabling technologies have not being provided by these institutions, but rather by independent initiatives such as the case of fracking [29]-[31].

In order to analyze the world of energy, we will study in Section 2 the current situation, which is fast evolving [32] [33]. The (re)evolution is catalyzed by technology, and the influence of higher efficiency of the energy conversion chain, from raw sources to end uses. Efficiency is the subject of Section 3. Section 4 is devoted to review technology, which is the main driver to change the Energy world. Section 5 deals with the different scenarios identified according to the priorities in the energy policies, which mean that some scope for evanescent variables is needed. Special attention will be paid to the case of unconventional gas (and oil).

The paper ends with a collection of prospects that could be derived from attempts to improve significantly the utilization of available energy sources, which can then open a new age for corporations prepared to meet these opportunities.

2. The Current Energy World: Crossroads or Road-Maps?

There are several studies and series of analysis [1]-[4] devoted to a better knowledge of Energy. The same facts are presented to argue in favor different points of views. The International Energy Agency tries to offer a more eclectic landscape of energy, without pushing it to extreme opinions. The European Union has to integrate different views on topics such as Nuclear Energy, but it is generally well aligned with environmental matters (*i.e.* Directive 20/20/20 [34]). Furthermore, the EU has a taste for technology and development, but lacks establishing strong priorities (*i.e.* the Alternative Fuel Directive).

The US has adopted recently a much liberalized position in the energy market, and has shown a lot of interest in energy information [3] (http://www.eia.gov/forecasts/ieo/). The US has demonstrated very limited interest in developing new energy technologies in recent years. However, several institutions have produced interesting analysis on the true values of energy research for the long run (*i.e.* the IEEE [4] with the document USA-NEPR-2014). This long-run advantage seems to be of secondary importance for the current Obama administration as compared to those related with lowering cost and providing higher energy independence (*i.e.* the unconventional hydrocarbon industry). The 20th Century ended with an international concern about the long term case, and a quest for sustainability was launched to avoid Global Warming [35]. As already quoted, an international body, the IPCC [36], was created to produce forecasts on climatic for the coming decades. This effort was not match to foresee problems in the Energy sector and much less to identify pathways to development markets and technologies.

It is true that an organization as the International Energy Agency is already making forecast about the evolution of the sector, but the IEA only has 29 members, most of them from the European Union, but unfortunately not the EU as such.

We are already aware that Climatic Change deserves attention at the highest level, but Energy is not a minor problem. Energy in turn is dominating the environmental impact at every scale.

Energy is at a manifold crossroads where several concepts open different lines, especially the following ones: sources, investment, technologies, environment, sectors, emerging countries demands, independence, markets, and efficiencies.

A minimum of $20 trillion investments is estimated by the IEA for the next 25 years to meet surging energy demand and to offset the declining reserves and production of the world's major oil fields. These numbers can be challenged; both as a whole and as an aggregation of sectorial figures. However, they are based on projections that are easier to accept. For example, the rise in the number of automobiles in the world goes from circa 1 billion to 1.7 billion.

There is, however, another set of unknowns behind these numbers. For instance, automobiles will not be the same ones as those of present days. Automobiles will be used in a slightly different pattern. People tend to live in macro-urban areas and the use of car mainly in that environment can be substituted by public transportation. As a matter of fact, as seen in **Table 1**, IEA [1] projections on total consumption in transportation for 2030 (and even for 2050) do not change.

Some of the hypotheses underlying in those numbers can be challenged. The forecast for Light Road values in 2050 is based on the following allocation based on energy:

- Oil products, 20
- Natural gas, 1
- Biofuels, 11
- Electricity, 7

This projection considers a very short penetration of Electric Vehicles, and this is even more relevant as it includes H_2 vehicles in this source type.

As presented later, the EV presents very appealing features in fuel consumption and environmental quality. The main limitation on range should not be considered a burden as 85% of the mileage done by private cars corresponds to daily round-trips under 150 km.

This subject will be dealt with later; it is at the very center of a potential revolution in town transportation modes and air quality. It seems to be extremely difficult to calculate the cost of environmental consequences, specifically in terms of health problems as a result of micro-particles aggression. This hazard can only be dramatically reduced by electric cars.

The Future of Energy will be the result of a confrontation among facts from Nature, political principles & programs; and the evolution of both technology and society. **Figure 1** presents a comparison between the current North-American model and the European model on the basis of a 4-corner ring where the energy battle takes place.

Nowadays, the North-American masterpiece is the new aggregation of reserves which appear in the inventory thanks to hydraulic fracturing in horizontal wells, an enabling technology which has changed at depth the oil and gas industry both in the US and at a worldwide scale.

Table 1. Total world final energy consumption in transportation (EJ) for several modes and years (source, IEA).

Mode	Year		
	2011	2030	2050
Heavy road	26	27	29
Light road	51	49	39
Rail	2	3	5
Air	11	12	13
Sea	10	12	13
TOTAL	100	103	99

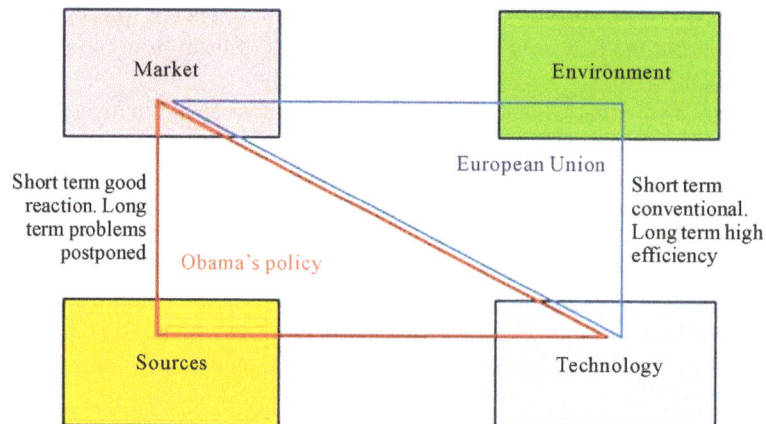

Figure 1. Sketch of the energy policy priorities in the US and the EU.

Breakthroughs in technology can be fundamental pieces in shaping the future. Nuclear Energy was an example in the past; and Shale hydrocarbons are an example right now. The Future of Energy will also be affected by changes in the demand side, but this is still harder to predict. Some of the new products and new activities, as widespread communication and digitalized information, can create new consumption modes (relying on electricity) and can save energy because they reduce the need to move for peoples and documents.

It is worth pointing out that IEA estimates predict a huge increase in energy consumption from the chemicals and petrochemicals industry. Increasing from 40 EJ in 2011 to 80 in 2050. The remaining industrial sectors do not undergo a similar rise. Furthermore, this industry has an important peculiarity: it uses hydrocarbons as primary matter, not only as fuel. This fact points out the importance of preservation of hydrocarbons for the highest added value end use, and it is an advice for looking to petro-chemistry with greater interest.

Such a dual role of fossil fuels is not shared by the rest of electricity energy sources, namely renewables and nuclear. In particular, the success of oil products for powering vehicles comes from one of their important properties; the high energy store capacity in terms of amounts of energy per unit mass and/or unit volume.

However, efficiency of Internal Combustion Engines (ICE) is very small as compared to other energy conversion machines (*i.e.* turbines). That fact was not important when oil was cheap and abundant, but is becoming a problem, from economical to environmental. ICE plus oil products were a very good tandem for powering vehicles, ships and airplanes, but such a good fitting is not enough for the problems of a sector which has to become more efficient and clean.

Figure 2 shows the strong changes produced in our 4-corner ring from the appearance of fracking as a new enabling technology. **Figure 3** has the EV as the technology driver.

Long term energy scenario will likely be more complex in its internal relations than the today's situation. Nowadays, as presented in **Figure 4**, the world of energy is rather specialized and segmented into compartments.

Close to 70% of coal currently goes to electricity generation, and 95% of oil goes to transport. **Figure 5** portrays the long distant future, Electric Vehicles could give much more flexibility to the energy system, and oil could be used for more expensive applications (*i.e.* higher added value). This includes new energy utilizations, as discussed in next section.

Understanding how new scenarios can be built based on new enabling technologies is useful for guiding energy policies and to guide corporations too. This double objective will be treated after an insight into energy enabling technologies.

3. Energy Efficiency Structure

More than 90% of the primary energy goes through the process of combustion; which, as presented in **Figure 6**, has a temperature. This is particularly relevant for thermodynamic cycles, limited in efficiency by Carnot's principle as presented in **Figure 7**.

An alternative to combustion is offered by electrochemistry, left side of **Figure 6**, which is not limited by Carnot's principle. However, it also has limitations [37] because of the generation of entropy. The theoretical

Figure 2. Four pillars of Energy and the sprout of shale gas and oil changing the scenario.

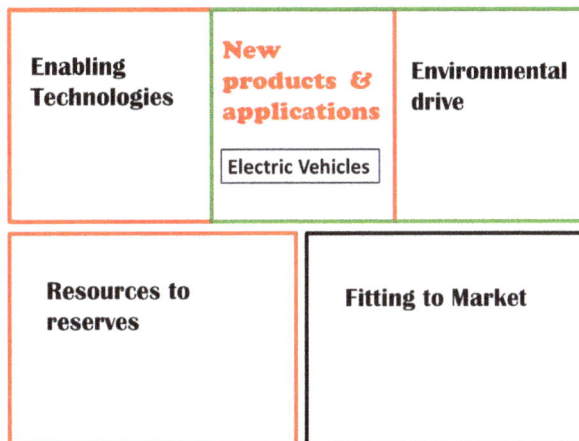

Figure 3. Four pillars of Energy and the sprout of Electric Vehicles, a key element in the future of transportation and environmental quality.

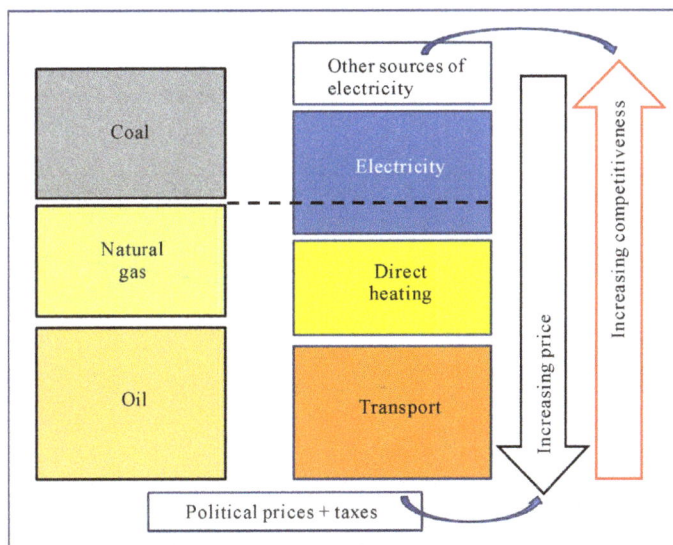

Figure 4. Fossil fuel coupling to end uses. In general, transportation presents lower competitiveness but support higher taxes.

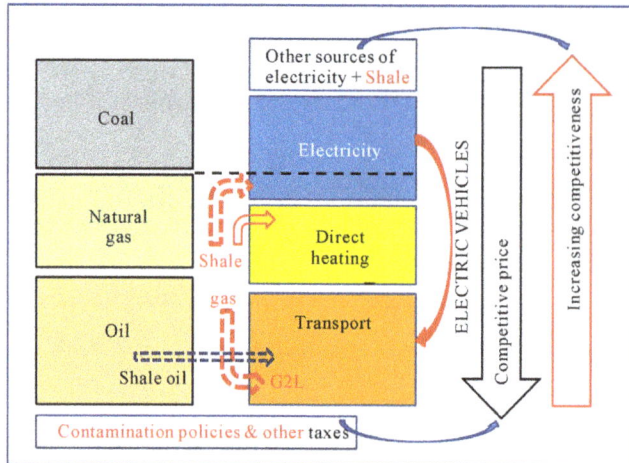

Figure 5. The deployment of Electric Vehicles will convey a deep change in the Energy sector. For instance, coal can contribute to transportation, through the electric grid, so increasing competitiveness in that field. The same can be said about electric renewable sources.

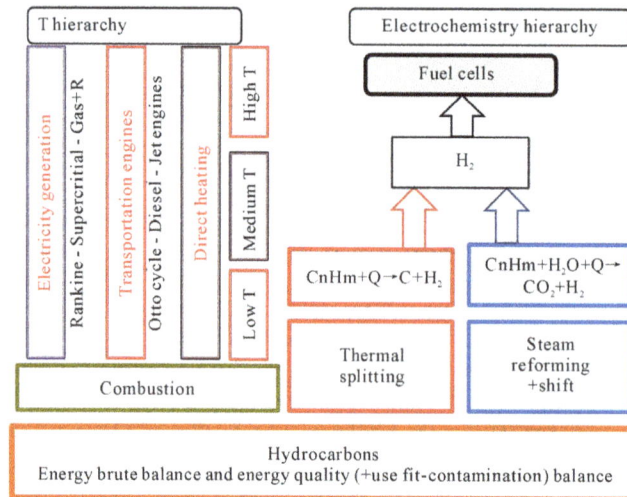

Figure 6. Diagram of the pathways for exploiting the energy of the hydrocarbons. The combustion branch at left, and electrochemistry at right.

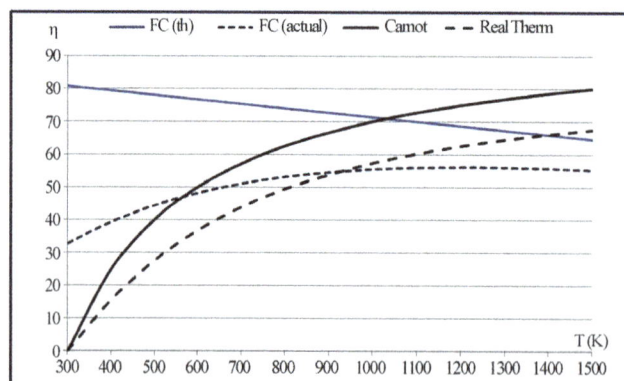

Figure 7. Fuel cell efficiency and thermal energy efficiency versus system temperature.

limit of efficiency in this case corresponds to the ratio between Gibbs' free enthalpy (ΔG), and enthalpy (ΔH), and it is a decreasing function of temperature (T), just the opposite of the thermal branch.

A first reaction to this figure is to make a complementary approach, using the brute force of combustion when high temperatures are achievable, and looking for electrochemistry devices when temperature must be low.

The latter, however, presents a hidden problem: electrochemistry needs that some electric charges (electrons and ions) move over an adequate substrate. This substrate can be a liquid electrolyte, as in lead batteries, or a solid. When moving fast, electric charges interact with those of the substrate, loss energy, and heat the substrate. Hence, if charges have to move fast because high power is needed, the system becomes very hot and efficiency decreases. This is an important rule for electrochemistry, which has been a drawback for its development [38].

Nonetheless, the development of new materials (*i.e.* polymer membranes) have open the way for extending the range of electrochemistry applications, including the famous case of "Hydrogen automobiles". Several studies and assessments have been conducted recently on this subject, notably by countries as Canada [39], Norway [40], and Australia [41] with very important and competitive electricity industries.

Fuel cells working at moderate T (*i.e.* below 100°C) have significant lower efficiencies than the theoretical values (*i.e.* PEFC devices) [38]. Solid Oxide Fuel Cells (SOFC) [42]-[43] need very high operational temperatures, but they can work directly with hydrocarbons. This simplifies the previous phases of the system, with the H2 generator.

They are also less sensitive to poisoning by carbon monoxide and other contaminants. They could be used for electricity generation wherever gas or oil is delivered (*i.e.* underground in the center of a city or any remote location). Proposals based on SOFC and other fuel cells are starting to emerge [44] driven by a better integral use of the energy stored in hydrocarbons.

A different branch of electrochemistry is related to batteries, which are becoming of paramount importance for powering electric cars [45]-[49]. Main problem presented by these cars (in plug-in fully electric type) is the range. For the moment, commercial cars have a range of 200 km with a specific consumption of 12 kWh/100 km (more expensive models can double this range).

Another concern is the time required to recharge (aka speed).Batteries don not cope well with high electric energy rate over a short time. A high fraction of the energy is transformed into heat. Therefore, electric cars appear as machines very well suited for daily traffic in cities, provided the charge can be done at night in a proper place (*i.e.* a private parking site).

Efficiency in this case rises quite a lot as compared to Internal Combustion Engines. Let us take the example of an average petrol car, in moderately congested street circuit, with real consumption of 10 liters per 100 km. This consumption is equivalent to circa 70 grams/km or 0.66 moles of octane per km; that would yield an emission of about 5 moles of CO_2/km. This is a mass flow of 220 g/km.

This is almost double the values required for new cars under EU legislation. However, these "administrative" values refer to ideal running conditions, for which consumption lowers to 6 liters per 100 km (or less), or approximately 130 g/km.

The value of the mechanical energy to drive an average car is 0.12 kWh (*i.e.* 0.432 MJ). The heat load in the combustion of 70 g of gasoline is 700 kcal, equivalent to 2.92 MJ. Therefore, the actual efficiency of the average automobile is about 15%.

The optimum engine performance is almost double the latter figure, circa 28%. Inefficiencies of the engine out of its optimum, accelerations, and idle periods make the optimum performance decline significantly.

An electric motor has the advantage of not having idle periods, as it rests when the car stops. Its performance is very high for most of the regimes of the system, and is exceptionally controlled by power electronics. On average it can support an 88% yield. However, a correction factor due to charge/discharge performance of the battery is required. A conservative estimate can be a correction of 75%. This results in an overall efficiency of 66%. So its consumption (from the electric grid) is 0.18 kWh electric per kilometer.

An additional adjustment is required before the final benchmark. We must add grid and generation losses to the latter figure. Again, a conservative estimate is a 10% of the total loss. This yields a need of 0.2 kWh per km for the electric system.

Octane burned in a combined cycle Brayton-Rankine, currently yields an efficiency of 60% and it is expected that this efficiency is boosted up to 70% in the near future. Taking the latter figure (which best represents the future), we obtain a primary (thermal) energy consumption of 0.28 kWh per km travelled (afterward) by car, which is equivalent to 1 MJ.

This means that we will need to burn 24 grams of octane in an electric power plant to produce the required energy in electric form. This is just over a third of the direct consumption of an ICE gasoline car.

Generation of CO_2 follows a parallel route. It is reduced from 220 to 75 g/km. This accounts for the energy chain after the gasoline is delivered; as neither process account for upstream emission of that delivery point.

It is inevitable to note that CO_2 emissions would be much lower if the batteries were recharged with electricity from nuclear or renewable sources. In this case, full-life cycle emissions are in the range of 30 to 40 g/kWh, or 10 g/km. This is one-twentieth of that emitted by the average car.

As for the cost to the end user, efficiency is not the only advantage. The price in electric domestic market are circa 20 c€/kWh (electricity), hence the cost per km would be 3.6 c€/km, or 3.6 euros per 100 km. With gasoline at 1.5 €/liter, and an average of 10 liters per 100, the cost would increase to € 15 per 100 km. This is a fourfold increase. ON the one side, this value includes the very high tax burden to gasoline, but on the other side it reflects the price of gasoline at either high and low international prices of oil as a commodity in the international market, *i.e.* Brent oil. At 100,000 km the electric car would save more than 11,000 €.This should compensate for the cost of the battery and the ancillary systems.

In summary, the largest environmental cleanup in both global and local pollutants, increased energy efficiency of automobile transportation and the positive economic impact to consumers, all point out that the Electric Car is a very powerful pathway to improve overall energy arena.

Additional economy considerations can be taken into account. If a car owner decides to optimize the investment and running costs for the total requirement of car performances in a given period of his/her life (*i.e.* 5 years) the owner will find that more about 85% of the total distance (number of km) traveled is done in daily roundtrips that are shorter than 100 km.

Owners might dive longer distances some long week-ends throughout the year, and could engaged in few (*i.e.* one or two) trips a year with distances longer than 2.000 km (forth and back). For the latter, an air flight plus a car renting strategy saves the owner a lot of money and a lot of time, and at the same time the total energy balance is also optimized (even for a car with 5 passenger this will not be a worst scenario). An Airbus 320 spends 5.000 liters of kerosene in a trip from London to the Spanish cost in the Mediterranean Sea, and carries around 125 passengers. This means 40 liter per passenger. If driven in a usual European car, the owner will need about 20 hours of effective drive (for 2.100 km), consuming about 200 liters of gasoline.

Similarly, if a depreciation of 0.5 €/km is applied to the car, the trip London-Murcia implies a roundtrip cost of 2.100 €. Adding the cost for fuel, the total cost of the round trip is circa 2.700 €.

Again, the comparison with the flight depends on the number of passengers, but one can find air fares for that flight at less than 250 € (excluding low cost airlines). Therefore, both from a personal and an energy efficiency perspective, the optimum solution is to accommodate the features of the trip to the characteristics of the machine.

It can be argued that the cost for renting a car must be added, but this cost is significantly lower to the associated cost of three driving days for a fifteen-day holidays.

The option of renting an ICE car for long domestic week-end trips is an option that will help optimize the energy consumption over the time-span under consideration (*i.e.* 5 years).

It is true that ordinary people are not used to this type of calculation, but this is rather simple and one can foresee an information campaign to illustrate these calculations. All electric car vendors already have calculation schemes to demonstrate the savings associated to a given type of daily trips and other journeys. And should everything fail an app for smartphone can help deliver this advantage.

Social and cultural changes are in many cases encouraged by technology, with or without government advertisement. There was no governmental campaign to foster mobile phones or the use of Internet.

Cars are much more expensive than a phone or a computer, and the reaction time constant for buying a car is much longer than the period to replace a cellular phone. This means that this "evanescent" variable of public perception and decision making will be not as prompt as in telecommunication. However, many hints point out in that direction.

Electric vehicles can become household goods on this basis faster that they can do so by local government policies to impose them to fight local contamination.

4. Technology Changes in the 21st CENTURY

It is not easy to generate prospects on Energy Technology. Previous experience points out that some promising lines grow little by little while other lines lagging behind for many decades burst suddenly and make an early market breakthrough.

An example of the former is nuclear fusion, which 40 years ago it was considered a matter to be controlled within the forthcoming 30 years. In reality, nuclear fusion has kept in the main road of tokamak reactors but it still considering that industrial maturity will arrive 30 years from now.

An example of the latter is photovoltaics, which was considered a very expensive and inefficient technology for electricity generation, and explode over the last decade. Investment costs have decreased from over 10 to under 2 US$/kW.

The key factor was that PV did not require a significant breakthrough in physics, but rather a development of technology. On the contrary, the control of magnetic fusion reactors was not mastered 40 years ago and there it still holds several unknowns nowadays. It is obvious that many advanced have been made, but research is still at the laboratory scale and with fundamental problems to be addressed.

The situation for renewable energies is just the opposite. These sources of energy rely on well-known physical mechanisms and the main problem has always been to reduce costs (both capital expenditures and operating costs). The best results have been achieved by improving designs and materials.

The learning curves of all renewable energy sources have been successful in the last 10 to 15 years. The first renewable energy to achieve a significant reduction of costs was wind power. It became possible to manufacture a machine (the 3 blade horizontal generator mounted on a tower) with a higher efficiency than the rest of the machines.

PV still presents many potential lines of development, although the best cost reduction step has been associated to cheap semiconductor material (*i.e.* poly-crystalline silicon).

Nevertheless, the new energy technology port-folio includes advancements in all levels or stages of the energy supply chain and the form of energy and its changes throughout the chain.

In terms of sources of energy, it is obvious that "fracking" represents one of the key milestones of the past years, and efforts should be done to make it cleaner and more effective in terms of the hydrocarbon recovery factor. Otherwise, some areas (*i.e.* Europe [50] or Quebec (Canada) [51]) will not join the fracking club.

In the energy conversion phases, electrochemistry [37] can give a boost to energy efficiency, particularly in the domain of distributed generation. Fuel cells working on very clean hydrogen [38] need a previous chemical double reactor, with one endothermic process for steam reforming and other exothermic process for shifting the reaction.

For the moment, technological advanced in this technology have been very limited. However, many possibilities can be identified in this technology option such as reactor configuration, materials, and temperature regime.

Two important aspects about fuel cells are the following: first, some fuel cells [42]-[44] run on hydrocarbons, which can easily be stored; and second, fuel cells do not have moving parts, which can an advantage for operation and maintenance.

On the other hand, experience with fuel cells is still limited if compared to the enormous potential of electrochemistry.

Electrochemistry has already been identified as a candidate for new energy technology alternatives. Electrochemistry can be considered as the most important one in terms of economic impact on environmental quality and integral energy efficiency, which is the case of the electric vehicle.

It is eloquent to revise the history of electric cars [52]-[54] which were proposed more than one century ago. The electrical vehicle has had peaks and troughs along these years. Even after the oil crisis, when some analysts and researchers propose electric cars fed by nuclear – generated electricity for coping with the energy problem. Troublesomeness for this proposal is that it took too long to develop the nuclear power park for supplying all the energy that was required, and beyond and above, the very limited range that an electrical car had at that days.

Batteries for electric cars and the cars themselves started to change 25 years ago, once electrochemistry was identified as a powerful tool for energy transformation with high efficiency. **Table 2** is a good example of the efforts needed to push technology ahead, and it is homage to those who believed in Electric Vehicles when they were not yet a promise.

However, upcoming energy technology developments will affect most (if not all) segments in the world of energy. This applies for the domain of sources, its conversion, and its applications. Of course, in the short term there will be a clear priority to make natural gas the center of many fundamental decisions, particularly because of the success of fracking and other new techniques, but also because of the vast amount of resources discovered over the past years.

Table 2. An important relic from the past: assessment on batteries for electric cars as foreseen in 1989. Projections were incomplete, but not out of the case [52].

	Characteristics of selected electric passenger vehicles						
							Projected
	ETV-1	ETV-2	ETV-20	ETV-I	1987 BMW	1990 BMW	VW Jetta
Top speed. mph	†	62	60	60	53	75	78
Urban range. Mi	Up to 75	66	<75‡	†	43-77§	62-124§	118
0 - 30 mph acceleration secs	†	8	†	7	14	7	6
Mi/kuh from battery city	3.41	3.14	†	3.61	†	3.73‖	†
Passenger capacity	4	4	2	2	4	4	4
Power train	dc	dc	ac	ac	dc	ac?	ac
Battery	Pb/acid	Pb/acid	Pb/acid	Pb/acid	Na/s	Na/s	Na/s
Approx year of tests	1980	1980	1986?	1985	1987	1987	†
Reference	Kurtz (1981)	AiResearch (1981)	Wyczalek (1987)	Ford & GE (1987)	Regar (1987)	Regar (1987)	Angelis & Sedgwick (1988)‖

Top speed is maximum continuous cruising speed. Range and efficiency data for ETV-1. ETV-2. and ETX-I refer to FUDS (Federal Urban Drive Schedule);for Jetta. They refer to ECE (European) urban cycle. †Not available. ‡Range is at constant speed of 25 mph. §Lower range estimate at top speed; higher at 30 mph. ‖Based on ABB Na/S battery projections of Table 3. with an improved powertrain.

Properly speaking, fracking had been known for decades, and it was pursued in the peculiar "Plowshare Project" [55] which was conceived for extending the use of Nuclear Explosives to peaceful applications, as ultra-large civil works and underground intrusions for getting minerals and above all, getting gas from source and/or trapping rocks. It could be said that "fracking gas" was known but remained unknown for many years.

Energy conversion and customization for final uses will surely be another domain for development and novelties. Notably in the process of becoming more dependent on electricity for many applications, including a substantial part of ground transportation in private vehicles.

Fuel cell, batteries and combustion will have to come closer to enable synergies that we are currently missing. Integral Energy Efficiency it is somehow dependent on those synergies.

Currently, oil products are mainly consumed in Internal Combustion Engines with a theoretical efficiency circa 30%, but with a real efficiency in the range 15 to 20%. This is in full contradiction with the inherent capabilities of a chemical fuel, which could burn at 1500 K and above, so creating a very hot energy source. Of course the main feature appreciated in liquid oil products is their capability to store energy in small volumes. This is just the end of the string respect electricity.

Electric engines have very high efficiencies (from plug to shaft) but electricity cannot be stored. A potential way to avoid this mismatch is to rely on hydrocarbon-fueled high temperature Fuel Cells in order to generate electricity with a very high efficiency that will later be used to recharge batteries. Of course it is not possible to assemble a car with a SOFC component, at is operates at 600°C and has a power in the order of few MW. However, it can be located in a secured facility close to consumers.

This leads to an integration of routes for energy conversion in order to optimize the Integral Energy Efficiency within the framework established by the applications. A sketch on this integration between the electrochemical hierarchy and the thermal hierarchy is shown in **Figure 8**.

Integrating all possible mechanisms to optimize the exploitation of energy does not disturb the energy markets. On the contrary, it aids sorting them out in a better manner. In some instances it is considered that the deployment of the electric vehicle at large scale will be harmful for the oil industry, and this is not the case. It is not so because oil has unique properties, that range from a well stablished global market to its own physical and chemical properties.

In particular it can be considered one of the best, if not the best way, to store energy. **Figure 9** depicts this characteristic in comparison with Natural Gas and major Renewable Energies (Solar ones plus Wind). Oil has the largest range for accessibility (because it can be transported easily across long distances) and also has the widest range of applications (including future generation of electricity with a much higher efficiency, and lower

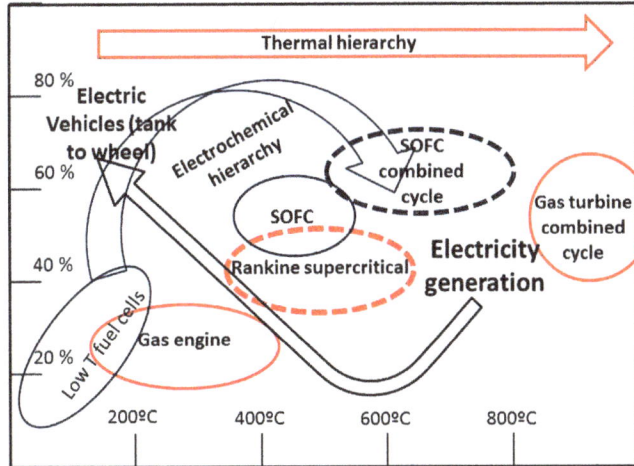

Figure 8. A map of efficiencies and temperatures for different combinations of thermal machines and Fuel cells.

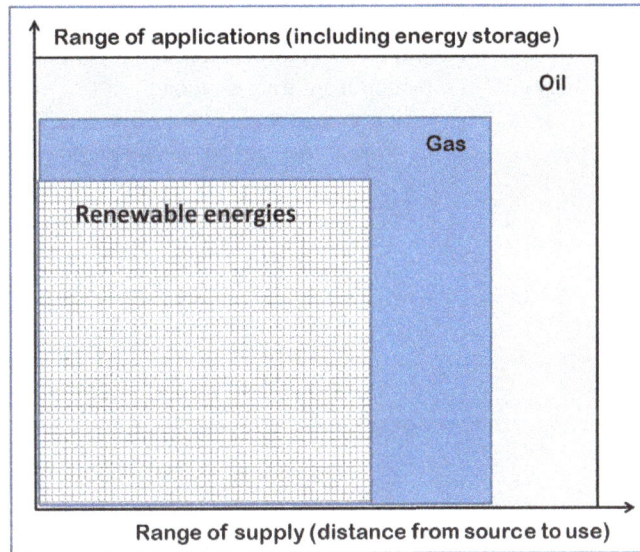

Figure 9. A qualitative graph of the ranges covered by different sources of energy in relation to supply and environment.

CO_2 emissions unit of energy per consumed).

5. Energy Policies and Corporation Strategies

The actual evolution of the Energy sector cannot be outlined with a high level of confidence. Even in centrally planned economies, because of technology leaps, market oscillations and environmental pressure, the task of forecasting evolution of energy is hard. What can be done is to analyze specific scenarios. Relying on coherent assumptions we can identify shortage of materials, or excessive financial support to materialize. Those scenarios can be defined presuming selected evolution in relevant variables, including evanescent ones, which can come from the past in the Energy sector.

This exercise is achievable in the case of automobiles, their engines and the energy infrastructure to power them. Good projections of automobiles have been carried out in the past using specific ratios as a basis for forecasting. This is based on the relationship between the number of automobiles and total population in a specific geographical area (*i.e.* continents or countries), as well as other variables in relation to the number of cars (*i.e.* as work force and gross domestic product). The fraction of total income devoted to vehicle acquisition and fuel is another fundamental ratio. Last but not least, sound judgment is needed to preserve necessary coherence of the study.

The penetration of EV in a given market can be studied as an example of this procedure. This will at least highlight the different requirements on the key variables of the problem. Each scenario includes different patterns in the energy conversion chain (from sources to market) depending in this case of the type of vehicle. A conversion chain is defined by a number of steps where some energy is lost, some contamination is generated and some investments or cost must be expensed.

Some values can be assigned to different steps and patterns to qualify their maturity or readiness, but credibility of the study will largely remain a personal decision. Energy technologies still needing very basic R&D programs as Nuclear Fusion should not be included in the picture.

The International Energy Agency, the US Energy Information Administration and the European Commission (through the Joint Research Center, mainly) have similar models for characterizing energy scenarios. The main difference among them is the role assigned to some of the new technologies.

The main deficiency for this methodology is that it cannot integrate innovations which are not yet in the agenda. The most visible example is happening now in the USA, where the "breakthrough" of fracking and shale resources has changed completely the Energy sector. **Figure 10** shows the impressive evolution of "shale gas" in the USA over the past six years. This revolution deserves a suitable analysis and some further comments.

The most important one is that "unconventional gas" (and oil) is not an interim explosion in the North-American industry but a long-standing trend with important consequences in geopolitics, environmental qualification and economics of Energy. Projections of gas production in the States are given in **Figure 11**, where it can be seen that "unconventional gas" will be dominant from now on.

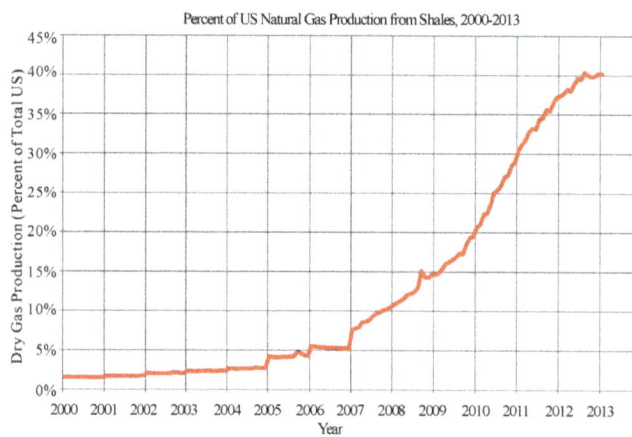

Figure 10. Percentage of shale gas production over total, in the USA.

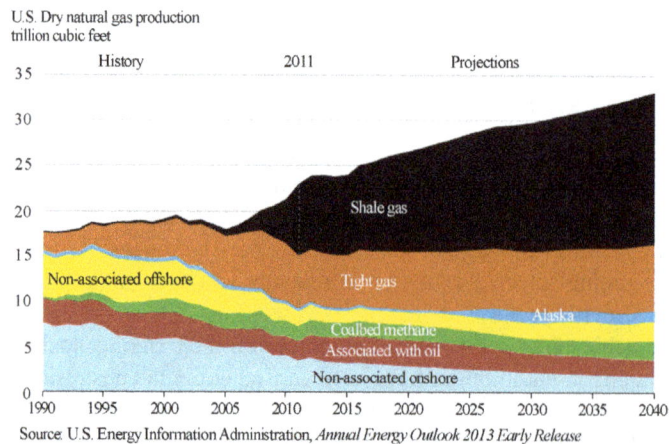

Figure 11. Share of different types of gas production in the USA.

"Shale" was not in any official agenda, and had not been previously evaluated in environmental terms. It was a "gold rush" created by technology pioneers, sprawling immediately. The full story of "fracking" is still to be told, but we need to highlight that the US DOE was promoting other lines of action. Most of these lines are currently frozen.

The rush has been so fast that potential cost associated with toxic waste or by products are not internalized yet. The main reason for this is a lack of reliable accounting on them. Moreover, shale hydrocarbons have not yet become an international commodity but rather locally traded.

A suitable infrastructure was not available when the gold rush started, and a sort of mismatch went on between drilling and full-scale commercialization. Drilling is not cheap, and completions are expensive. However, the biggest threshold was to develop an infrastructure for exporting goods throughout the country. The core of the infrastructure was for importing crude oil and refined products, and that shift implied a change in mentality and big capital investments.

The development of that network will only take a couple of years (partially aided by the Obama's Administration support and commitment) and will start a new age (one of uncertain time duration) where oil will not be linked only to critical domains.

Important countries as United States and Canada, with huge shale reserves, can convey a lot of stability to the oil market in the mid-term, mainly in terms of supply.

This process to achieve the international market cannot be deployed faster. "Fracking" was born and grew up without asking permission, and without it being present in the energy agenda or planning.

This delay is neither good for the energy system nor goof for shale-hydrocarbons producers. Potential revenues are not realized because they do not have yet access to the global market.

Figure 12 shows the differences in price among three major gas markets. They are paramount, and they have several root causes. Access to cheap reserves is one of them, and this includes a long term of guaranteed supply.

Short term profit has been in this case the driver planning tool. The process was somehow capped; but it has not experienced a strong change, due to a change in priority in President Obama's Administration. At first, this Administration seemed very committed to global environmental issues. For example, it created in 2009 a new entity called "White House Office of Energy and Climate Change Policy". However, it this entity did not quite deliver on its expectations as it was integrated in the "Domestic Policy Council" after two years.

Anyhow, first presidential statements clearly indicated direction in policies related to energy. In the vice-president's memorandum dated December 15, 2009, under the title "Progress report: The transformation to a clean energy economy", the driving force for those policies are defined as "jumpstarting a major transformation of our energy system including unprecedented growth in the generation of renewable sources of energy, enhanced manufacturing capacity for clean energy technology, advanced vehicle and fuel technologies, and a bigger, better, smarter electric grid."

In fact the report is divided in the following sections: Renewable Energy, Vehicles and Fuels of the Future, Grid Modernization, Energy Efficiency, Carbon Capture, Nuclear Power, and Science and Innovation. Neither the word "shale" nor "fracking" appear in the text.

• In current times, "unconventional gas" has already been embodied in Obama's Administration. Efforts have been made to justify it not only by private profit, but according to a philosophy established on the following points: "Unconventional gas" can be made "sustainable" by selecting the regions where it can be exploited and establishing the corresponding regulations.

• It is a very good element to support freedom from Russian energy dependence and freedom from Persian Gulf oil dependence.

• Unconventional gas profits will have to be invested, in a sizeable fraction, into Clean Energy Projects.

Industrial and commercial corporations, as anyone else, live in a certain social and legal environment, including technical regulations. A specific feature of corporations is their strong dependence of being profitably fitted to that environment. Adaption is important for any living being, but it is even more critical in companies. Companies can easily be created or discontinued according to decisions made by investors, who act as watchdogs for corporation profitability. Although each corporation has its own internal structure, coherent to its objective, a generic description of its functional articulation can be depicted as illustrated in **Figure 13**.

Money (and Profit) is usually the criterion to assess the health of a corporation. However, some intangible assets (such as know-how and expertise) should also be evaluated in terms of potential profit for the future.

If money is the outermost shell which is indeed connected to the general financial system, the Physical Assets

Trends in natural gas spot prices at major global markets
U.S. Dollars per million British thermal units (MMBtu)

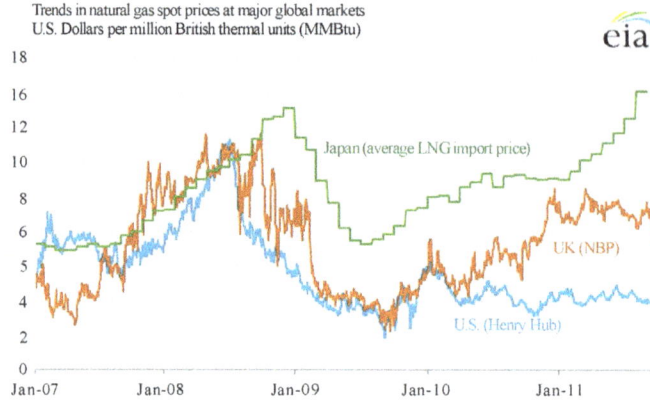

Figure 12. Differences in price among three major gas markets.

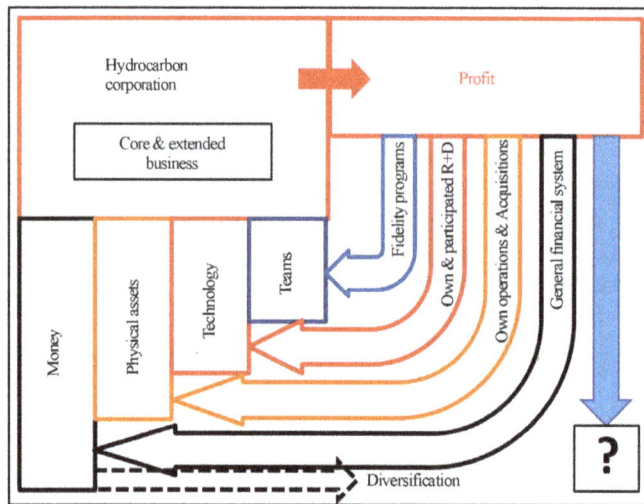

Figure 13. A schematic flowchart of activities in a corporation.

of the company, acquired along its history, constitute the bones of such particular body.

Technology is the third shell of the company; and a clear distinction can be seen between this shell and the Physical Assets: these ones are private property of the corporation, and Technology is in many cases a "cloud" of knowledge, which can be partially protected by patents and other rights, but it is created out of scientific and commercial knowledge, and it can therefore evolve and be improved.

In the innermost shell we find the people, or teams, of the company. They can go out as they can come in, and properly speaking are not assets of the corporation, regardless of confidentiality clauses.

People however, constitute the nervous system and the brain, and this is why the so called "human factor" is so important. This is none the less outside the scope of this paper.

The objective we had when setting the foundations of this work was to analyze the main features of the evolution of Energy macro-policies, and to analyze how Energy corporations should do for being successful in the new Energy scenario where it will have to compete. , which includes a long series of influences characterized booth by physical and evanescent variables.

The answer to this question is very complex, and in some cases will not exist as such, and it will be a collection of considerations and advises coming from the general analysis of the Energy sector.

A Directive from the European Union "European alternative fuels strategy" (COM) 2013, 17) requires Member States to adopt national policy frameworks to development markets for alternative fuels and develop their infrastructure. The Directive sets binding targets for the build-up of alternative fuel infrastructure. An accompanying Impact Assessment (SWD (2013) 5/2) evaluates cost and benefits of different policy options and identifies conditions for a comprehensive coverage of the main alternative fuel options.

The assessment is aimed at being totally neutral in both economics and technology terms. This boundary condition limits drastically the scope of the assessment. The Directive in opens all possible ways (different from the classical ones) and does not attempts to define a priority list of actions based on a given number of criteria's. The Directive seeks insight of long-term potential, environmental impact, security and flexibility of energy supply.

It does not mean that a corporation has to deal with all possibilities. On the contrary, a corporation must identify the ways, means, and objectives to address the future utilizing all the knowledge and judgment acquired in the past to succeed in its quest to perform in the future world of energy.

6. Summary and Expectations

From any known analysis, it must be understood that there is no possibility to define and develop a unified and integrally coherent Global Energy Policy for the entire planet.

Even in the world covered by the International Energy Agency, a unique policy seems out of reach. Nevertheless, corporations can be oriented by analytical approaches to the scenarios envisioned by the prospective work done by several agencies and the proper central team of the corporation and external specialized consultants.

Scenarios identified in this analysis as potential integrations of coherent choices are depicted in **Figure 14**. Business as usual should not be considered as an option, because it is impossible to ignore the strong changes in technologies and markets. It is a BUS going nowhere.

Environmentally dominated scenarios seem rather unlikely, because they will be very expensive and would only be meaningful if all countries accept the same procedures and standards. This is very unlikely based on the aftermath of the Kyoto Protocol.

Integral optimization of energy seems the most sensible scenario for a more efficient future. It has the very important appeal of advanced technology. At the same time, it has the drawback of expensive investments to deploy new technologies. The key advantage is that the system will be more effective in the long run, reduces fuel consumption, and reduction collateral side effects (as those produced by contamination of combustion by-products) All these will represent a major benefit which will likely overwhelm the initial drawback of higher expenses. This scenario follows the typical law of economics of engineering systems: the higher the investment cost (a proxy to higher quality) the smaller the running cost, especially fuel costs.

Last scenario shown in the picture is the most aggressive one, and corresponds to a general deployment of fracking. Local and global opposition to fracking can disturb this course of action. It is certainly not a well-accepted option within the European Union. It has been briefly explained that this is mainly a North-American story, and a rather fast one. The "fracking" technology has got such momentum that Obama's Administration has changed its general vision of Energy.

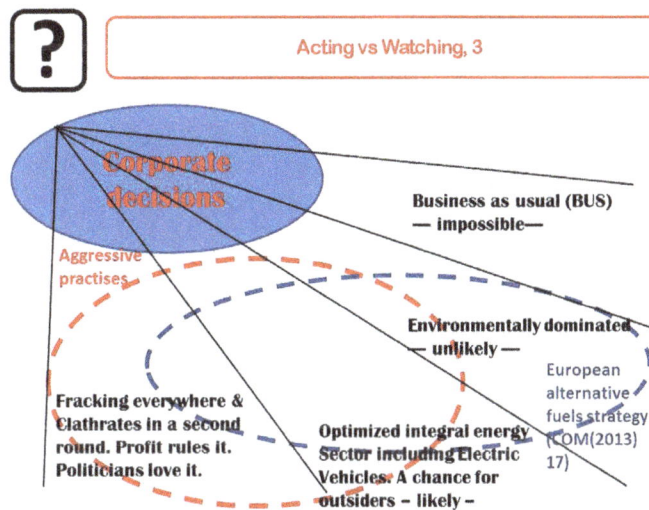

Figure 14. A map of pathways for arriving to different futures.

After considering that fourfold pathway towards a profitable future for the Energy corporation, what advice could be drawn from the contents of this paper? It seems the advice should be to invest and rely on technology. Further, to focus on the effects that the full development of an array of well coupled technologies would have on the market where the corporation works and the corporation itself.

From all papers studies under this scope, technologies that seem mandatory in this advice are: Electric Vehicles; High temperature fuel cells; Distributed and controlled electricity generation; and energy storage.

This can be counter intuitive and somehow a paradox, but oil can contribute towards those objectives (as well as natural gas), with higher efficiencies and lower CO_2 emissions per unit of energy applied to any given end use.

In other words, hydrocarbons can play a unique role in the context of Integral Energy Efficiency. Petroleum products can become the tools that guarantee the energy supply chain thanks to their suitable properties for storage and for feeding new electrochemistry devices.

This means that it is possible to make a corporation that is based on hydrocarbons operation compatible with the requirements for lowering the greenhouse gases emissions. In fact, it is possible because there are many roads for improving synergies among energy conversion mechanisms.

Hence, there seems to be a good agreement between the overall mandate of increasing the Integral Energy Efficiency as a general priority at global scale and the corporation objective of maximizing profits within a framework of commercial activity.

Corporations should be prepared for the higher complexity of future energy sector, with more degrees of freedom in a less segmented fueling structure.

Acknowledgements

This paper is part of the Ph. D work of J.M "Chema" Martinez-Val, at Madrid Technical University (UPM) in the School of Mines. The Ph. D work is carried out under the direction of professors Maldonado and Rodriguez-Pons. He also studied at The French Institute of Petroleum and the Colorado School of Mines. He is a Registered Engineer at Texas Board of Engineering.

References

[1]   International Energy Agency (2014) http://www.iea.org/etp/explore/

[2]   International Energy Agency (2014) World Energy Outlook 2014. www.worldenergyoutlook.com

[3]   U.S. Energy Information Administration (2014) Annual Energy Outlook 2014 with Projections to 2040. http://www.eia.gov/

[4]   IEEE (2014) IEEE National Energy Policy Recommendations. www.ieeeusa.org

[5]   International Monetary Fund (2012) Coping with High Debt and Sluggish Growth. http://www.imf.org/external/pubs/ft/weo/2012/02/

[6]   Chema Martínez-Val, J.M., Maldonado-Zamora, A. and Ramon Rodríguez Pons-Esparver, R.R. (2013) Adapting Business of Energy Corporations to Macro-Policies Aiming at a Sustainable Economy: The Case for New Powering of Automobiles. *Energy and Power Engineering*, **5**, 92-108. http://dx.doi.org/10.4236/epe.2013.51010

[7]   Chema Martínez-Val, J.M. (2013) Improving the Global Energy Industry by Integrating Macro-Technologies: Challenges and Opportunities for Corporations. *Energy and Power Engineering*, **5**, 604-621. http://dx.doi.org/10.4236/epe.2013.510067

[8]   Red Eléctrica de Espa—A Website. http://www.ree.es/sites/default/files/downloadable/sintesis_ree_2013_v1.pdf

[9]   Perez, Y. and Ramos-Real, F.J. (2009) The Public Promotion of Wind Energy in Spain from the Transaction Costs Perspective 1986-2007. *Renewable and Sustainable Energy Reviews*, **13**, 1058-1066. http://dx.doi.org/10.1016/j.rser.2008.03.010

[10] Martínez Montes, G., Serrano López, M.M., Rubio Gámez, M.C., Menéndez Ondina, A. (2005) An Overview of Renewable Energy in Spain. The Small Hydro-Power Case. *Renewable and Sustainable Energy Reviews*, **9**, 521-524.

[11] Hernández, F., Gual, M.A., Del Río, P. and Caparrós, A. (2004) Energy Sustainability and Global Warming in Spain. *Energy Policy*, **32**, 383-394. http://dx.doi.org/10.1016/S0301-4215(02)00308-7

[12] Foidart, F., Oliver-Solá, J., Gasol, C.M., Gabarrell, X. and Rieradevall, J. (2010) How Important Are Current Energy Mix Choices on Future Sustainability? Case Study: Belgium and Spain—Projections towards 2020-2030. *Energy Policy*, **38**, 5028-5037. http://dx.doi.org/10.1016/j.enpol.2010.04.028

[13] Batlle, C. and Rodilla, P. (2010) A Critical Assessment of the Different Approaches Aimed to Secure Electricity Generation Supply. *Energy Policy*, **38**, 7169-7179. http://dx.doi.org/10.1016/j.enpol.2010.07.039

[14] Moreno, F. and Martinez-Val, J.M. (2011) Collateral Effects of Renewable Energies Deployment in Spain: Impact on Thermal Power Plants Performance and Management. *Energy Policy*, **39**, 6561-6574. http://dx.doi.org/10.1016/j.enpol.2011.07.061

[15] European Union (2011) Materials Roadmap Enabling Low Carbon Energy Technologies. Commission Staff Working Paper, SEC (2011) 1609 Final.

[16] European Union (2013) Guidelines for Financial Incentives for Clean and Energy Efficient Vehicles. Commission Staff Working Document, SWD (2013) 27 Final.

[17] Klaassen, G. and Riahi, K. (2007) Internalizing Externalities of Electricity Generation: An Analysis with Message-Macro. *Energy Policy*, **35**, 815-827. http://dx.doi.org/10.1016/j.enpol.2006.03.007

[18] Intergovernmental Panel on Climate Change (2007) Fourth Assessment Report, 2007. www.ipcc.ch

[19] Terrell, H. (2011) US Gas Reserves Estimated at Record High. *World Oil*, **232**, 13.

[20] Hydrocarbon Processing (2011) Natural Gas Enters a New Era of Abundance. *Hydrocarbon Processing*, **90**.

[21] McIlvaine, R. and James, A. (2010) The Potential of Shale Gas. *World Pumps*, **7**, 16-18. http://dx.doi.org/10.1016/S0262-1762(10)70195-4

[22] Lebre, E., Borghetti, J., Basto, L. and Lauria, T. (2010) Sustainable Expansion of Electricity Sector: Sustainability Indicators as an Instrument to Support Decision Making. *Renewable and Sustainable Energy Reviews*, **14**, 422-429.

[23] Holsapple, C.V. and Singh, M. (2001) The Knowledge Chain Model: Activities for Competitiveness. *Expert Systems with Applications*, **20**, 77-98. http://dx.doi.org/10.1016/S0957-4174(00)00050-6

[24] Shina, M., Holden, R. and Schmidt, R.A. (2001) From Knowledge Theory to Management Practice: Towards an Integrated Approach. *Information Processing & Management*, **37**, 335-355. http://dx.doi.org/10.1016/S0306-4573(00)00031-5

[25] Liao, W., Heijungs, R. and Huppes, G. (2011) Is Bioethanol a Sustainable Energy Source? An Energy-, Exergy-, and Emergy-Based Thermodynamic System Analysis. *Renewable Energy*, **36**, 3479-3487. http://dx.doi.org/10.1016/j.renene.2011.05.030

[26] Luo, L., Van Der Voet, E. and Huppes, G. (2009) Life Cycle Assessment and Life Cycle Costing of Bioethanol from Sugarcane in Brazil. *Renewable and Sustainable Energy Reviews*, **13**, 1613-1619. http://dx.doi.org/10.1016/j.rser.2008.09.024

[27] Hanff, E., Dabat, M.-H. and Blin, J. (2011) Are Biofuels an Efficient Technology for Generating Sustainable Development in Oil-Dependent African Nations? A Macroeconomic Assessment of the Opportunities and Impacts in Burkina Faso. *Renewable and Sustainable Energy Reviews*, **15**, 2199-2209.

[28] Koh, M.Y. and Ghazi, T.I.M. (2011) A Review of Biodiesel Production from *Jatropha curcas* L. Oil. *Renewable and Sustainable Energy Reviews*, **15**, 2240-2251. http://dx.doi.org/10.1016/j.rser.2011.02.013

[29] Kinnaman, T.C. (2011) The Economic Impact of Shale Gas Extraction: A Review of Existing Studies. *Ecological Economics*, **70**, 1243-1249. http://dx.doi.org/10.1016/j.ecolecon.2011.02.005

[30] Arthur, J.D., Hochheiser, H.W. and Coughlin, B.J. (2011) State and Federal Regulation of Hydraulic Fracturing: A Comparative Analysis. *Proceedings of the SPE Hydraulic Fracturing Technology Conference*, The Woodlands, Texas, 24-26 January 2011.

[31] Lafollette, R.F. and Holcomb, W.D. (2011) Practical Data Mining: Lessons Learned from the Barnett Shale of North Texas". *Proceedings of the SPE Hydraulic Fracturing Technology Conference*, The Woodlands, Texas, 24-26 January 2011.

[32] BP Statistical Review of World Energy 2014. http://www.bp.com/en/global/corporate/about-bp/energy-economics/statistical-review-of-world-energy.html

[33] Eurostats (2014) Energy Price Statistics. http://epp.eurostat.ec.europa.eu/statistics_explained/index.php/Energy_price_statistics

[34] European Union Directive 2009/28 CE. http://eur-lex.europa.eu/legal-content/EN/ALL/?uri=CELEX:32009L0028

[35] Brundtland, G. (1987) Our Common Future. Report, United Nations World Commission on Environment and Development. Oxford University Press, UK.

[36] IPCC, Mitigation of Climate Change. http://report.mitigation2014.org/spm/ipcc_wg3_ar5_summary-for-policymakers_approved.pdf

[37] Gellings, P.J. and Bouwmeester, H.J. (1997) Handbook of Solid State Electrochemistry. CRC Press, Boca Raton.

[38] Brandon, N.P. and Thompsett, D. (2005) Fuel Cells Compendium. Elsevier, Oxford.

[39] Ference Weiker & Company Ltd. (2010) Assessment of the Economic Impact of the Canadian Hydrogen and Fuel Cell Sector. British Columbia Ministry for Technology and Economic Development.

[40] Godo, H., Nedrum, L., Rapmund, A. and Nygaard, S. (2003) Innovations in Fuel Cells and Related Hydrogen Technology in Norway—OECD Case Study in the Energy Sector. NIFU Report 35/2003.

[41] Wyld Group Pty Ltd. (2008) Hydrogen Technology Roadmap. Australian Government, Department of Resources, Energy and Tourism, ABN:53099078485.

[42] Ormerod, M. (2003) Solid Oxide Fuel Cells. *Chemical Society Reviews*, **32**, 17-28. http://dx.doi.org/10.1039/b105764m

[43] Park, S., Vohs, J.M. and Gorte, R.J. (2000) Direct Oxidation in a Solid-Oxide Fuel Cell. *Nature*, **404**, 265-267.

[44] Tse, L.K.C., Wilkins, S., McGlashan, N., Urban, B. and Martinez-Botas, R. (2011) Solid Oxide Fuel Cell/Gas Turbine Trigeneration System for Marine Applications. *Journal of Power Sources*, **196**, 3149-3162. http://dx.doi.org/10.1016/j.jpowsour.2010.11.099

[45] Turton, H. (2006) Sustainable Global Automobile Transport in the 21st Century: An Integrated Scenario Analysis. *Technological Forecasting & Social Change*, **73**, 607-629. http://dx.doi.org/10.1016/j.techfore.2005.10.001

[46] Thomas, C.E. (2009) Fuel Cell and Battery Electric Vehicles Compared. *International Journal of Hydrogen Energy*, **34**, 6005-6020. http://dx.doi.org/10.1016/j.ijhydene.2009.06.003

[47] Van Mierlo, J., Maggetto, G. and Lataire, Ph. (2006) Which Energy Source for Road Transport in the Future? A Comparison of Battery, Hybrid and Fuel Cell Vehicles. *Energy Conversion and Management*, **47**, 2748-2760.

[48] Moriarty, P. and Honnery, D. (2008) Low-Mobility: The Future of Transport. *Futures*, **40**, 865-872. http://dx.doi.org/10.1016/j.futures.2008.07.021

[49] Moriarty, P. and Honnery, D. (2008) The Prospects for Global Green Car Mobility. *Journal of Cleaner Production*, **16**, 1717-1726. http://dx.doi.org/10.1016/j.jclepro.2007.10.025

[50] Mullen, M. (2010) The State of Shale Plays in Europe. *World Oil*, **231**, D-79.

[51] Editorial News (2011) Quebec's Shale-Gas Moratorium. *Petroleum Economist*, **78**.

[52] Deluchi, M.A., Wang, Q. and Sperling, D. (1989) Electric Vehicles: Performance, Life Cycle Costs, Emissions and Recharging Requirements. *Transportation Research*, **23A**, 255-270.

[53] Schot, J., Hoogma, R. and Elzen, B. (1994) Strategies for Shifting Technological Systems: The Case of the Automobile System. *Futures*, **26**, 1060-1076. http://dx.doi.org/10.1016/0016-3287(94)90073-6

[54] Ford, A. (1994) Electric Vehicles and the Electric Utility Company. *Energy Policy*, **22**, 555-570. http://dx.doi.org/10.1016/0301-4215(94)90075-2

[55] Plowshare Project. https://www.osti.gov/opennet/reports/plowshar.pdf

16

The Climate Change Impact on Russia's Wind Energy Resource: Current Areas of Research

Sergei Soldatenko[1], Lev Karlin[2]

[1]A.M. Obukhov Institute of the Atmospheric Physics of the Russian Academy of Sciences, Moscow, Russia
[2]Russian State Hydrometeorological University, St. Petersburg, Russia
Email: soldatenko@ifaran.ru

Abstract

Exploration of the climate change impact on wind energy is a focus of scientific analysis and research in many countries around the world. Previous studies have demonstrated that over the last three decades measured wind in the boundary and surface layer of the atmosphere has changed all over the globe. However, effects of climate change on the wind energy sector of Russia are not well explored. Therefore, the Russian climate change research needs to focus on improving the analysis and prediction of wind characteristics that are most relevant to Russia's wind energy development. This paper analyzes the effects of global climate change on the patterns of the general circulation of the atmosphere, large-scale atmospheric temperature field and dynamics, as well as wind speed in the planetary boundary layer and, in particular, in the atmospheric surface layer, with regards to Russia's geographical location and its climatic characteristics. This paper also explores and discusses current areas of climate change research relevant for estimating the wind energy potential in Russia. Two areas of research are emphasized: study of the impact of global warming on poleward shifts of the large-scale synoptic eddies which strongly affect the weather patterns and wind field over large territories; and the study of the effects of ice melting in Arctic seas which significantly alter the properties of the underlying surface and, thus, speed and direction of wind in the surface layer.

Keywords

Climate Change, Wind Resources, Baroclinic Instability, Synoptic-Scale Eddies, Surface Layer

1. Introduction

Wind energy is one of the well-established sources of renewable energy that is currently used in 103 countries around the world. According to the key statistics provided by the World Wind Energy Association (WWEA), the world wind energy capacity by the end of 2013 reached 318,529 MW, after 282,275 MW in 2012 [1]. The world total installed wind energy capacity since 1997 is shown in **Figure 1** [1]. In 2013 only 35,550 MW of new wind power capacity was added, compared to 44,609 MW added in 2012. **Figure 2** illustrates newly installed wind energy capacity from the beginning of 1998 [1]. In 2013 the growth rate of wind power installations has been only 12.8%, the smallest since 1998. To date, wind power contributes about 4% to the global electricity demand. WWEA projects that the wind power capacity in 2020 could realistically reach more than 700,000 MW. Further growth of the wind energy has potential to significantly reduce the short-term (2020) and long-term (2050) emissions of greenhouse gases (GHGs) thereby mitigating the climate change and its negative consequences on nature and human society.

For the last few years, the wind energy markets in South America and Eastern Europe have been the most dynamic and experienced the highest growth rates. Russian Federation (RF), being one of the largest by land-mass countries in the world, has very high wind energy potential. However, the wind energy development is inhibited by the vast reserves of conventional energy resources and lack of strong political interests and legislative framework to promote renewable energy development. As a result, in 2013 Russia's total installed wind capacity was only 16.8 MW, putting it in 69-th place among all countries. By contrast, in 2013 the installed capacities in China, the USA and Germany, who are the leaders in the development of wind energy, were 91,324, 75,324 and 34,669 MW respectively. Nevertheless, in 2013 the Russian government approved new measures to support renewable energy. In particular, up to 16 new wind farms are to be built across regions of the country. Among these are four wind farms in the Northwestern Federal District: the 300 MW wind farm in the St. Petersburg re-

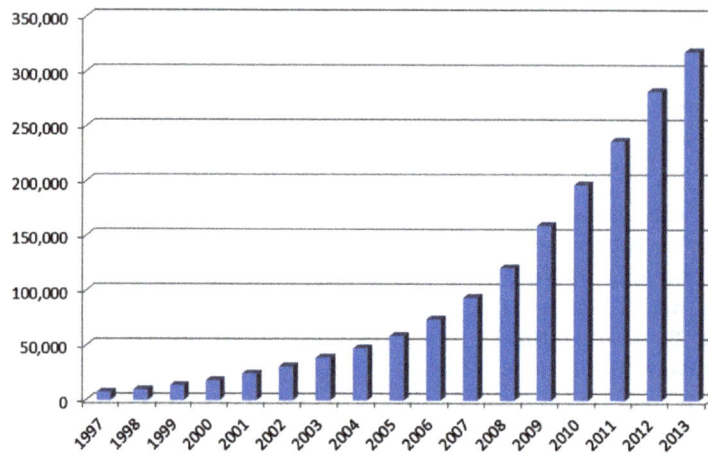

Figure 1. Total installed wind capacity in the world (MW) [1].

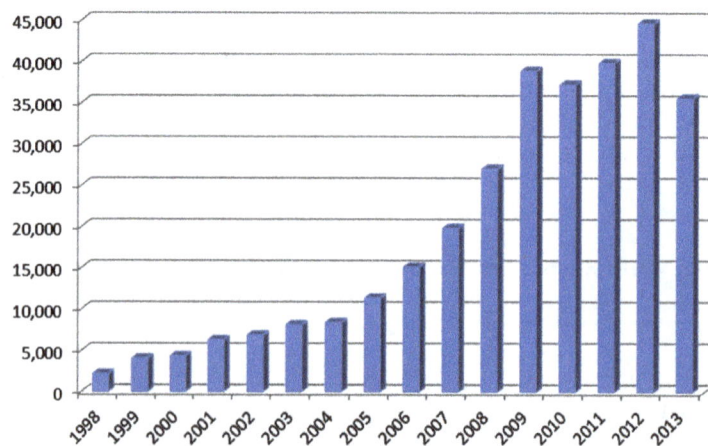

Figure 2. Newly installed capacity (MW) [1].

gion, the 200 MW project in the Kaliningrad region, as well as 300 MW and 500 MW projects in the Murmansk region.

A number of previous studies have estimated the global wind power resource. The theoretical wind potential in terms of the global annual flux has been estimated at 6000 EJ/year [2], but the global technical potential for wind energy strongly depends on the technology and assumptions made with respect to various constraints for development of wind farms. Unfortunately, no standardized and approved method has been proposed to estimate the global technical potential of wind energy. In addition, different studies used different meteorological and technical data, approaches and assumptions. Consequently, the obtained estimates differ significantly from one another [3]-[8] and range from 70 EJ/year (19,400 TWh/year) to 3050 EJ/year (840,000 TWh/year) [9].

Two approaches can be used to estimate wind power resources. The first method is based on available climatic wind data. In this approach, average historical wind speed measurements are interpolated on a high-resolution grid to obtain a surface wind distribution. The second approach involves the use of models of the general circulation of the atmosphere, with subsequent applications of high-resolution numerical weather forecasting models and/or downscaling technique. Combinations of these two methods can also be used to estimate the global, regional and local wind energy potential. It is very important, that the estimates of the technical potential for wind energy take into account not only the wind energy technologies but also the state and variability of the Earth's climate system because the driving force of wind energy, the wind speed, is determined by the wide-range spectrum of atmospheric motions, which, in turn, are impacted by many factors such as external radiative forcing.

Detailed analysis of wind energy resource in Russia is presented in [10]. An estimate of the technical wind potential in Russia is more than 50 EJ/year (14,000 TWh/year). The most suitable regions for wind energy development within Russia, as shown in **Figure 3** [11], are situated in the western part of the country, western Siberia, the South Ural area, and on the seacoasts of the Arctic and Pacific Oceans. However, in many of these areas population density is less than one person per square kilometer and the existing electricity infrastructure is not sufficiently developed. Opportunities for wind energy development for both small facilities and large wind farms in different parts of Russia were discussed in [12].

Russia's territory covers different climate zones. The global climate change, which is currently observed on our planet, affects different geographical regions differently. Although climate change represents a global phenomenon, its effects are heterogeneous across geographical regions of the world and, in particular, across Russia's territory. The Earth's climate system (ECS) is a complex, interactive, nonlinear dynamical system consisting of the atmosphere, hydrosphere, cryosphere, lithosphere and biosphere. The state of the ECS at a given time and place with respect to variables such as temperature, barometric pressure, wind velocity, moisture, precipitations is known as the weather. Climate is usually defined as "average weather" or, in other words, as an ensemble of states traversed by the climate system over a sufficiently long period of time. Commonly, this period corresponds to ~30 years, as defined by the World Meteorological Organization. The evolution of the ECS in time is due to both its own internal oscillating processes, such as El Niño-Southern Oscillation (ENSO), and changes in external factors that influence climate system. However, the solar radiation is the fundamental energy source that drives the whole ECS [13]. Thus, the thermal structure of the climate system, its dynamics, as well as tem-

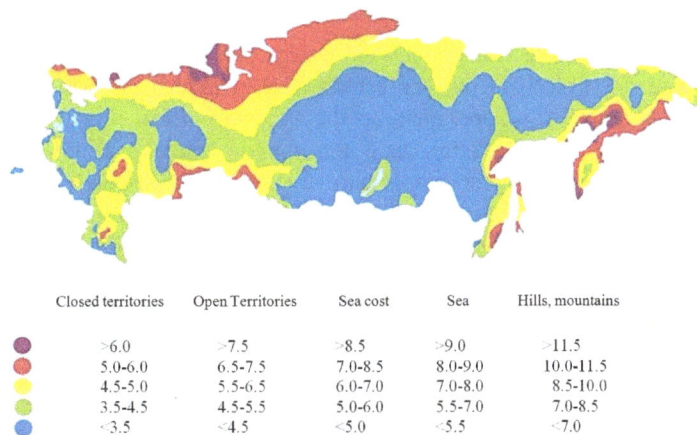

	Closed territories	Open Territories	Sea cost	Sea	Hills, mountains
(purple)	>6.0	>7.5	>8.5	>9.0	>11.5
(red)	5.0-6.0	6.5-7.5	7.0-8.5	8.0-9.0	10.0-11.5
(yellow)	4.5-5.0	5.5-6.5	6.0-7.0	7.0-8.0	8.5-10.0
(green)	3.5-4.5	4.5-5.5	5.0-6.0	5.5-7.0	7.0-8.5
(blue)	<3.5	<4.5	<5.0	<5.5	<7.0

Figure 3. Wind speed at the height of 50 m above the surface [11].

poral and special behavior are significantly influenced by natural and anthropogenic factors that affect the radiation energy balance (energy budget) of the ECS. The Earth's radiation balance represents the accounting of the balance between incoming shortwave radiation, which is almost entirely solar radiation, and outgoing longwave radiation, which is partly reflected solar radiation and partly radiation emitted from the Earth system, including the atmosphere. Changes in the radiation balance of the ECS are governed by:

- variations in the incoming short-wave solar radiation, caused, for instance, by changes in the Earth's orbit;
- fluctuation of the Earth's albedo (reflectivity) due to variations in cloud cover, atmospheric aerosol composition, vegetation, etc.; and
- alteration of the long-wave radiation emitted by the Earth back into the space, caused, for example, by changing greenhouse gas (GHG) concentrations.

Climate reacts directly to such perturbations and also indirectly through positive and negative feedback mechanisms that exist in ECS. According to climate theory, observational data and research findings, the global warming is a man-made phenomenon caused by the increase of concentrations of GHGs such as carbon dioxide (CO_2) and methane (CH_4). Observations show that since the beginning of the 20-th century, the Earth's global average surface temperature has increased by almost 0.8°C, with about two-thirds of the increase occurring since 1980. Presently, global warming is acknowledged by both the scientific community and majority of policymakers. The IPCC Fifth Assessment Report (AR5) provides a distinct view of the up-to-date state of scientific knowledge regarding climate change. It is recognized, that mankind is causing global warming by anthropogenic CO_2 emissions generated by human activities through combustion of fossil fuel, mainly coal, oil, natural gas and wood. Due to the burning of fossil fuels and destruction of native forests, the concentration of carbon dioxide is increased from 280 to more than 390 parts per million (ppm) since the beginning of the so-called Industrial Revolution (~1750).

Generally speaking, Earth's atmosphere is currently overburdened with carbon dioxide, which significantly affects the Earth climate with potentially disastrous consequences. Therefore, the humanity must act now to prevent further increases in the concentration of GHGs and therefore global warming by adopting clean and renewable energy resources such as wind energy, solar energy and bioenergy.

As mentioned above, renewable energy sources depend on the state and variability of the Earth's climate. Thus, the wind energy production is really a function of the ECS' state since surface wind speed is one of the state variables of the ECS. Therefore, if the future behavior of the ECS will be different from how the ECS has evolved in the past, the estimates of renewable energy resources must be periodically updated. Changes in measured surface wind speeds over the last three decades have been discussed in [14] [15]. Since the climate is not stationary, the assessment of resources of renewable energy, including wind energy, based on historical climate data is no longer valid. It is expected that GHG concentrations in the atmosphere will continue to grow. Climate models predict that, depending on the amount of GHG emitted into the atmosphere by human activities, global temperature could further rise by 1°C - 5°C over the next several decades. Increase GHG concentrations are also expected to reduce ice cover, snow cover and permafrost, raise sea levels as well as change the amounts and patterns of precipitation. Changes in the thermal regime of the climate system and alterations of the physical properties of the underlying surface, certainly affect Earth's climate, and therefore also affect the characteristics of the wind in the atmospheric boundary layer.

The impact of global climate change on wind energy resources has been previously studied for select geographical regions (e.g. [16]-[25]). However, effects of climate change on the wind energy sector in Russia are presently not well explored. This paper analyzes the effects of global climate change on the patterns of the general circulation of the atmosphere, large-scale atmospheric temperature field and dynamics, as well as wind speed in the planetary boundary layer and, in particular, in the atmospheric surface layer, with regards to Russia's geographical location and its climatic characteristics. This paper also explores and discusses current areas of climate change research relevant for estimating the wind energy potential in Russia. Two areas of research are emphasized: study of the impact of global warming on poleward shifts of the large-scale synoptic eddies, which strongly affect the weather patterns and wind field over large territories; and the study of the effects of ice melting in Arctic seas, which significantly alter the properties of the underlying surface and, thus, speed and direction of wind in the surface layer.

This topic is chosen because climate change research needs to focus on improving the analysis and prediction of wind parameters that are most important for wind energy production. It is recommended, that wind farm developers have to take into consideration the effects of climate change on the wind characteristics in the atmospheric surface layer, so that wind energy production remains beneficial and profitable in the future.

2. Wind and Main Factors Determining Its Variability

2.1. Wind and Its Characteristics

Wind is the movement of air masses relative to the earth's surface. Wind is caused by differences in atmospheric pressure (pressure gradient): air masses move from regions of high pressure towards to regions of low pressure. The pressure gradient is generated by heterogeneity of heating and cooling of air masses under the influence of radiation, latent heat of condensation of water vapor, as well as turbulent and convective heat transfer. The following forces influence the moving air masses: the Coriolis force due to the Earth's rotation, inertial forces, gravity, and, in the atmospheric boundary layer, the surface force of friction.

For wind energy the practical interest represents the horizontal component of the wind velocity vector, which is usually much larger than the vertical component of the wind speed. The horizontal wind is characterized by two measured parameters: speed and direction from which the wind blows. Wind speed and direction vary in time and space because of turbulence of the air flow and external factors resulting from spatial and temporal heterogeneity of atmospheric temperature and pressure. Variability of wind parameters is characterized by spatial and temporal variations of various scales. Wind speed and direction are measured and defined as values averaged over the certain time interval (usually 2 - 10 min).

The wind velocity in the lowest layer of the atmosphere, the atmospheric surface layer, is the most important parameter for wind power generation. The characteristic height of the surface layer is about 100 meters. In this layer turbulent fluxes of momentum, sensible and latent heat are nearly constant with height. The wind speed in the surface layer at a given location may vary within a wide range: from zero to 100 m·s^{-1} and even more.

To determine the efficiency of wind energy, a set of parameters is used which includes up to two dozen characteristics. These characteristics can be divided into three groups. The first group includes climatic characteristics that assess wind energy potential of the region in the atmospheric boundary layer such as the average annual and monthly wind speed and direction. The second group includes parameters that define the performance of wind turbines and the selection of their optimum operational regime, for example continuous duration of wind speed and direction within the specified intervals, the frequency distribution of wind speed, information on the duration of calms and wind speeds above a given value. The third group includes wind characteristics used in the design and calculation of the strength and stability of wind turbines: the intensity of atmospheric turbulence, maximum wind speed and gusts, estimates of their duration and frequency.

For wind power energy, the set of important indicators includes also long-term average values of wind characteristics calculated for different time of the day for given month, season and year based on long-term time series of observations. However, in reality the wind parameter variations in a given location can significantly differ from statistical data.

2.2. Temporal and Spatial Variability of Wind Characteristics

Wind variability is characterized by the wide-range temporal spectrum: from decadal (climatic variability) and several years (interannual variability) to a few seconds (micro-turbulent variability). For wind power energy, of greatest interest are the temporal variations of wind characteristics (e.g., mean speed) caused by regular natural processes that determine sustained weather and climate variability. For example, Earth's rotation around its axis leads to diurnal variations in the radiation balance of the earth's surface and therefore leads to diurnal variations in temperature, wind characteristics and other meteorological parameters. By averaging the long-term observational data one can obtain the diurnal variation of wind characteristics for each month, season and the whole year. Typically, the diurnal variation of wind characteristics is a periodic function and is characterized by some maximum and some minimum that together determine the amplitude of the diurnal variation. The relative amplitude of the diurnal wind variation depends essentially on the geographic region. In most parts of Russia the relative amplitude of diurnal wind variation is highest in the summer time and can reach 30% - 40% with respect to daily average values at the anemometer height [10]. It is important to note that the amplitude of the diurnal variation of the wind characteristics decreases with height.

Inclination of the axis of the Earth's rotation and the rotation in an elliptical orbit around the sun lead to annual variations in solar energy received by the planet, which in turn lead to the substantial changes in the atmospheric general circulation and therefore wind field around the globe. The relative annual changes of wind characteristics on the territory of Russia can reach 40% - 50% with respect to mean annual values. There are also long-term and decadal oscillations of the wind climate regime. Interannual oscillations can be regular, generated by periodic changes of radiation regime and processes of large-scale atmosphere-ocean interaction, but also quasi-periodic (e.g., ENSO). The amplitude of interannual wind variations usually reaches 15% - 20% with respect to the amplitude of the seasonal and irregular oscillations [10].

Spatial variability of wind is also multiscale: from global and synoptic scale down to the centimeter and millimeter for turbulent fluctuations. The theory of atmospheric general circulation is well-developed and well-described in the literature (e.g., [13] [26]). It is very important to underline, that the main drivers of the earth's atmosphere are the temperature gradient between equator and poles and the earth's rotation which produce the zonal air flow. The equator-to-poles temperature gradient is formed due to the relative position of the earth with respect to the sun, since much more radiation is received by the earth near the equator than in other areas, with the least radiation being received at or near the poles.

2.3. Local Winds

An important factor in the formation of the wind climate in the boundary layer of the atmosphere and the surface layer are local winds. Local winds represent the air flows in a certain relatively small geographical region. These winds are created and affected by topographic features, vegetation, water bodies, heterogeneity of the underlying surface etc. Characteristic temporal and spatial scales of local winds and their intensity are determined by the essential features of external forcing that generate local winds. Usually local winds change very often and can move from mild to extreme winds in just hours. These winds have usually names unique to the area where they occur.

Local winds can be subdivided into three general classes. The first contains diurnally varying airflows that are driven by diurnal heating or cooling of the ground surface or by local gradients of surface heat flux (e.g., land and sea breezes, mountain-valley circulations, drainage and slope winds). The second class includes winds generated by the interaction of large-scale airflows with orography. The third group consists of those winds accompanying thermal convective activity.

3. Climate Change Impacts on Surface Wind Velocity

3.1. Climate Change Impacts on the Atmospheric Dynamics

The sphericity of Earth and the resulting spatially non-uniform distribution of solar heating significantly influence the formation of general circulation of the atmosphere and global three-dimensional configuration of atmospheric winds. It is known, that tropical areas absorb about twice the solar short-wave energy than the polar regions, producing a meridional temperature gradient (MTG) and available potential energy. Some of this energy can be transformed into kinetic energy, which is manifested in the wind. Rotation of the Earth deflects the winds, creating the westerly quasi-zonal airflow which under some conditions can be unstable. Instability of the zonal-flow produces the wave-like atmospheric motions and creates very complicated air flow patterns. The general circulation of the atmosphere plays very important role in redistributing energy, water vapour, latent and sensible heat within the ECS.

There is a large difference between tropical and extratropical atmosphere. In the high and middle latitudes, large-scale atmospheric motions are govern by the nearly perfect balance between the Coriolis force and the pressure gradient. The baroclinic instability of the westerly quasi-zonal flow generates large-scale eddies, cyclones and anticyclones, which dominate the extratropical circulation. The typical length of these eddies is about 4000 km. On the surface weather maps, cyclones (low pressure systems) are located at the ridges of pressure-waves, and anticyclones (high pressure systems), at the troughs of pressure-waves. In the tropics diabatic heating and surface friction affect the circulation since the Coriolis force is weak. One of the most prominent tropical circulation attribute is the Hadley cell (HC). The HC's features include rising motion near the equator, poleward flow 10 - 15 kilometers above the surface, descending motion in the subtropics, and equator-ward flow near the surface. The trade winds and subtropical jet streams are closely related to the HC.

Climate change causes alterations of the radiative balance of the ECS which, in turn, transforms the global and regional temperature structure. It is observed, that the warming is more significant in higher latitudes that in lower latitudes. Thus, the temperature difference between the poles and the equator decreases affecting the intensity of global circulation. In particular, it is observed that there is poleward expansion of the tropical circulation (about 0.7° latitude per decade), which weakens the strength of the mean tropical circulation (e.g. [27] [28]). However, for the extratropical atmospheric dynamics, the intensity changes in the tropical circulation are less important than the poleward movement of the Hadley cell that characterizes the tropical zonal circulation and its width.

The extratropical circulation is also modified under global warming. The most prominent examples of this are the changes in the annual modes, which are the dominant modes of large-scale extratropical variability, manifesting themselves in poleward shifts of the jet streams and storm tracks, increases in the length scale of extratropical eddies, changes in frequency of eddy formation and their intensity [29]. Recent studies show that over the past few decades surface winds have declined in many geographical areas around the world (e.g. [14] [30]-[33]). However, the precise cause of the stilling is not well understood. The comprehensive study of potential cause of changes in surface wind speeds over the mid-latitudes of the Northern Hemisphere between 1979 and 2008, using data from 822 surface weather stations, is represented in [15]. It was observed that surface wind speeds have declined by 5% - 15% over almost all continental areas in the northern mid-latitudes, and that strong winds have slowed faster than weak winds. An increase of surface roughness could explain between 25% and 60% of the surface wind slowdown [15].

3.2. Storm Tracks Shifts under Climate Change

Large-scale extratropical eddies, cyclones and anticyclones, are instrumental for meridional and vertical transporting heat, water vapor and momentum and, thus, they exert strong influence on the weather and climatic conditions over large geographical regions. Cyclones and anticyclones play a significant role in the general circulation of the atmosphere smoothing and leveling temperature contrasts between high and low latitudes [13] [26] [34]. The frequency and intensity of cyclones have a marked influence on the variability of the surface wind velocity, a factor that is very important for wind energy.

Low-pressure systems, cyclones, form the so-called storm tracks, the predominant trajectories of moving cyclonic eddies. Storm tracks are defined as the regions of strong baroclinicity (maximum meridional temperature gradient), which are determined on the basis of eddy characteristics like eddy fluxes of angular momentum, energy, and water vapour. Generally, the climate of Europe and the Western areas of Russia is dominated by weather systems of the extratropical storm tracks. In the Northern Hemisphere, there are two major storms tracks along which most extratropical cyclones travel: Atlantic and Pacific. The behaviour of the storm tracks is highly variable and has a profound impact on the climate of Russian territories. Climate and atmospheric general circulation models, supported by a growing quantity of observational data, have demonstrated that storm tracks shift poleward while the climate becomes warmer. In addition, one can observe the changes in frequency and intensity of cyclones and their length scale. These changes have significant effects on the wind field, in general, and on the surface wind, in particular, over large territories.

The dominant physical mechanism for generating large-scale atmospheric eddies in the high- and mid-latitudes is baroclinic instability of the westerly quasi-zonal atmospheric flow [34]-[36]. Atmospheric static stability and the meridional temperature gradient are the most important fundamental parameters characterizing the development of baroclinic instability and, therefore, the formation and dynamics of extratropical large-scale synoptic eddies. There is evidence of an increase over the past decades of the static stability in the extratropics [37] and a poleward movement of zones with strong MTG-baroclinic zones (e.g. [29] [38] [39]). The MTG produces a vertical shear of the geostrophic wind that forms the westerly quasi-zonal atmospheric air jet. The speed of this flow and its vertical distribution can serve as an indicator of the meridional thermal structure of the atmosphere. Therefore the global warming can be indirectly identified through the speed and vertical shear of the westerly quasi-zonal atmospheric flow. To study how the patterns of this flow and the baroclinic instability have changed in recent decades, a comparative analysis of climates for periods 1949-1968 and 1975-1994 has been carried out using National Center for Environmental Prediction (NCEP) Reanalysis and the European Centre for Medium-Range Weather Forecasting (ECMWF) Reanalysis (ERA40) data [40] [41]. **Figure 4** shows the height (in pressure units) and latitudinal cross-section of the zonal wind for the periods 1949-1968 and 1975-1994 as well their difference. One can see that there are significant differences between the two periods. Alterations in the zonal wind caused by changes in the MTG strongly affect the development of baroclinic instability, generation

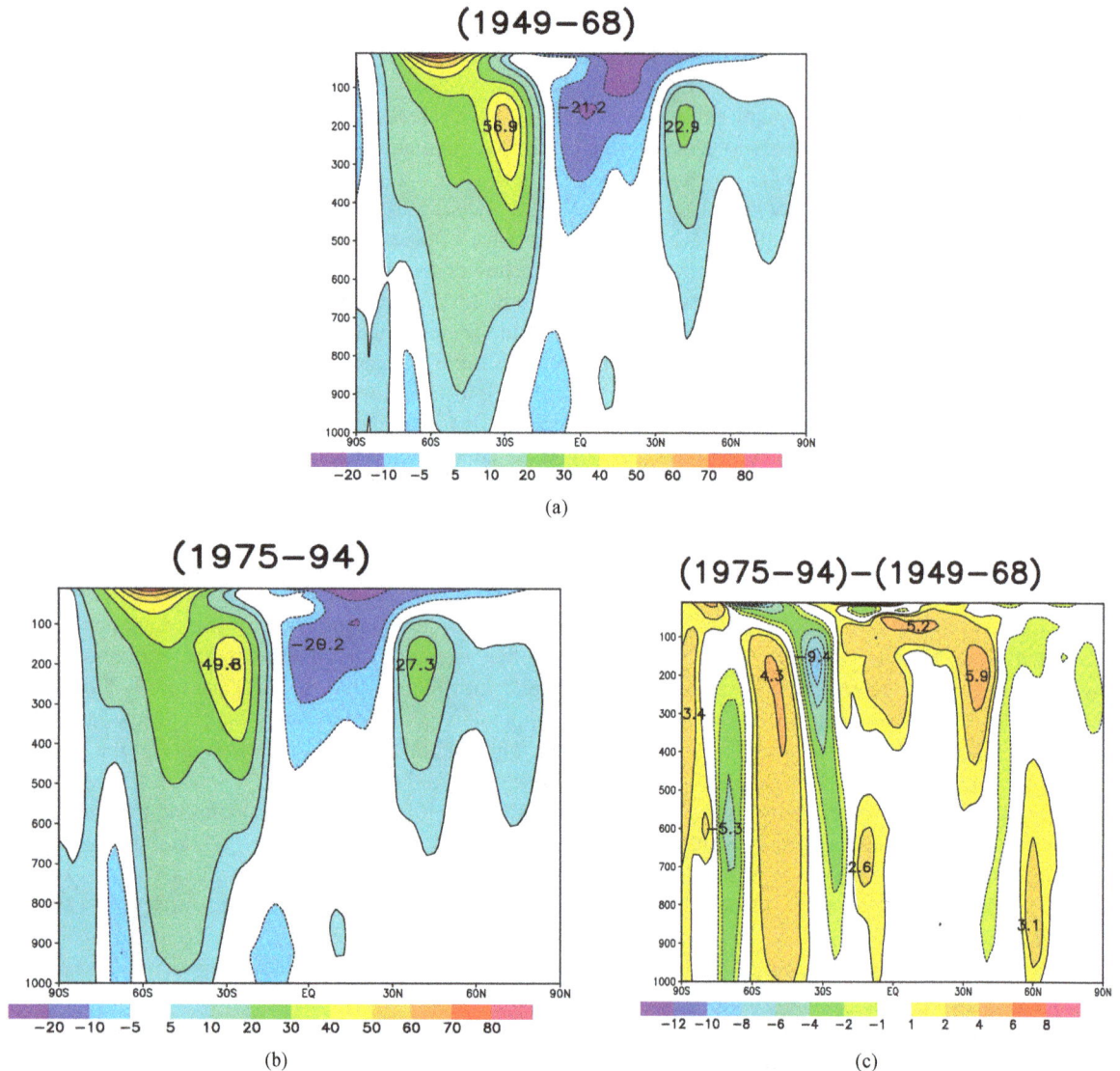

Figure 4. Vertical cross-section of July zonal wind (m·s^{-1}) as a function of latitude and pressure (in hPa) for the 1949-1968 basic state (a); the 1975-1994 basic state (b and their difference (1975-1994)-(1949-1968) [40].

of large-scale atmospheric eddies and, therefore, surface wind field and its distribution. The impact of variations in the atmospheric static stability and vertical shear of the zonal wind on the main characteristics of baroclinically unstable waves of synoptic scales (e.g. the growth rates of unstable waves as function of wavelength) is explored in [42]. The sensitivity functions obtained in [42] allow estimation, using first-order approximation, of the influence of changes in fundamental atmospheric parameters on the baroclinic instability and formation of synoptic-scale eddies. As an example, **Figure 5** shows the sensitivity function S_σ versus zonal wavenumbers k_z for different values of static stability parameter σ_0. The function S_σ shows changes in the growth rate of unstable waves due to variations in σ_0. The results obtained in [42] are consistent with observations: an increase in static stability and a decrease of the MTG, which occurred over the past few decades in some areas of the globe, have led to the decrease in the growth rate of baroclinic unstable waves, a shift of the spectrum of unstable waves in the long wavelength part of spectrum, and weakened intensity of cyclogenesis. Naturally, these changes affect the formation and development of synoptic eddies as well as the essential features of weather patterns and surface wind velocity over large territories, particularly over Western Russia.

Under global warming, the horizontal length of synoptic eddies is also affected [43]. Cyclones and anticyclones, during the onset of their temporal evolution, are strongly impacted by baroclinic factors. However, after 5

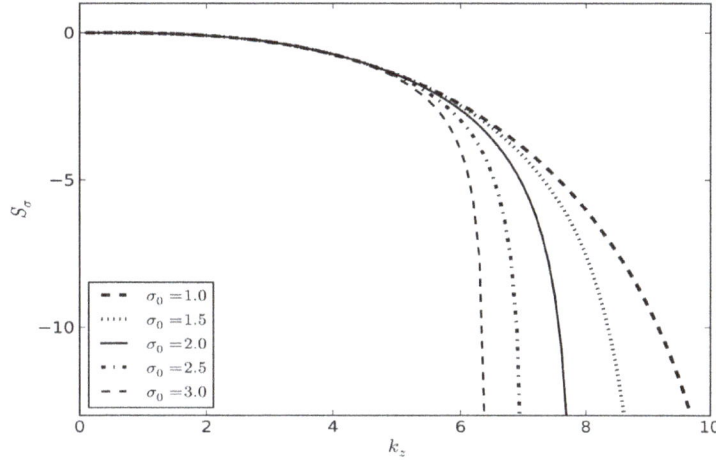

Figure 5. Sensitivity function S_σ vs. zonal wavenumber k_z for different values of the static stability parameter σ_0 ($\times 10^6$ m^2·Pa^{-2}·s^{-1}) [42].

- 7 days of their evolution, large-scale eddies are converted into quasi-vertical quasi-barotropic synoptic objects. A characteristic length of these objects can be estimated by the Obukhov scale

$$L_0 = \frac{(gH)^{1/2}}{f} = \frac{(RT)^{1/2}}{f},$$ (1)

where g is the gravity acceleration, H is the height of the homogeneous atmosphere, f is the Coriolis parameter, R is the gas constant, and T is temperature. Thus, for barotropic atmosphere, the temperature growth causes the increase in L_0:

$$\frac{dL_0}{dT_0} > 0.$$ (2)

Here, T_0 is the surface temperature. Typical values of H and f in the extratropical atmosphere are 8.5 km and 1.1×10^{-4} s^{-1}, respectively, so the Obukhov scale is $L_0 \approx 2600$ km. For baroclinic atmosphere the characteristic length scale of synoptic eddies can be estimated by the Rossby radius of deformation

$$L_R = \frac{(NH)^{1/2}}{f} = \frac{1}{f}\left(\frac{T(\gamma_d - \gamma)}{g}\right)^{1/2},$$ (3)

where N is the buoyancy frequency, γ_d is the dry adiabatic lapse rate, γ is the reference state lapse rate. Typical value of N in the extratropical atmosphere is 0.01 s^{-1}, so the Rossby radius of deformation is $L_R \approx 850$ km. In baroclinic atmosphere the influence of the temperature growth on the horizontal size of synoptic-scale eddies is not as clear:

$$\frac{1}{L_R}\frac{dL_R}{dT} = \frac{1}{2}\left[\frac{1}{T} - \frac{d\gamma/dT}{(\gamma_d - \gamma)}\right].$$ (4)

Depending on trends in the lapse rate under the influence of global warming, two regimes are possible:

$$\frac{dL_R}{dT_0} > 0 \quad \text{and} \quad \frac{dL_R}{dT_0} < 0.$$

Within the middle latitudes, static stability demonstrates a robust increase [3] with the upper tropospheric warming outpacing the lower tropospheric warming by about 2°C. Therefore, the atmospheric lapse rate decreases under global warming:

$$\frac{d\gamma}{dT_0} < 0.$$ (5)

Taking into account (5), from (4) one can conclude that in baroclinic atmosphere the characteristic length of synoptic-scale eddies also increases under global warming.

Observational data and results from numerical modeling show that the poleward shift of the storm tracks during the last several decades is not the only observable trend. Additional changes due to climate change include a decrease in occurrence frequency of synoptic-scale eddies and a rise in the intensity of more extreme cyclones [28] [44]. This is believed to be due to the enhanced surface warming in higher latitudes and weaker surface warming in the tropical zone, which leads to a decreasing MTG and the associated atmospheric baroclinicity. The impact of climate change on the trend in the large-scale wave activity of the extratropical atmosphere has been explored in [45] based on the theory of baroclinic instability developed in [46]. The rate of synoptic-scale eddy formation in the extratropical atmosphere can be estimated by the following expression [45]

$$p^c \left(\Delta S \right) = \frac{1}{\tau} \frac{4 \cos \varphi \Delta S \left(\varphi \right)}{\pi L_R^2 \cos \varphi_0}, \tag{6}$$

where p^c is the number of large-scale eddies formed per unit of time, $\Delta S(\varphi)$ is the area of the Earth's surface between latitudes φ and $\varphi + \Delta \varphi$, and τ is the increment of growing unstable modes. Expression (6) was obtained under the assumption that the number of eddies formed per unit of time and the intensity of wave activity reach their maxima [47]. Another asymptotic model with the assumption that the number of eddies generated per unit of time p^f and the intensity of wave activity reach their minima [47] provides the following formula for p^f [45]

$$p^f \left(\Delta S \right) = \left(\frac{L_R}{L_0} \right)^2 p^c \left(\Delta S \right). \tag{7}$$

The global warming causes the increase in both L_R and L_0. Therefore, qualitatively analyzing the equations (6) and (7), one can conclude that, with the increase in global atmospheric temperature, the generation rate of large-scale eddies for both asymptotic models decreases. Moreover, since the growth rate of unstable baroclinic waves under global warming decreases, the characteristic time of eddy development τ increases, so that, according to (6) and (7), the eddy generation rates p^c and p^f are further reduced. However, with respect to Russian territory, this problem is poorly studied and therefore represents the new area of research. On weather and climatic maps storm tracks indicate geographical areas with significant changes in wind speed and direction in the atmospheric surface layer. The exploration of the storm tracks behavior under global warming and the improvement in our knowledge of the effects of climate change on the extratropical large-scale dynamics, spatial structure and temporal evolution of eddies can be very helpful for estimating the wind energy resources in any particular region of the world.

3.3. Changes in the Physical Properties of the Underlying Surface

Wind energy potential is primarily determined by the wind speed in the atmospheric boundary layer (ABL) and, in particular, in the surface layer, which is the lowest part of the ABL. The characteristic height of the atmospheric surface layer, formerly known as the constant flux layer, is about a tenth of the height of the ABL ($\sim 10^2$ m). The behavior of the ABL and surface layer is strongly affected by Earth's surface drag and other physical properties of the surface. Thus, in the ABL and the surface layer, physical variables such as wind velocity and direction, temperature and moisture demonstrate quite rapid changes with height and with time. The surface layer dynamics is strongly affected by solar radiation and thus related to the diurnal cycle.

Monin-Obukov similarity theory is usually used to estimate the mean wind speed u at height z above the ground as follows:

$$u \left(z \right) = \frac{u^*}{\kappa} \left[\ln \frac{z}{z_0} + \Psi \left(\frac{z}{L} \right) \right], \tag{8}$$

where u^* is the friction velocity, $\kappa \approx 0.4$ is the von Karman constant, z_0 is the surface roughness, Ψ is an empirical function and L is the Monin-Obukhov stability parameter. Function Ψ is the correction to the logarithmic wind profile resulting from the deviation from neutral stratification. Under neutral atmospheric stability conditions, the ratio z/L tends to zero since $L \to \infty$ and, thus, Ψ drops out. Obviously, the surface roughness is one the most important parameters that affect the vertical wind velocity profile in the surface layer. **Table 1** provides

Table 1. Surface roughness length for various surface types [48].

Surface type	z_0 m
Smooth sea	0.00001
Rough sea	0.000015 - 0.0015
Ice	0.00001
Snow	0.00005 - 0.0001
Short grass	0.04 - 0.1
Coniferous forest	0.28 - 3.9
Broadleaf deciduous forest	2.7

typical surface roughness lengths for different surface types [48].

The impact of topography on large-scale atmospheric motions has been previously explored in the context of numerical modeling (e.g. [49] [50]). The effect of local roughness variations at the Earth's surface on the small-scale atmospheric dynamics has only recently attracted the growing interest from around the world thanks to the rapid development of wind energy generation industry [51]. As mentioned above, the coasts of Russia's Arctic seas offer good opportunities for wind farm development. However, there is evidence of a rapid decline, over the past three decades, of the Arctic sea ice cover during the summer months [9]. Temperature in the Arctic has increased at twice the rate compared to the rest of the globe, and likely will increase by another 8°C in the 21st century. Climate simulation results show that the rate of ice cover decline will accelerate (e.g. [52]-[54]) under global warming. The average rate of the Arctic sea ice melting is about 12% per decade, which suggests the Arctic will be ice-free by 2030.

Continued decline of the Arctic sea ice at a disastrous rate may dramatically alter regional and global climate and weather patterns in the decades to come. For instance, storm tracks and jet stream will possibly move further north in response to warmer temperatures near the pole, altering the general circulation of the atmosphere and weather patterns around the globe and especially in the Arctic [55]. In addition, melting of the Arctic ice significantly transforms the sea surface, especially in the coastal areas, from snow-covered ice to complex mixtures of ice and melt ponds. Moreover, as a result of global warming the large ice-free areas (polynyas) occur close to the coast. Many polynyas may freeze during the peak of winter. The large polynyas, however, can stay open all winter. Consequently, these lead to an increase in the sea surface roughness length z_0. Changes in the surface roughness in the Arctic sea coastal areas decrease the wind speed in the ABL, influencing the exchanges of heat, moisture and momentum between the atmosphere and surface, affecting the synoptic scale pressure field, and resulting in changes in regional circulation patterns. The surface roughness z_0 of sea ice is also affected by ridges, floe edges and patterns formed of drifting snow. Observations show that the parameter z_0 has wide range of variations: from 0.3 mm up to 30 mm [56]. Thus, under global warming, the surface of the Arctic sea in coastal areas becomes very heterogeneous and rapidly changing, which, in turn, has a strong influence on the wind speed in the ABL and surface layer of the atmosphere.

To estimate the influence of variations in z_0 on the wind speed in the ABL u, one can use the sensitivity function S, which is the derivative of u with respect to z_0. The expression for S can be obtained by differentiating (8) with respect to z_0:

$$S = \frac{\partial u}{\partial z_0} = -\frac{u^*}{\kappa z} z_0^{-1}. \tag{9}$$

Sensitivity function S shows changes in u due to variations in z_0 and is evaluated in the vicinity of some basic value of the parameter z_0. In general, several basic values can be selected to cover some range of changes in z_0. Let's calculate the sensitivity function for $z = 50$ m. Typical value of u^* is 0.2 m·s^{-1}, therefore

$$S_{50} \approx -10^{-2} \times z_0^{-1}. \tag{10}$$

Assuming the surface roughness parameter for snow-covered ice is $z_0 = 10^{-4}$ m, the sensitivity is $S_{z=50} \approx 10^2$ s^{-1}. If δz_0 is a variation in the parameter z_0, then the change in the wind speed δu at a given height z due to δz_0 can be estimated as follows

$$\delta u(z) = S \times \delta z_0 + \mathcal{O}\left[\left(\delta z_0\right)^2 \right]. \tag{11}$$

For example, if $\delta z_0 \approx 10^{-2}$ m, then $\delta u \approx 1$ m·s^{-1}. This result has an illustrative character, showing strong dependence of the surface layer wind speed on the surface roughness.

Therefore, melting of the Arctic sea ice has a wide range of impacts on the global and regional atmospheric dynamics. The study of how global warming affects the sea surface roughness and, as a result, the wind speed in the Arctic coastal regions represents a current area of research interest, and call for both field experiments as well as theoretical explorations.

4. Conclusions

Observational data and climate simulation results indicate that the Earth's global temperature has been consistently rising over past several decades with more rapid rising seen over last three decades. Global temperature will likely continue to grow in the next decades. Climate change influences not only the temperature regime of the ECS globally and regionally, but also the atmospheric dynamics—the general circulation of the atmosphere, regional and local winds. The global warming is a human-made phenomenon caused by the increase of GHG concentrations in the atmosphere. Using renewable energy, wind energy in particular, to reduce the negative consequences of the global warming should be considered as a possible option. However, the wind energy production is a function of the state and variability of the ECS. Previous studies have shown that over the last three decades measured wind in the boundary and surface layer of the atmosphere has changed significantly around the world. At the same time, estimates of climate change on the wind energy sector of Russia are not well explored.

Developers of wind farm projects should take into consideration potential impact of climate change on the operation, financial expectations and potential benefits of wind generation facilities. Therefore, exploration of impact of climate change on wind changes in the areas suitable for wind farm development is a focus of current feasibility studies and research. This paper discussed the analysis of the effects of climate change on the general circulation of the atmosphere, large-scale atmospheric dynamics and wind speeds in the ABL and surface layer, and considered geographical regions suitable for potential development of wind farm projects. Two current areas of research were emphasized: First, the study of the global warming impact on the formation and evolution of the synoptic-scale eddies and storm tracks associated with prevailing trajectories of the movement of cyclones; Second, the exploration of the influence of melting ice in Arctic coastal areas on changes in the surface roughness length. This parameter significantly influences the wind speed in the atmospheric surface layer.

Traditionally, feasibility studies and meteorological assessment conducted at potential areas for the development of wind farm projects are based on the wind climatology. For the most part, these studies ignore changes in weather patterns that may be connected with the global warming, such as alterations of wind patterns, poleward shifts of storm tracks, changing length scale of cyclones, increasing number of intense cyclones with high wind velocities, and modified wind speeds. Consequently, the results of feasibility studies can contain uncertainty, which increases with the life cycle period of wind farms (20 - 30 years). In conclusion, to improve the accuracy of feasibility studies, it is essential that potential effects of global warming are fully taken into account.

References

[1] WWEA (2014) Key Statistics of World Wind Energy Report 2013. WWEA, Shanghai, 7 April 2013, 13 p.

[2] Rogner, H.H., Barthel, F., Cabrera, M., Faaij, A., Giroux, M., Hall, D., Kagramanian, V., Kononov, S., Lefevre, T., Moreira, R., Notstaller, R., Odell, P. and Taylor, M. (2000) Energy Resources. In: *World Energy Assessment, Energy and the Challenge of Sustainability*, United Nations Development Program United Nations Department of Economics and Social Affairs, and World Energy Council, New York, 135-171.

[3] Grubb, M.J. and Meyer, N.I. (1993) Wind Energy: Resources, Systems and Regional Strategies. In: Johansson, T.B., Kelly, H., Reddy, A.K.N. and Williams, R.H., Eds., *Renewable Energy: Sources for Fuels and Electricity*, Island Press, Washington DC, 157-212.

[4] Archer, C.L. and Jacobson, M.Z. (2005) Evaluation of Global Wind Power. *Journal of Geophysical Research*, **110**, Article ID: D12110.

[5] Lu, X., McElroy, M.B. and Kiviluoma, J. (2009) Global Potential for Wind-Generated Electricity. *Proceedings of the*

National Academy of Sciences of the USA, **106**, 10933-10938. http://dx.doi.org/10.1073/pnas.0904101106

[6]   Jacobson, M.Z. and Archer, C.L. (2012) Saturation Wind Power Potential and Its Implications for Wind Energy. *Proceedings of the National Academy of Sciences of the USA*, **109**, 15678-84.

[7]   Marvel, K., Kravitz, B. and Caldera, K. (2012) Geophysical Limits to Global Wind Power. *Nature Climate Change*, **3**, 118-121. http://dx.doi.org/10.1038/nclimate1683

[8]   Adams, A.S. and Keith D.W. (2013) Are Global Wind Power Resource Estimates Overstated? *Environmental Research Letters*, **8**, Article ID: 015021.

[9]   Wiser, R., Yang, Z., Hand, M., Hohmeyer, D., Infield, D., Jensen, P.H., Nikolarv, V., O'Malley, M., Sinden, G. and Zervos, A. (2011) Wind Energy. In: Edenhofer, R., Pichs-Madruga, R., Sokona, Y., Seyboth, K., Matschoss, P., Kadner, S., Zwickel, T., Eickemeier, P., Hansen, G., Schlomer, S., and von Stechow, C., Eds., *IPCC Special Report on Renewable Energy Sources and Climate Change Mitigation*, Cambridge University Press, Cambridge, UK and New York. http://dx.doi.org/10.1017/CBO9781139151153.011

[10]   Nikolaev, V.G., Ganaga, S.V. and Kudriashiv, K.I. (2008) National Inventory of Wind Resources of Russia and Methodological Foundations for Their Determination. Atmograph, Moscow, 590 p.

[11]   Starkov, A.N., Landberg, L., Bezroukikh, P.P. and Borisenko, M.M. (2000) Russian Wind Atlas. Russian-Danish Institute for Energy Efficiency, Moscow: Risø National Laboratory, Roskilde, 551 p.

[12]   Dmitriev, G. (2001) Wind Energy in Russia. VetrEnergo for Gaia Apatity and INFORSE-Europe. http://www.inforse.org/europe/word_docs/ruswind2.doc

[13]   Lorenz, E.N. (1967) The Nature and Theory of the General Circulation of the Atmosphere. World Meteorological Organization.

[14]   Pryor, S.C., Barthelmie, R.J., Toung, D.T., Takle, E.S., Arrit, R.W., Flory, D., Gutovsky, W.J. and Roads, J. (2009) Wind Speed Trends over the Contiguous United States. *Journal of Geophysical Research: Atmospheres*, **114**, Published Online. http://dx.doi.org/10.1029/2008JD011416

[15]   Vautard, R., Cattiaux, J., Yiou, P., Thepaut, J.N. and Ciais, P. (2010) Northern Hemisphere Atmospheric Stilling Partly Attributed to an Increase in Surface Roughness. *Nature Geoscience*, **3**, 756-761. http://dx.doi.org/10.1038/ngeo979

[16]   Pryor, S.C., Barthelmie, R.J. and Kjellström, E. (2005) Potential Climate Change Impact on Wind Energy Resources in Northern Europe: Analyses Using a Regional Climate Model. *Climate Dynamics*, **25**, 815-835. http://dx.doi.org/10.1007/s00382-005-0072-x

[17]   Harrison, G.P., Cradden, L.C. and Chick, J.P. (2008) Preliminary Assessment of Climate Change Impacts on the UK Onshore Wind Energy Resource. *Energy Sources, Part A: Recovery, Utilization, and Environmental Effects*, **30**, 1286-1299. http://dx.doi.org/10.1080/15567030701839326

[18]   Pryor, S.C. and Barthelmie, R.J. (2002) Comparison of Potential Power Production at On- and Off-Shore Sites. *Wind Energy*, **4**, 173-181. http://dx.doi.org/10.1002/we.54

[19]   Pryor, S.C. and Barthelmie, R.J. (2003) Long-Term Trends in Near-Surface Flow over the Baltic. *International Journal of Climatology*, **23**, 271-289. http://dx.doi.org/10.1002/joc.878

[20]   Pryor, S.C. and Barthelmie, R.J. (2010) Climate Change Impacts on Wind Energy: A Review. *Renewable and Sustainable Energy Reviews*, **14**, 430-437. http://dx.doi.org/10.1016/j.rser.2009.07.028

[21]   Breslow, P.B. and Sailor, D.J. (2002) Vulnerability of Wind Power Resources to Climate Change in the Continental United States. *Renewable Energy*, **27**, 585-598. http://dx.doi.org/10.1016/S0960-1481(01)00110-0

[22]   Sailor, D.J., Smith, M. and Hart, M. (2008) Climate Change Implications for Wind Power Resources in the Northwest United States. *Renewable Energy*, **33**, 2393-2406. http://dx.doi.org/10.1016/j.renene.2008.01.007

[23]   Wang, C. and Prinn, R.G. (2009) Potential Climatic Impacts and Reliability of Very Large-Scale Wind Farms. Report No. 175, MIT, Cambridge.

[24]   Nolan, P., Lynch, P., McGrath, R., Semmler, T. and Wang, S. (2012) Simulating Climate Change and Its Effects on the Wind Energy Resource of Ireland. *Wind Energy*, **15**, 593-608. http://dx.doi.org/10.1002/we.489

[25]   Hueging, H., Haas, R., Born, K., Jacob, D. and Pinto, J.G. (2013) Regional Changes in Wind Energy Potential over Europe Using Regional Climate Model Ensemble Projections. *Journal of Applied Meteorology and Climatology*, **52**, 903-917. http://dx.doi.org/10.1175/JAMC-D-12-086.1

[26]   Masaki, S. (2014) Atmospheric Circulation Dynamics and General Circulation Models. Springer-Verlag, New York.

[27]   Lu, J., Vechhi, G.A. and Reichler, T. (2007) Expansion of the Hadley Cell under Global Warming. *Geophysical Research Letters*, **34**, Published Online.

[28]   Reichler, T. (2009) Changes in the Atmospheric Circulations as Indicator of Climate Change. In: Letcher, T.M., Ed., *Climate Change: Observed Impacts on Planet Earth*, Elsevier, Amsterdam, 145-164. http://dx.doi.org/10.1016/B978-0-444-53301-2.00007-5

[29] Bengtsson, L., Hodges, K.I. and Roeckner, E. (2006) Storm Tracks and Climate Change. *Journal of Climate*, **19**, 3518-3543. http://dx.doi.org/10.1175/JCLI3815.1

[30] Kink, K. (1999) Trends in Monthly Maximum and Minimum Surface Wind Speeds in the Coterminous United States, 1961 to 1990. *Climate Research*, **13**, 193-205. http://dx.doi.org/10.3354/cr013193

[31] Pirazzoli, P.A. and Tomasin, A. (2003) Recent Near-Surface Wind Changes in the Central Mediterranean and Adriatic Areas. *International Journal of Climatology*, **23**, 963-973. http://dx.doi.org/10.1002/joc.925

[32] Pryor, S.C., Barthelemie, R.J. and Schoof, J.T. (2005) Inter-Annual Variability of Wind Indices across Europe. *Wind Energy*, **9**, 27-38. http://dx.doi.org/10.1002/we.178

[33] Smits, A., Klein-Tank, A.M.G. and Können, G.P. (2005) Trends in Storminess over the Netherlands, 1962-2002. *International Journal of Climatology*, **25**, 1331-1344. http://dx.doi.org/10.1002/joc.1195

[34] Charney, J.G. (1947) The Dynamics of Long Waves in Baroclinic Westerly Current. *Journal of Meteorology*, **4**, 136-162. http://dx.doi.org/10.1175/1520-0469(1947)004<0136:TDOLWI>2.0.CO;2

[35] Holton, J.R. (1992) An Introduction to Dynamic Meteorology. 3rd Edition, Academic Press, New York.

[36] Eady, E.T. (1949) Long Waves and Cyclone Waves. *Tellus*, **1**, 33-52. http://dx.doi.org/10.1111/j.2153-3490.1949.tb01265.x

[37] Frierson, D.M.W. (2006) Robust Increases in Midlatitude Static Stability in Simulations of Global Warming. *Geophysical Research Letters*, **33**, Published Online. http://dx.doi.org/10.1029/2006GL027504

[38] Hall, N.M.J., Hoskins, B.J., Valdes, P.J. and Senior, C.A. (1994) Storm Tracks in a High-Resolution GCM with Doubled Carbon Dioxide. *Quarterly Journal of the Royal Meteorological Society*, **120**, 1209-1230. http://dx.doi.org/10.1002/qj.49712051905

[39] Yin, H. (2005) A Consistent Poleward Shift of the Storm Tracks in Simulations of 21st Century Climate. *Geophysical Research Letters*, **32**, Published Online. http://dx.doi.org/10.1029/2005GL023684

[40] Frederiksen, J.S. and Frederiksen, C.S. (2007) Interdecadal Changes in Southern Hemisphere Winter Storm Track Modes. *Tellus*, **59**, 599-617. http://dx.doi.org/10.1111/j.1600-0870.2007.00264.x

[41] Frederiksen, C.S., Frederiksen, J.S., Sison, J.N. and Williams, R.H. (2011) Observed and Projected Changes in the Annual Cycle of Southern Hemisphere Baroclinicity for Storm Formation. *Proceedings of the* 19*th International Congress on MODSIM*, Perth, 12-16 December 2011, 2719-2725

[42] Soldatenko, S.A. and Tingwell, C. (2013) The Sensitivity of Characteristics of Large Scale Baroclinic Unstable Waves in Southern Hemisphere to the Underlying Climate. *Advances in Meteorology*, **2013**, Article ID: 9812711. http://dx.doi.org/10.1155/2013/981271

[43] Kidston, J., Dean, S.M., Renwick, J.A. and Vallis, G.K. (2010) A Robust Increase in the Eddy Length Scale in the Simulation of Future Changes. *Geophysical Research Letters*, **37**, Published Online. http://dx.doi.org/10.1029/2009GL041615

[44] Mizuta, R., Matsueda, M., Endo, H. and Yukimoto, S. (2011) Future Changes in Extratropical Cyclones Associated with Change in the Upper Troposphere. *Journal of Climate*, **24**, 6456-6470. http://dx.doi.org/10.1175/2011JCLI3969.1

[45] Mokhov, I.I., Mokhov, O.I., Petukhov, V.K. and Khairullin, R.R. (1992) On Trends of Atmospheric Cyclogenesis Activity under Climate Change. *Atmospheric and Oceanic Physics*, **28**, 11-26.

[46] Phillips, N.A. (1954) Energy Transformation and Meridional Circulation Associated with Simple Baroclinic Waves in a Two-Level, Quasi-Geostrophic Model. *Tellus*, **6**, 273-286. http://dx.doi.org/10.1111/j.2153-3490.1954.tb01123.x

[47] Palmen, E. and Newton, C.W. (1969) Atmospheric Circulation System. Academic Press, New York.

[48] Jacobson, M.Z. (2005) Fundamental of Atmospheric Modeling. Cambridge University Press, Cambridge. http://dx.doi.org/10.1017/CBO9781139165389

[49] Hoskins, B.J. (1983) Dynamical Processes in the Atmosphere and the Use of Models. *Quarterly Journal of the Royal Meteorological Society*, **109**, 1-21. http://dx.doi.org/10.1002/qj.49710945902

[50] Cook, K.H. and Held, I.M. (1992) The Stationary Response to Large-Scale Orography in a General Circulation Model and a Linear Model. *Journal of the Atmospheric Sciences*, **49**, 525-539. http://dx.doi.org/10.1175/1520-0469(1992)049<0525:TSRTLS>2.0.CO;2

[51] Kirk-Davidoff, D.B. and Keith, D.W. (2008) On the Climate Impact of Surface Roughness Anomalies. *Journal of the Atmospheric Sciences*, **65**, 2215-2234. http://dx.doi.org/10.1175/2007JAS2509.1

[52] Comiso, J., Parkinson, C., Gertsen, R. and Stock, L. (2008) Accelerated Decline in the Arctic Sea Ice Cover. *Geophysical Research Letters*, **35**, Published Online. http://dx.doi.org/10.1029/2007GL031972

[53] Kumar, A., Perlwitz, J., Eischeid, J., Quan, X., Xu, T., Zhang, T., Hoerling, M., Jha, B. and Wang, W. (2010) Contribution of Sea Ice Loss to Arctic Amplification. *Geophysical Research Letters*, **37**, Published Online.

http://dx.doi.org/10.1029/2010GL045022

[54] Petoukhov, V. and Semenov, V.A. (2010) A Link between Reduced Barents-Kara Sea Ice and Cold Winter Extremes over Northern Continents. *Journal of Geophysical Research*: *Atmospheres*, **115**, Published Online. http://dx.doi.org/10.1029/2009JD013568

[55] Jaiser, R., Dethloff, K., Handorf, D., Rinke, A. and Cohen, J. (2012) Impact of Sea Ice Cover Changes on the Northern Hemisphere Atmospheric Winter Circulation. *Tellus*, **64**, 11595. http://dx.doi.org/10.3402/tellusa.v64i0.11595

[56] Guest, P.S. and Davidson, K.L. (1991) The Aerodynamic Roughness of Different Types of Sea Ice. *Journal of Geophysical Research*: *Oceans*, **96**, 4709-4721. http://dx.doi.org/10.1029/90JC02261

17

Power Producing Preheaters—An Approach to Generate Clean Energy in Cement Plants

Amitesh Pandey

Jaypee Cement Limited, Rewa, India
Email: amitesh.mech17@gmail.com

Abstract

Demand of cement in developing countries is directly proportional to the development rate of that country. But increasing input cost of cement manufacturing, decreasing margin of profit, scarcity of raw coal availability and emission of greenhouse gases are some constraints, which restrict the growth of cement industry. Hence to combat with all these adverse situations simultaneously, this project report introduces and efforts to generate clean and green energy with the help of combination of preheater tower, which is available in all integrated cement plants and an augmented wind turbine. Hence, the technology is named as "Power Producing Preheaters" or 3P.H. Introduction of 3P.H. in cement industry, generates a definite amount of clean and green energy (as per site conditions), which is directly used in cement production to avoid grid connectivity cost of wind turbine output. Calculations are done to show the overall cost of project, its payback period and reduction in emission of greenhouse gases along with its benefits in cement industry.

Keywords

Wind Energy, Augmented Wind Turbine, Power Producing Preheaters, Clean Energy in Cement Plants

1. Introduction

The manufacturing process of cement consists of mixing, drying and grinding of limestone, clay and silica into a composite mass. The mixture is then heated and burnt into a preheater and kiln, to be cooled in an air cooling system to form clinker, which is the semi finished product. This clinker is cooled by air and subsequently ground with gypsum to form cement [1]. There are three types of process to form cement-wet, semi dry and dry process. In the wet/semi dry process, raw material is produced by mixing limestone and water (called slurry) and blending it with soft clay. In the dry process technology, crushed limestone and raw materials are ground and mixed together without addition of water. The dry and semi wet processes are more fuel efficient. In the wet process, 0.28 tons of coal and 110 kWh of power consume to manufacture one ton of cement, where as in the dry process only 0.18 tons of coal and 100 kWh of power consume for the same [1].

Cement consumption rates per capita of different leading countries as per 2008-09 data are—1245 kg for Saudi Arabia, 1040 kg for China, 491 kg for Japan, 378 kg for Russia, 285 kg for United States 271 kg for Brazil and 147 kg for India [1]. These data clearly indicate that growth of per capita consumption of cement indicates the growth of country, which proves that growth of cement industry in any country is directly related to the growth of that country. Hence, encouragement of cement industry is highly needed for development of any country. Growth of per capita cement consumption rate indicates the growth of infrastructure, growth of steel sector along with growth in financial sector also. But the restrictions for cement industry growth are—high input cost of raw material, high power and fuel cost, scarcity of raw coal and emission of green house gases.

Breakup of different costs associated with the production of cement is as follows:

1) Power and Fuel Cost: This cost accounts for nearly 30% of the total production cost. Hence, power and fuel costs have major impact on total operating margin [2].

2) Raw Material Cost: It is the second major part of total cost, after power and fuel cost. Different raw materials used are—limestone, gypsum, silica and fly ash. Raw materials account for 30% - 40% of the cost of sales [2].

3) Transportation Cost: It is the third major part of total cost. Increasing price of crude oil affects the transportation cost. It constitutes more than 10% of the cost of sale [2].

4) Other Miscellaneous Cost: It includes the maintenance cost, manpower cost, inventory cost, marketing and selling cost, etc. In overall, it accounts for 15% - 20% of the total cost [2].

As per above cost break up, it is clear that power and fuel costs have major role in total cost and that this cost drives the total cost of cement production. Fuel is used to heat the kiln of cement plant which is a type of furnace used to burn the material, and amount of fuel depends on the optimization of the process and the type of process used, while power cost is the cost of total power consumed in production of per ton clinker/cement. This cost is also dependent on process optimization and on equipment condition that is maintenance of equipments. Most of the cement plants use thermal power for cement production (burning of coal) due to which green house gases generate which is responsible for the global warming *i.e.* depletion of ozone layer and hence responsible for climate change.

Hence to combat with above mentioned restrictions *i.e.* to reduce the power cost and reduction in emission of green house gases, in production of cement, this project introduces a new technology named as "Power Producing Preheaters" *i.e.* "3P.H.", which introduces generation of clean and green energy up to a definite amount (as per site conditions) for cement production with the combination of preheater tower, available in all integrated cement plants and augmented wind turbine.

2. Introduction of Wind Power

Wind is air in motion. It is generated due to earth's rotation due to which there is uneven heating of earth's surface by sun rays. The sun rays cover a much greater area at the equator and smaller area at the poles. Hence the hot air rises from the equator and expands towards the poles, which causes wind. Wind has mass and mass in motion has a momentum, which is a form of energy that can be harvested through a wind turbine. A wind turbine is a system which transforms the kinetic energy available in the wind into mechanical or electrical energy that can be used for any required application.

2.1. Advantages of Wind Power

Lots of advantages of wind turbine or wind power are clearly visible. Some of them are—one time installation cost, low operational and maintenance cost, no fuel cost, environment friendly and pollution free, lowest gestation period, limited use of land. In all above advantages the most important is zero emission of green house gases which is the most needed step for current scenario to avoid climate change.

Power (theoretical) in the wind can be expressed as [3]

$$P_{wind} = \frac{1}{2} \times \rho \times A \times V^3$$

where: P_{wind} = Theoretical power in the wind (w/m^2).

ρ = air density (kg/m^3).

A = swept area or projected area (m^2) (Wind turbine rotor area).

V = average wind speed towards turbine blades (m/s).

In 1919, the physicist, Albert Betz showed that from an ideal wind turbine, maximum kinetic energy conversion limit is 16/27 *i.e.* 59.3% of the kinetic energy of wind to be capture. This is known as Betz limit or Betz law and can be used in any wind turbine. Since this is the maximum limit of kinetic energy conversion hence in any case of modern wind turbine design, the practical limit may reach up to 70% to 80% of the theoretical limit.

Hence in actual power obtained from the wind turbine can be expressed as

$$P_{wind} = \frac{1}{2} \times \rho \times A \times V^3 \times Cp \quad [3]$$

where: P_{wind} = Power in the wind (w/m^2).

ρ = Air density (kg/m^3) = 1.225 kg/m^3.

A = Projected area or wind turbine rotor are (m^2).

V = Average wind speed (m/s).

Cp = Coefficient of efficiency.

$$Cp = \eta_m \times \eta_e \times \eta_{aero} = (0.57 - 0.45)$$

where: η_m = Mechanical efficiency.

η_e = Electrical efficiency.

η_{aero} = Aerodynamic efficiency.

2.2. Typical Arrangement of a Wind Turbine

The conversion of wind energy to useful energy involves two processes. The primary process of extracting kinetic energy is from the wind and conversion of mechanical energy at the rotor axis. In the secondary process, the conversion of rotor's mechanical energy into useful energy carried [3].

$$ROTOR \rightarrow GEAR\ BOX \rightarrow GENERATOR \rightarrow OUTPUT$$

Practically the power obtained by a wind turbine [3] is given as

$$P_{wind} = k \times \frac{1}{2} \rho \times A \times V^3$$

where: K = C$_P$ N$_G$ N$_B$.

C$_P$ = Coefficient of performance of kinetic energy extraction = 0.593 = $\dfrac{16}{27}$ = Betz limit.

N$_g$ = Generator efficiency.

N$_b$ = Gear box/bearing efficiency.

- The torque generated by the wind turbine is

$$Ts = \frac{P}{W_s}$$

Ts = Mechanical torque at the turbine side.

P = Power output of the turbine.

W$_s$ = Rotor's speed of the wind turbine.

- The power coefficient Cp is the percentage of power in the wind that can be converted into mechanical power and the ratio of the blade tip speed to the wind speed is referred as the TIP-SPEED RATIO (TSR)

$$TSR = \frac{W_s}{V} \times R \quad [3]$$

R = Radius of the wind turbine rotor. Since

$$P_{wind} = k \times \frac{1}{2} \rho \times A \times V^3 \quad [3]$$

where k is a constant. Hence, wind power output depends on.

2.3. Density of Air

Density of wind varies with temperature as

$$\rho\left(kg/m^3\right) = \frac{353.12}{273.15 + T} \quad [4]$$

Generally it is counted as 1.225 kg/m³ at sea level. Since it is directly proportional to wind power, hence play a role in output power. But the value of density is as low as it does not important (impact) on the total power generated by wind turbine. Density depends on temperature of air and elevation or height of air from sea level. From the above relationship it is clear that as the temperature increases air density reduces and wind power output also reduces. The maximum air density is at the earth's surface. Air density decreases with height away from the surfaces of earth as the pull of the earth's gravity is less.

2.4. Wind Power Density

Variation in wind power density also affects the wind power output. It is directly proportional to the wind power output. Wind power density is dependent of air density and height of location as

$$P_2/P_1 = \left[h_2/h_1\right]^{3/7} \quad [4]$$

From above, it is clear that wind power density increases with height of the turbine rotor.

2.5. Turbine Swept Area (TSA)

Turbine swept area or diameter of the rotor blades is directly proportional with wind power output. Wind power output is directly proportional to T.S.A which is directly proportional to (Rotor Diameter)². TSA is a part of rotor design which includes following considerations Blade length, Blade number, Blade pitch, Blade shape, Blade material, and Blade weight etc

The design of the blades used is based on blade element theory and on Betz equation. Drag powered wind turbines are characterized by slow rotational speed and high torque capabilities, while the lift powered wind turbines have much higher rotational speed than drag types and therefore are well suited for electricity generation.

2.6. Angle of Attack (Blade Angle)

Practically blade angle can vary in between 1.0 to 15.0 degrees.

2.7. Blade Number

Generally aerodynamic efficiency increases with the number of blades but with diminishing return. The fewer the number of blades, the lower the material and manufacturing cost will be. Higher rotational speed reduces the torques in the drive train, resulting in lower gear box and generation costs. Turbine with many blades or very wide blades will be subjective to very large forces when the wind blows at a critical fast speed.

2.8. Tip-Speed Ratio (TSR)

The Tip-Speed ratio is the ratio of the rotational speed of the blade to the wind speed. The larger this ratio, the faster the rotation of the wind turbine rotor at a given wind speed. Electricity generation requires high rotational speed. Lift type wind turbines have maximum TSR of around 10 while in drag type turbine the TSR is approximately 1.0.

2.9. Wind Velocity

Wind power output is directly proportional to third power of velocity of wind striking on rotor. Hence doubling the wind velocity will result in eight times power output.

2.10. Blade Shape

The speed with which the tip of the rotor blade moves through the air is known as tip speed. At the tip of the blades the speed is some 8 times higher than the speed of the wind, hitting the front of the turbine. Hence rotor blades for wind turbine are always twisted.

2.11. Rotor's Height

Since wind power output is directly proportional to third power of wind velocity, and wind velocity increases as per $1/7^{th}$ law at height, hence height of turbine rotor is another important parameter for turbine output. As per $1/7^{th}$ law of wind velocity, which is also known as $1/7^{th}$ power law to get the wind velocity at any altitude

$$\frac{V_2}{V_1} = \left[\frac{Z_2}{Z_1}\right]^{\frac{1}{7}} \quad [4]$$

The energy in the wind increases with the cube of the wind speed (P directly proportional to V^3) and wind speed increases with height. An increase of just 26% in wind speed means twice as much power available in the wind, and the turbine till produce almost twice as much. Double the wind speed and almost eight times power output we will get.

2.12. Conclusion for Wind Power

As we discussed above, and from wind power output equation

$$P_{wind} = k \times \frac{1}{2}\rho \times A \times V^3 \quad [3]$$

We can conclude that wind velocity and turbine rotor's area are the two important parameters to increase the wind power output. Since wind velocity increases as per $1/7^{th}$ law with respect to height, hence height of turbine rotor is also a critical parameter to increase the wind power output.

3. An Emerging Technology for Modern Wind Turbines (Augmented Wind Turbines)

Energy obtained from wind is proportional to the cubic of wind speed and hence a small increase in wind speed will result in a large magnification or augmentation of the wind energy. Hence the modern wind turbines are augmented in such a way to increase the wind speed of turbine's rotor to a required value. Now we will discuss about the same design of wind turbine.

To increase the wind turbine output, by increasing the wind velocity, is considered by introducing the ducted wind turbine, which will use the equation of continuity as per **Figure 1**.

$$A_1V_1 = A_2V_2 \quad \text{or} \quad V_2 = \frac{A_1 \times V_1}{A_2} \quad [5]$$

where A_1 = Area of intake.

V_1 = Velocity of air flow at intake.

A_2 = Cross sectional area at venturi.

V_2 = Velocity of air at venturi.

Construction of such a wind turbine will be such as turbine's rotor will be placed in an augmented or ducted path at venturi so that to utilize the augmentation in velocity (V_2).

As the power extracted from wind has a cubic relationship to wind velocity and linear relationship to pressure, this is exploited in the ducted turbine and given an advantage of a factor nearly 17 (improvement factor) over the conventional turbine in theoretical calculations, not including coefficients of power transformations.

To reduce complexity of design in the augmented wind turbine, controlling angle of attack is not built in to the turbine blades , as with conventional wind turbines . The ducted turbine uses Variable Inlet Guide vanes (VIGVS) mounted in the air stream prior to the turbine rotor, which controls angle of attack maintaining optimum performance, while the mechanism do not have to be mounted in confines of a hub. An annular arrangement is proposed that houses the pitch change mechanism in the nacelle as inner ducting reducing inertia on the rotating mechanism.

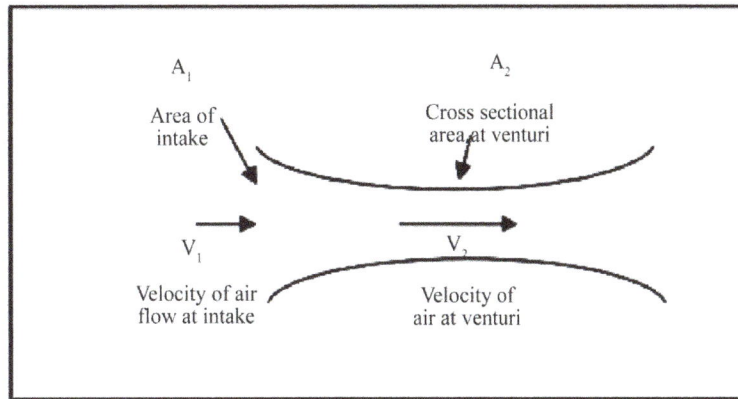

Figure 1. Wind turbine augmentation [5].

$$\text{Since} \quad V_2 = \frac{A_1 \times V_1}{A_2} \quad [5]$$

Hence from above it is clear that by placing the wind turbine inside a convergent duct, we should see increased efficiency as the air flow is accelerated through the venturi. Gains will also be made with reduction of span wise flow and the elimination of blade tip vortices. Accordingly six percent reduction in drag can be gained by the elimination of blade tip vortices.

3.1. Comparison of Wind Turbine Output with and without Augmentation

3.1.1. Without Augmentation

No ducting is used in such condition, as shown in **Figure 2**. Turbine diameter is set as 3 meter with an unusable hub diameter as 250 mm and wind velocity as 6 m/sec and considering the un useful hub diameter in both cases we get

$$A = \frac{\pi}{4} = (3.14/4) \times (3 - 0.25)^2 = 5.936 \text{ m}^2$$

Hence Power obtained will be

$$P = \frac{1}{2} \times \rho \times A \times V^3$$

$$= \frac{1}{2} \times 1.293 \times 5.936 \times 6^3 \left(\text{air density} = 1.293 \text{ kg/m}^3 \right)$$

$$= 828.93 \text{ Watt}$$

3.1.2. With Augmentation

A ducted wind turbine with acceleration of air flow due to the venturi effect aligned with Bernoulli's equation of continuity is shown in **Figure 3**.

All the set parameters are fixed as in previous case while dimensions of ducting at venturi and intake are set as 3.0 meter and 4.5 meter.

Now

$$A_1 = \frac{\pi}{4} d_1^2 = 3.14/4 \times (4.5)^2 = 15.896 \text{ m}^2$$

$$A_2 = \frac{\pi}{4} d_2^2 = 3.14/4 \times (3 - 0.25)^2 = 5.936 \text{ m}$$

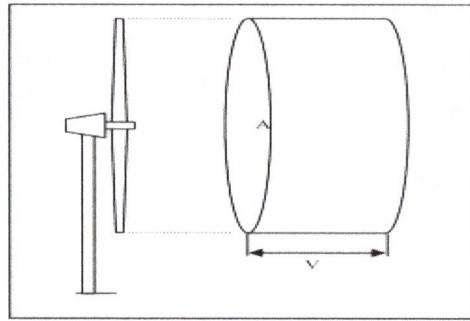

Figure 2. Wind turbine without augmentation [5]. (A ducted horizontal wind turbine for efficient generation. I. H. Al-Bahadly and A. F. T. Petersen, Massy University New Zealand).

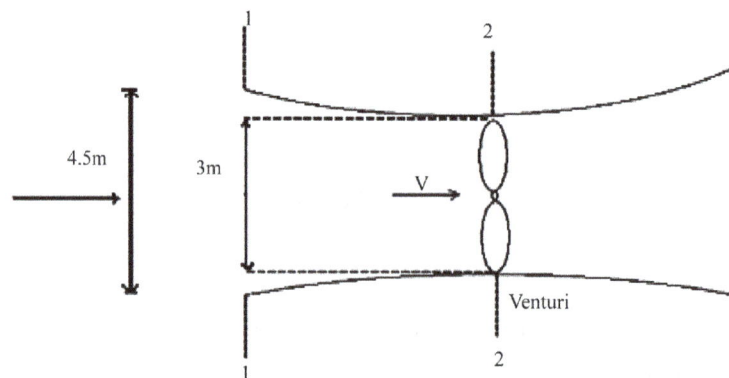

Figure 3. Wind turbine with augmentation [5].

Using

$$A_1 V_1 = A_2 V_2 \quad \text{or} \quad V_2 = \frac{A_1 \times V_1}{A_2}$$

Or,

$$V_2 = \frac{15.896 \times 6}{5.936} = 16.066 \text{ m/sec.}$$

If we look at the relationships between the factors of the power eq^n, we notice a linear relation between density and power, where as there exists a cubic relation between velocity and power. This will be exploited with the ducted turbine design. The density will be affected by the acceleration of the air flow due to the venturi-effect. To investigate the change in density to the air flow, we must assume dry as being an ideal gas, then apply the ideal gas law equation:

$$P_a \times V_g = nRT \quad \text{(Ideal gas equation)}$$

where

P_a = Pressure,
V_g = volume of gas,
N = number of kilo moles,
R = gas constant,
T = Temp. In Kelvin,
m = mass.
If we consider the density of air as:

$$\rho = \frac{m}{Vg}$$

Then from above equations, we can calculate the density of air in the turbine:

$$\rho = \frac{n \times Pa}{R \times T}$$

If we consider, the minimum pressure gradient as 101,221 Pa then from above equation we can calculate the density drop in air flow through the duct. Using equation as:

$$n = 29 \text{ kg/kilo mol.}$$

$$P_a = 101,212 \text{ Pascal}$$

$$R = 8.314 \text{ J/K/mol}$$

$$T = 293 \text{ K}$$

Hence, $\rho = \dfrac{29 \times 101221}{8.314 \times 293}$ or $\rho = 1.205 \text{ kg/m}^3$

While density of air at standard temperature and pressure is $\rho = 1.293 \text{ kg/m}^3$

$$\text{Hence \% change in density} = \frac{1.293 - 1.205}{1.293} \times 100 = 6.80\%$$

Hence considering the change in air density (1.205 kg/m^3) and calculating the power output as

$$P = \frac{1}{2} \times \rho \times A_2 \times V_2^3$$

$$P = \frac{1}{2} \times 1.205 \times 5.936 \times (16.06)^3 = 14,821.3362 \text{ Watt}$$

Hence augmentation obtained = 14,821.3362/828.93 = 17.88

This proves the increase in efficiency from a ducted wind turbine. The calculations proves an increase in efficiency of a factor of 17, from the same ambient wind velocity of 6 m/sec and includes the reduction in density as the air accelerates through the venturi, and idol diameter of hub too.

This significant increase in efficiency can be exploited in a number of ways, either to make more money with the same size diameter turbine or to make the turbine smaller which is more desired for our purpose.

4. Introduction of Power Producing Preheaters (P.P.P.H.) in Cement Industry

As we discussed in previous topics that power generated by a modern ducted wind turbine, mainly depends on
- Wind velocity.
- Turbine swept area.
- Air density.
- Augmentation used.

Since wind velocity increases with height or altitude from ground or sea level, hence it is the function of height from the ground. Turbine swept area or rotor diameter is a design parameter and depends on site condition and installation and fabrication cost. Air density has very less effect on wind turbine output. Augmentation used is also a design parameter which helps to reduce the rotor size and to increase the wind velocity to a desired value. From all these facts we can conclude that the favorable conditions for better wind power output are as follows:
- Obstacle and turbulence free wind flow.
- High altitude.
- Optimum augmentation.
- High average wind speed.

Now considering about a cement plant. Generally in all integrated (Complete) plants, following properties, which are very much favorable to install a ducted wind turbine, are observed
- Situated far away from urban area.
- A preheater cyclone tower of reinforced cement concrete (R.C.C) with overall height range 80 - 130 meters as shown in **Figure 4**.
- Necessary supporting structure to install a ducted wind turbine for augmented performance of turbine.

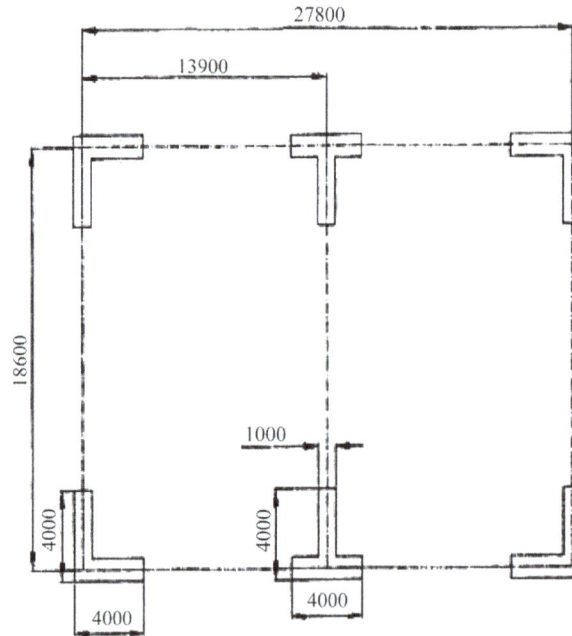

PREHEATER TOWER PLAN

Figure 4. Typical plan (top view) of R.C.C preheater tower in cement plant. (All dimensions are in mm).

These entire favorable conditions offer us to introduce the concept of Power Producing Preheaters *i.e.* 3P.H.in cements plants. The structure of such a ducted wind turbine will be installed on preheater tower top roof i.e. at 80 - 130 meter height above from the ground to utilize $1/7^{th}$ law of wind velocity. Augmentation used increases the average wind velocity near rotor hub i.e. near venturi and r.c.c structure of 80 - 130 meter height will reduce the fabrication cost of wind turbine tower structure. Use of such an augmented turbine will have following visible advantages.

4.1. Utilization of 1/7th Law of Wind Velocity

As per this law

$$V_1/V_2 = \left(H_1/H_2\right)^{1/7} \quad [4]$$

Generally in all integrated cement plants, we have an r.c.c (reinforced cement concrete) cyclone preheater tower of 80 - 130 meter height, of material composition M20/M30 while top slab thickness is 200 - 300 mm, is available. Now considering the average standard wind speed at any site as 4 m/sec (which is measured at 10 meter height above the ground) and height of rotor above the ground level from top slab is 100 meter, then

$$V_2/V_1 = \left(H_2/H_1\right)^{1/7}$$

Or,

$$V_{100}/V_{10} = \left(H_{100}/H_{10}\right)^{1/7} = \left(100/10\right)^{1/7}$$

Or,

$$V_{100}/4 = \left(10\right)^{1/7} = 1.3895$$

Hence

$$V_{100} = 5.558 \text{ m/sec}$$

Hence, the 4 m/sec wind velocity converts into 5.558 m/sec wind velocity. This magnification is very important as it reduces the requirement of big diameter rotor and hence reduces the total cost and load on preheater tower also. Similarly we can calculate the power density amplification as

$$P_2/P_1 = \left(H_2/H_1\right)^{3/7} \quad [4]$$

4.2. Introduction of Augmented Wind Turbines

As we saw that ducted wind turbine increases the wind power output nearly 17 times, hence this magnification in total output power is very important to reduce the rotor size and hence its fabrication cost. Ducted turbine works on the principle of equation of continuity as shown in **Figure 1**

$$A_1V_1 = A_2V_2 \quad [5]$$

where A_1 and A_2 are area of duct at any two places while V_1 and V_2 are the velocities of wind at these two points respectively **Figure 1**. Due to this principle we can change or get the desired wind speed at turbine's rotor which in turn gives the opportunity to reduce the rotor diameter because installing a large diameter rotor will not be suitable for tower top in cement plants. Hence using ducted wind turbine the rotor diameter can be kept as minimum as required. Along with this, another important advantage of using ducted wind turbine is availability of wind in a unique direction. As we know that ducts of augmented wind turbines use the variable inlet guide vanes to guide the outside wind to rotor's direction, hence any change in the direction of wind will cause no effect on the turbine.

Inlet guide vanes and stators have been incorporated into the design of ducted wind turbine to ensure that air flow is offered to the turbine at an optimum angle of attack. For any airfoil cross section to be efficient, it has to be offered to the air flow at the optimum angle of attack.

4.3. Low Cost Wind Turbines

Another and equally important advantage of installing a wind turbine on preheater tower top is low cost wind energy. As we know that installing a ducted wind turbine at a high altitude is always favorable, but the fabrication cost of wind turbine tower will increase so much on high altitude that it will not be economical to install. Hence using the r.c.c preheater tower's height for wind turbine tower installation on its top, will reduce almost 60% - 70% cost of turbine tower fabrication cost. On the other hand since nearly 11% cost of total cost belongs to grid connection cost, hence using wind turbine output for cement production will eliminate this 11% cost of total cost [6]. The cost break up of total installation cost wind turbine is as follows

Wind turbine fabrication cost (64%)

Grid connection cost (11%)

Planning/Miscellaneous cost (9%)

Foundation cost (16%)

From the above mentioned table of break up cost for installation of wind turbine, it is clear that 11% of total cost includes grid connection cost while 26.3% of total turbine fabrication cost includes in wind turbine tower fabrication [6]. Hence installing a ducted wind turbine on preheater tower top will reduce this 26.3% cost of turbine tower fabrication and 11% of total cost is reduced by using the generated power in production of cement itself. Both these cost will make a major relief in installation cost per MW.

5. A Typical Case Study for Introducing Power Producing Preheaters in Cement Plant

Consider an r.c.c cyclone tower of preheater with certain dimensions as per a particular site. The plan *i.e.* top view and the dimensions this preheater tower is shown in **Figure 4**. Typical parameters and dimensions of this preheater tower are as follows

- Breadth of preheater tower = 18.6 meter.
- Height of top slab from ground = 100 meter.
- Width of preheater tower = 27.8 meter.
- Average wind speed recorded at site = 4 m/sec = V_{10}.

  Considering 30 meter hub height from tower top slab, hence overall height of wind turbine rotor from ground = 100 + 30 = 130 meter.

  Considering dimensions of prepared duct, as shown in **Figure 6**.
- Diameter of duct at venturi = d_1 = 15 meter.
- Diameter of duct at opening = d_2 = 30 meter.

  Hence, the final structure of such wind turbine is shown in **Figure 5** as below.
- Average wind density = 1.25 kg/m^3 (considering reduction in density at venturi).

  Now using 1/7 th law of wind velocity at high altitudes, we have

$$V_{10} = 4 \text{ m/sec}, H_{10} = 10 \text{ meter}, H_{130} = 130 \text{ meter}, V_{130} = ?$$

Hence using 1/7 th law

$$V_{130}/V_{10} = \left(H_{130}/H_{10}\right)^{1/7}$$

Or,

$$V_{130}/4 = \left(130/10\right)^{1/7} = \left(13\right)^{1/7} = 1.442$$

Hence

$$V_{130} = 4 \times 1.442 = 5.768 \text{ m/sec.}$$

Hence, wind speed at rotor's height = 5.768 m/sec

  Now using equation of continuity for prepared duct as in **Figure 6**

$$A_1 V_1 = A_2 V_2$$

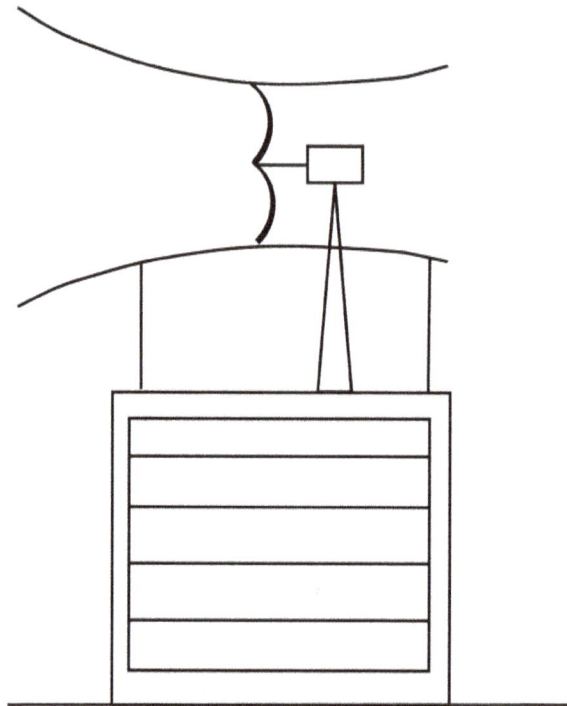

Figure 5. Complete technology of 3P.H.

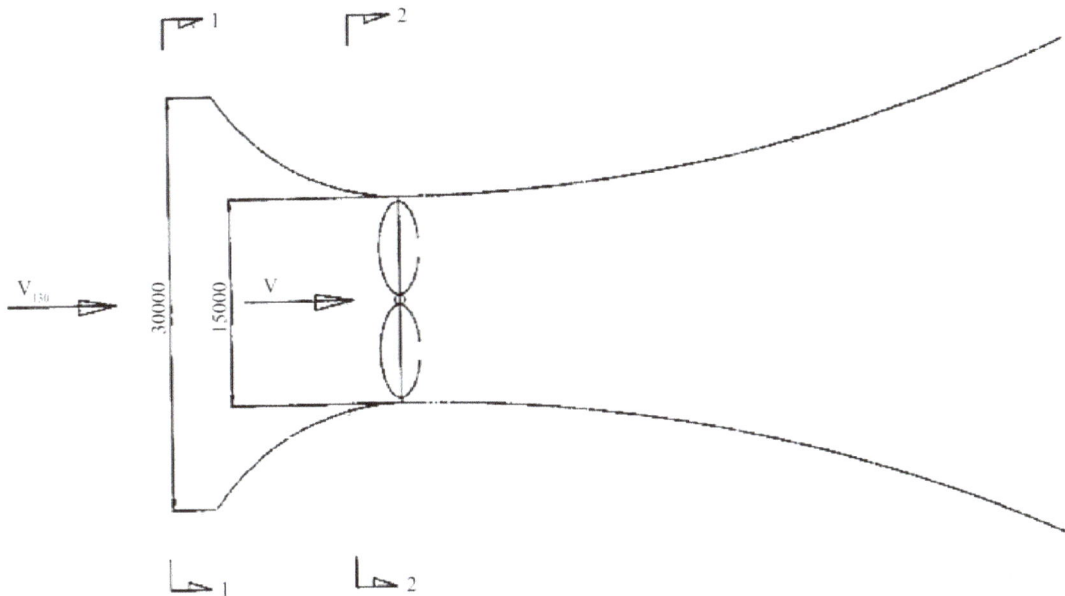

Figure 6. Dimensions of ducting (all dimensions are in mm).

Here

$$A_1 = \frac{\pi}{4} d_1^2, \ d_1 = 30 \text{ meter}$$

$$V_1 = V_{130} = 5.768 \text{ m/sec}$$

$$A_2 = \frac{\pi}{4} d_2^2, \ d_2 = 15 \text{ meter}$$

$$V_2 = \text{velocity at venture} = ?$$

Hence

$$A_1 V_1 = A_2 V_2 \quad \text{or} \quad \frac{\pi}{4} d_1^2 \times V_1 = \frac{\pi}{4} d_2^2 \times V_2$$

Or,

$$(30)^2 \times 5.768 = (15)^2 \times V_2$$

Or,

$$V_2 = 23.072 \text{ m/sec.} = \text{Wind speed at rotor's hub}$$

Now we will calculate the maximum power obtained from these parameters and dimensions of wind turbine, taking considerations of Betz limit of maximum power conversion as 16/27 and considering all efficiencies including mechanical, electrical, and aerodynamic efficiency. Hence overall efficiency

$$C_p = \text{Betz limit} \times C_{mech} \times C_{elec} \times C_{aero} = 16/27 \times 0.96 \times 0.95 \times 0.96 = 0.5188$$

For safety consideration taking this overall efficiency = 0.50.

Hence, maximum power obtained from such an augmented wind turbine is

$$P_{max} = \frac{1}{2} \times \rho_{air} \times A_{rotor} \times V_{rotor}^3 \times Cp$$

$$= 1/2 \times 1.25 \times 3.14 \times (15)^2 \times (23.072)^3 \times 0.50$$

$$= 671561.367 \text{ Watt} = 671.562 \text{ kW}$$

Hence

$$P_{max} = 671.562 \text{ kW}$$

6. Some Results and Discussion

As per the study report of CEA (Central Electricity Authority of India) [7], based on the study of 105 thermal plants in India of more than 100 Mw capacity each with total installed generation capacity of 93,172 Mw, the combustion technology in these 86 plants is based on pulverized coal burning but the type of furnace technology design of the boiler, forced draught fans etc differ with plants. Based on CEA data (2010), specific coal usage at 03 plants is less than 0.6 kg/kWh, at 19 plants usage is between 0.6 - 0.7 kg/kWh, at 07 plants usage is between 0.9 - 1.0 kg/kWh and at 03 plants usage exceeds 1.0 kg/kWh. In most general cases, specific fuel usage lies in between 0.7 - 0.9 kg/kWh *i.e.* on average 0.8 kg/kWh. In the same report of CEA, we know that emissions per unit of electricity are estimated to be in the range of 0.91 - 0.95 kg/kWh for CO_2, 6.94 to 7.20 gm/kWh foe SO_2 and 4.22 to 4.38 gm/kWh for NO, during coal consumption for electricity generation [7].

Hence keeping in mind about these facts, the effects and advantages of introducing Power Producing Preheaters technology in cement industry can be conclude as.

6.1. In Terms of Cost

As we know that cement industry is continuously struggling by high input cost of raw material and high power cost. With the use of 3P.H. we are trying to reduce the power cost for cement production.

We know that in modern cement plants, power consumption of per ton cement manufacturing lies in the range of 90 - 100 units per ton of cement production [1], while through grid it costs nearly 5.50 INR/unit [7], (INR = INDIAN NATIONAL RUPEE) and CPP costs nearly 4.0 INR/unit. Considering lowest cost per unit *i.e.* INR 4.0 per unit, then we are investing INR.380 per ton of cement. Let's take the example of a cement plant of capacity 2.0 MTPA delivering 5556 M.T of cement per day. Hence total investment in power will be around 5556 × 380 = 2,111,280 INR per day (USD 32481 per day) (USD = US Dollar, 1 USD = 65 INR).

Now the wind turbine installation cost in present scenario is nearly INR.70 million per Mw (USD 1,076,923 per mw) While our output is 671.562 kW. Further we are saving our cost by eliminating grid connection cost and tower fabrication cost by 11% and 23% respectively. Hence accounting all these reduction our cost of installation comes at 42.35 million INR (USD 651,538.45) for installing 672 kW (671.562 = 672 approx) capacity ducted wind turbine on preheater tower top, which will be one time investment for company.

Hence in terms of money—672 kWh means 672 units per hour means 672 × 4.0 INR per hour = 2688 INR per hour = 64,512 INR per day =1,935,360 INR per month = 23.224 million INR per year (USD 357,292.30).

Hence the payback period of the project will be = 42.35 million INR/23.22 million INR = 22 Months Approx.

After this payback period we are getting 672 kWh or 672 Units per hours as totally free which will save 1.935 million INR (USD 297,692.30) per month or 23.224 million INR (USD 357,292.30) per year.

6.2. In Terms of Power Consumption in Cement Production

Total power consumption per day in a 2.0 MTPA cement plant is = 5556 × 95 units = 527,820 units/day.

While we are generating 672 × 24 units = 16,128 units/day.

Hence actual consumption will be= 527,820 − 16,128 = 511,692 which is equal to 92 units per ton of cement.

Hence total 03 units per ton of cement production will be reduced. Hence total reduction per year will be = 5556 ton × 30 days × 12 months × 03 units = 600,480 *i.e.* 6.0 million units per year.

6.3. In Terms of Environment

Total generated output power = 672 kWh = 672 units/hr.

Hence total units generated per year = 672 × 24 × 30 × 12 = 5806080 units = 5.8 million units per year.

This generated energy is absolutely green energy and pollution free in nature.

Hence total reduction in green house gases will be

- 5,806,080 × 0.91 Kg of CO_2= 5,283,532.8 Kg CO_2.
- 5,806,080 × 6.94/1000 Kg of SO_2 = 40,294.1952 Kg of SO_2.

- 5,806,080 × 4.22/1000 Kg of NO = 24,501.65 Kg NO.

Hence, we are saving emission of 5283.532 ton of CO_2 per year, 40.294 ton of SO_2 per year and 24.501 ton of NO per year from a single cement plant of 2.0 MTPA capacities, using 3P.H. technology.

Reduction of emission in such amount will definitely result and effect up to a great extant in reduction of green house effect by reducing temperature and hence global warming to. This will help our society in a positive direction. Reduction in such amount of emission of green house gases is very important as per I.P.C.C study and warning.

6.4. In Terms of Raw Coal Availability

$$\text{Total generated output power} = 5,806,080 \text{ units per year.?}$$
$$= 5,806,080 \times 0.8 \text{ Kg of coal per unit}$$
$$= 4,644,864 \text{ Kg of coal per year}$$
$$= 4644.86 \text{ ton of coal per year}$$

Hence we are saving 4644.86 ton of coal per year from a single cement plant of 2.0 MTPA capacities by installing 3P.H. technology in plant.

All these benefits are only from a single plant. Applying 3.P.H. technology in whole cement industry will multiply the results up to a significant level.

7. Variations Expected on Results

So far we have discussed till now about the technical data, output and different parameters etc. All these will vary up to some extent on account of the following facts.

7.1. Due to seasonal Variations

Average wind speed varies with season to season. Generally it is recorded that wind speed in the month of May, June, July and August, is higher side while in the months of October, November, December, and January it is on lower side. In remaining period it remains in middle range. Due to these variations of wind speed with season total output power may vary up to certain extant.

7.2. Variations Due to Ducting Size

Calculations for wind power output are based on turbine rotor diameter and on size of ducting used that is on amount of augmentation used. These parameters are dependent on various factors like availability of space, feasibility in fabrication, fabrication cost, maintenance cost, load bearing capacity of preheater tower and finally on designed/desired output. So all the dimensions are site dependent and also depend on a particular preheater tower. Project feasibility and economics are also important to decide all parameters.

7.3. Variation in Emission Calculations

Amount of green house gases depends on type and quality of coal used for power generation and hence amount of emission vary with the type of coal used.

7.4. Variation in Wind Speed Due to Climate Change

In recent years it is observed that due to global climate changes the average annual wind speed is increased, which is an advantageous factor for installing 3P.H. concept in cement plants. This increment is continuously going up and hence encouraging us to install 3P.H. in cement industry.

8. Conclusion

The research paper shows the reduction in emission of green house gases during production of cement in cement industry. Simultaneously, implementation of this technology in cement industry reduces the power cost in cement production and hence reduces coal consumption through which cost saving in monetary terms is also possible. Since raw coal reserves are fixed in the world hence this technology is very useful to save raw coal of cement producing countries. Since availability of good average wind speed in cement plant preheater top reduces the installation of augmented wind turbine, hence the payback period is low which in turns increases the profit of the industry.

References

[1]   Cement Industry in India—Trade Perspective by Chamber of Indian Industry (CII). www.newsletters.cii.in

[2]   The Cost Elements of Cement by Armando Pabon. www.marketrealist.com

[3]   Calculation of Theoretical and Experimental Power Output of a Small 3 Bladed Horizontal Axis Wind Turbine by K. R. Ajao. www.jofamericanscience.org

[4]   Wind Power Calculator—Energy vs. Turbine Size vs. Speed. www.windpower.generatorguide.net

[5]   Al-Bahadly, I.H. and Petersen, A.F.T. A Ducted Horizontal Wind Turbine for Efficient Generation. Massy University, Palmerston North. www.cdn.intechweb.org

[6]   Renewable Energy Technologies—Cost Analysis Series of Wind Power—June 2012 by IRENA (International Renewable Energy Agency). www.irena.org

[7]   Estimates of Emissions from Coal Fired Thermal Power Plants in India by Moti. L. Mittal Department of Environmental and occupational Health, University of South Florida, Tamp, Florida, USA, Chhemendra Sharma and Richa Singh. www.epa.gov/ttn/chief/conference/ei20/session5/mmittal

18

A Statistical Analysis of Wind Speed and Power Density Based on Weibull and Rayleigh Models of Jumla, Nepal

Ayush Parajuli

Department of Mechanical Engineering, Pulchowk Campus, Tribhuwan University, Lalitpur, Nepal
Email: parajuliyush@gmail.com

Abstract

In the present study, wind speed data of Jumla, Nepal have been statistically analyzed. For this purpose, the daily averaged wind speed data for 10 year period (2004-2014: 2012 excluded) provided by Department of Hydrology and Meteorology (DHM) was analyzed to estimate wind power density. Wind speed as high as 18 m/s was recorded at height of 10 m. Annual mean wind speed was ascertained to be decreasing from 7.35 m/s in 2004 to 5.13 m/s in 2014 as a consequence of Global Climate Change. This is a subject of concern looking at government's plan to harness wind energy. Monthly wind speed plot shows that the fastest wind speed is generally in month of June (Monsoon Season) and slowest in December/January (Winter Season). Results presented Weibull distribution to fit measured probability distribution better than the Rayleigh distribution for whole years in High altitude region of Nepal. Average value of wind power density based on mean and root mean cube seed approaches were 131.31 W/m²/year and 184.93 W/m²/year respectively indicating that Jumla stands in class III. Weibull distribution shows a good approximation for estimation of power density with maximum error of 3.68% when root mean cube speed is taken as reference.

Keywords

Mean Wind Speed, Rayleigh Distribution, Weibull Distribution, Wind Power Density

1. Introduction

The world energy demand and consumption have increased seriously over last decades and most of this demand is met from fossil fuels. The use of fossil fuels and non-renewable form of energy has major impact on environment and our ecosystem through increasing pollution rate. In simple sense, energy and environment are major crisis of today. Many countries in the world are taking a step towards renewable form of energy to solve this challenge. Renewable energy sources like wind, solar, geothermal, hydro, biomass and ocean thermal energy have drawn attention from all over the world due to their almost inexhaustible and non-polluting characteristics. Wind energy being one of the important source for electricity production. It is vigorously pursued in many countries [1].

Like other kind of energy, wind energy is ultimately a solar resource. Wind systems are created mainly due to two main causes: 1) temperature differences between the equator and the poles (the earth's latitudes), 2) the rotation of the earth. Dry air in the vicinity of 30°N and 30°S flows towards the equator where it replaces rising hot air [2]. Wind is highly variable in space and time [3]. The energy potential of wind turbines can be calculated by toting up the energy corresponding to possible wind speeds in a certain period of time; so the probability distribution of different wind speeds of region is an important aspect in calculations. Thus, wind speed frequency distribution plays a critical role for predicting the energy output of a wind energy conversion system. During the last two decades, ample attentions were paid towards the development of a better statistical model for describing wind speed frequency distribution.

As seen from the literature, much concentration has been given to Weibull function because it is found to give fit to the observed wind speed data both at the surface [4] [5] and in the upper air [6]. Wentink compared the Weibull functions with other distributions like Plank's frequency distribution, Rayleigh Distribution and Gamma distribution [7]. Fyrippis *et al.* have analysed wind data and calculated wind power density in Naxos Island Greece [8]. Weibull distribution was useful for them to distribute huge statistical data and present it as a continuous distribution for further analysis. They presented the site falls under class 7, indicating possibility of large scale electricity generation. Oner *et al.* have studied about Weibull distribution, Rayleigh distribution & normal distribution and used them to study potential turbine locations [9]. For similar analysis, Odo *et al.* [10] in Enugu, Nigeria; Ahmed *et al.* [11] in Halabja, Iraq; Islam *et al.* [12] in Kudat and Labuan, Malaysia; Safari *et al.* [13] in Rwanda; Oyedepo *et al.* [14] in south-east Nigeria; Abbas *et al.* [15] in Pakistan and many other researchers have used Weibull distribution & Rayleigh Distribution in different places for analysing wind speed data. Such work have been done whole over the world except for Nepal. Thus, we also use weibull and Rayleigh distribution among others to understand wind potential in high altitude region in Nepal.

Weibull distribution is a two parameter function, namely, shape factor k (*dimensionless*) and scale parameter c (*dimensional*). It is used in describing the wind speed frequency distribution. Several methods have been proposed to estimate Weibull parameters. Graphic method, maximum likelihood method and moment methods are commonly used to estimate Weibull parameters. The purposes of estimation are: a) To retrospectively characterize past conditions; b) to predict future power generation at one location; c) to predict power generation within a grid of turbines; d) to calibrate meteorological data.

Nepal is landlocked country with diversity in its climate from Himalayan region to Terai (Plain Lands) within short range of distance. Nepal's total energy consumption in the fiscal year of 2008/09 was 400.5 million GJ. Traditional sources such as fuel wood, crop residues, and animal dung shared 87.1% of total energy consumption with commercial sources like petroleum products, coal and electricity, and other renewable energy sources contributing only 12.2% and 0.7% of the total energy consumption, respectively [16]. In context of Nepal, few researches regarding wind potential have been done. Studies made for the World Bank in 1977 indicate that the Khumbu area is a high potential area. The study by DANGRID, a Danish consulting firm, in 1992 reported a potential to generate 200 MW of electrical power with an annual energy production of 500 GWh from the wind resources along the 12 km valley between Kagbeni and Chusang in the Mustang district. The Kagbeni wind power project was one of the biggest projects with installed capacity of 20 kW built in 1987 under the support of the Danish Government. Although the government declared a plan to generate 20 MW electricity by wind energy in the Three Year Interim Plan (2007/08-2009/10), the lack of sufficient research data, and complicated geographical landscape of the country hindered its successful implementation [17] [18]. Ghimire *et al.* studied about wind energy resource assessment and feasibility study of wind farm in Mustang albeit predictive model of probability distributions were not applied. This article will look forward to provide information regarding theoretical wind harnessing potential in Himalayan region which is supposed to be region of high wind speed in Nepal and project to verify mostly accepted weibull probability distribution in Terrain of Nepal.

2. Materials and Methods

2.1. Site Location and Data Collection

Jumla is centre of Chandannath municipality in Jumla district and is located in Karnali zone of Nepal. The primary observation have shown that the region has wind potential. Since there was no similar study for this region, this study aimed to examine the wind energy potential of Jumla by finding Weibull and Rayleigh distribution parameters & determining the available power density. Besides applying mean wind speed, a root cube wind speed was applied to calculate the wind power and energy density. Since the wind power is proportional to cube of wind speed, it is a better representation of wind speed to be considered in calculations [19]. The average annual temperature in Jomsom is 13.5°C. The rainfall here averages 766 mm. Department of Hydrology and Meteorology has setup a synoptic station in Jumla which is at 2300 m above sea level. Wind speed was measured at height of 10 m from ground level and average daily wind speed was available. Wind speed from 2004 to 2014 (2012 excluded) was used for analysis.

2.2. Vertical Extrapolation of Wind Speed

In real measurement, the wind speed tends to increase with height in most locations and depends mainly on atmospheric mixing and terrain roughness. Therefore, to calculate the total wind energy potential, the measured surface wind speed must be modified for an altitude different (40 m in this literature) from the normalized height (*i.e.* 10 m). For this reason the following equation was used: [1] [20]

$$v = v_{mes} \left(z/z_{mes} \right)^m \tag{1}$$

where, v_{mes} is the wind speed at normalized height (m/s), z_{mes} is the normalized height (m) and Z is the turbine height (m). The exponent m depends on factors as surface roughness and atmospheric stability. Numerically, it lies in the range of 0.05 - 0.5. Surface roughness (m) which is dependent on the terrain condition varies from 0.128 to 0.160 even in a very homogenous surface as flat or farm land. A typical value for surface roughness is 0.14 (for low roughness surface) and varies from less than 0.1 (for very flat land, water or ice surfaces) to more than 0.25 (for forest and woodlands). According to the literature, for neutral stable condition, m is approximately 0.143, which is commonly assumed to be constant in wind resource assessments. In this research, the surface roughness (m) is taken as 0.143 [19].

2.3. Wind Speed Probability Distribution

To investigate the feasibility of the wind energy resource at any site, the best method is to calculate the wind power density based on the measured data of the meteorological station. Another method is to calculate the wind power density using frequency distribution functions like Weibull distribution, Rayleigh distribution, chi-squared distribution, generalized normal, log normal-distribution, three parameter log-normal, gamma distribution, inverse Gaussian distribution, kappa, wakeby, normal two variable distributions, normal square root of wind speed distribution, as well as hybrid distribution [19] [21]. Researches have shown that Weibull function fits the wind probability distribution more accurately compared to others [22]. Here the author uses Weibull and Rayleigh distribution to fit the time series data.

As wind speed changes regularly, frequency distribution of wind speed based on time series data can be calculated. Exact probability density function describing the speed data is difficult to find. Weibull distribution is a two parameter function characterized by scale parameter c (m/s) and shape parameter k (dimensionless). When Probability of occurrence of certain velocity is given by [11] [23]:

$$f_w \left(v \right) = \left(\frac{k}{c} \right) \left(\frac{v}{c} \right)^{k-1} \exp \left[- \left(\frac{v}{c} \right)^k \right] \tag{2}$$

The corresponding weibull cumulative density function (CDF) is given by

$$F_w \left(v \right) = 1 - \exp \left[- \left(\frac{v}{c} \right)^k \right] \tag{3}$$

Rayleigh function is special case of Weibull function. When shape parameter $k = 2$, Weibull distribution becomes Rayleigh distribution.

$$f_R \left(v \right) = \left(\frac{2v}{c^2} \right) \exp \left[- \left(\frac{v}{c} \right)^2 \right] \tag{4}$$

Shape parameter k and scale parameter c can be calculated using many methods as shown by previous researches. Graphical method (GM), Method of moments (MOM), Standard deviation method (STDM), Maximum likelihood method (MLM), Power density method (PDM), Modified maximum likelihood method (MMLM), Equivalent energy method (EEM) are widely used. In literature about wind energy, these methods are compared several times however results and recommendations of the previous studies are different from each other. For this reason, according to the results of the studies, it might be concluded that suitability of the method may vary with the sample data size, sample data distribution, sample data format and goodness of fit tests [24]. Research of Aazad *et al.* shows MOMs to be the most efficient method for determining the value of k and c to fit the Weibull distribution curves at any altitude [25]. This research proves to provide better information than others as 7 methods for parameter evaluation are compared at different altitudes. Researches must be conducted in terrains of Nepal because wind speed pattern shows different behaviour here. In absence of such studies, we choose work done by Azad *et al.* as our base, and choose MOMs to predict weibull parameters (k and c). Mean wind speed and variance of data shall be calculated beforehand then value of k & c can be as:

$$k = \left(\frac{0.9874}{\dfrac{\sigma}{\overline{v}}} \right)^{1.0983} \tag{5}$$

$$c = \frac{\overline{v}}{\Gamma\left(1+1/k\right)} \tag{6}$$

where Γ is the gamma function.

$$\Gamma(x) = \int_0^\infty e^{-t} t^{x-1} dx \tag{7}$$

2.4. Evaluation of Weibull and Rayleigh Distributions

In order to check how accurately a theoretical probability density function fits with observation data, in this paper, four types of statistical errors are considered as judgement criterion. To evaluate the performance of considered distribution, the mean percentage error (MPE), mean absolute percentage error (MAPE), root mean square error (RMSE) parameter, and the chi-square test are performed [25]. MPE shows the average of percentage deviation between calculated value from weibull & Rayleigh distribution from the observed value whereas MAPE shows average absolute percentage deviation. Best results are obtained when these values are close to zero. The Chi-square goodness-of-fit test judges the adequacy of a given theoretical distribution to a data sample. The size of class intervals chosen in this study is 1 m/s [22] [26].

$$\text{MPE} = \frac{1}{N} \sum_{i=1}^{N} \left(\frac{x_{i,w} - y_{i,m}}{y_{i,m}} \right) \times 100 \tag{8}$$

$$\text{MAPE} = \frac{1}{N} \sum_{i=1}^{N} \left| \frac{x_{i,w} - y_{i,m}}{y_{i,m}} \right| \times 100 \tag{9}$$

$$\text{RMSE} = \left[\frac{1}{N} \sum_{i=1}^{N} \left(y_{i,m} - x_{i,w} \right)^2 \right]^{1/2} \tag{10}$$

$$\chi^2 = \sum_{i=1}^{N} \left(\frac{\left(x_{i,w} - y_{i,m} \right)^2}{x_{i,w}} \right) \tag{11}$$

where N is number of observations, $y_{i,m}$ is frequency of observation or i^{th} calculated value from measured data, $x_{i,w}$ is frequency of weibull or i^{th} calculated value from the weibull distribution and same set of formulas can be used when subscript w is replaced by r representing Rayleigh distribution.

2.5. Wind Power Density (WPD)

Wind power density is measure of capacity of wind resources in specified site. Wind Power density can be measured based on many approaches [1] [19] [27]. It is well known that the power of wind that flows at (v) through a blade swept area ($A = 1$) increases as the cube of its velocity and is given by:

$$P = 1/2 \times \rho v^3 \tag{12}$$

Many researches have used mean velocity to calculate wind power density. Mean power can be calculated by:

$$P_m = 1/2 \times \rho v^{-3} \tag{13}$$

Because the wind power is proportional to cube of velocity, root mean cube of wind speed gives better result and is defined as [28]:

$$V_{rmc} = \sqrt[3]{\frac{1}{n} \sum_{i=1}^{n} V_i^3} \tag{14}$$

From Weibull distribution, power density can be calculated by:

$$P_W = \frac{1}{2} \rho c^3 \Gamma\left(\frac{k+3}{k}\right) \tag{15}$$

From Rayleigh distribution, power density can be calculated by:

$$P_R = \frac{3}{\pi} \rho c^3 \left(\frac{\pi}{4}\right)^{3/2} \tag{16}$$

where P represents Wind Power Density (W/m^2) and ρ is density (kg/m^3) at studied region. A typical value used in all the literature consulted is average air density 1.225 kg/m^3 corresponding to standard conditions (sea level, 15°C) [29]. However, air density is function of temperature T & Pressure P, both of which vary with altitude above z. The corresponding air density ρ could be evaluated using [8]:

$$\rho = \rho_o \frac{T_o}{T}\left(1 - \frac{\Gamma z}{T_o}\right)^{\frac{g}{\Gamma R}} \tag{17}$$

where g is the gravitational acceleration (9.81 m/s^2); T represents the average air temperature (K); $T_o = 288$ K $(273 + 15)$; R is the gas constant (287 J/Kg/k) for air; and Γ is vertical temperature gradient usually taken as 6.51 K/Km. Based on calculations, value of 1.231 was chosen as the air density.

However there is always an error in predicted value and measured value. Calculated wind power density by root mean cube speed or mean speed for the measured probability density distribution serves as reference power density (P_m). Power density predicted using Weibull and Rayleigh distribution (P_W) & (P_R) can be calculated using eqns. (15) & (16) respectively. Error in calculating the power density using distribution compared to measured value can be calculated as [15]:

$$\text{Error}\% = \frac{P_{W,R} - P_m}{P_m} \times 100 \tag{18}$$

2.6. Useful Wind Speeds

Knowing scale parameter (c) and shape parameter (k) of Weibull distribution function, average velocity can be predicted by Weibull and Rayleigh distribution [30].

$$\overline{V_w} = c\Gamma\left(1 + 1/k\right) \tag{19}$$

$$\overline{V_R} = c\sqrt{\frac{\pi}{4}} \tag{20}$$

Similarly, most probable wind speed (V_{mp}) and maximum energy carrying wind speed (V_{op}) also can be calculated using following formulas [12] [30]:

$$V_{mp,w} = c\left(1 - \frac{1}{k}\right)^{1/k} \tag{21}$$

$$V_{op,w} = c\left(1 + \frac{2}{k}\right)^{1/k} \tag{22}$$

Wind direction is important parameter for selection of wind turbines. For this purpose, a wind rose plot is needed which shows dominant wind direction. Wind rose can be done from 4 point, 8 point, 16 point and 32 points. To some users, the 8-point rose is sufficient for their needs. To another user, yet for same purpose, a 16-point rose is absolutely necessary [31]. Mainly two types of wind rose: wind frequency rose and wind speed rose can be drawn [32]. Unavailability of data on wind direction in stations of DHM barred us from plotting wind rose and thus is not presented in this article.

3. Results and Discussions

3.1. Monthly Mean Wind Speed and Seasonal Variations

Wind speeds are different as months and seasons vary. **Figure 1** shows mean wind speed for different months in different sets of years. Average of 2 consecutive years for same month is taken as a data point and plotted in graph. Figure shows a similar trend for all five curves where wind speed increases from January and reaches its peak value in May/June. Then, wind speed decreases in July/August as these months are susceptible to heavy rainfall in Nepal. But, wind speed again rises in October & decreases gradually till December. July & October are two months when wind speed rises in Jumla region. The same scenario is experienced in whole Nepal. The maximum & minimum mean wind speed in different months belonged to July 2004/05 and December 2013/14. For better analysis, a single year was divided in two season as cold season (November-April) and Warm Season (May-October). **Table 1** shows yearly mean wind speed for cold season & warm season. 80% of years analyzed have high mean wind speed in warm season. Average mean wind speed in warm season & cold season for 10 years are 6.17 m/s and 5.76 m/s respectively. Results clearly shows that warmer season has higher mean wind speed compared to colder season. The pattern is different compared to other countries seasonal variation where wind speed will be greater in colder season. In Nepal, lower wind speed coupled with colder dense air will give similar power as in warmer season where wind speed is higher but air density will be lower. So, the power

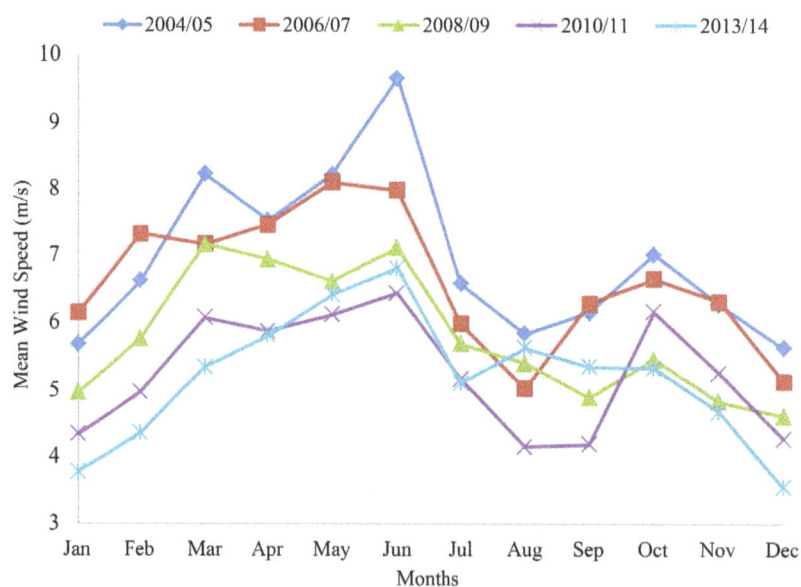

Figure 1. Monthly mean wind speed in Jumla (2004-2014).

Table 1. Yearly mean wind speed for cold and warm seasons in Jumla.

Year	Cold Season (November-April)	Warm Season (May-October)
2004	7.18	7.52
2005	6.14	6.89
2006	6.56	7.09
2007	6.59	6.26
2008	6.10	5.25
2009	5.34	6.46
2010	5.23	5.66
2011	5.07	5.06
2013	5.12	5.56
2014	4.29	5.95
Average	5.76	6.17

available will be constant throughout.

Table 2 shows yearly mean wind speed and corresponding standard deviation. Maximum mean wind speed & minimum mean wind speed was calculated to be 7.35 m/s and 5.07 m/s respectively. Yearly mean wind speed for 2014 was almost same as for 2011. The general trend in yearly mean wind speed seems to be decreasing gradually from 2004 to 2014. Especially, mean wind speed in December/January seems to be sharply decreasing as year progresses which lead to decrease in yearly mean wind speed. In other words, drop of wind speed on cold season is more severe than warm season. Several Processes on local, regional and global scales are likely contributing to this decrease. Increasing forest density can't alone explain this phenomenon described by Iacono [33]. On other hand, several researches have shown using both climate model simulations [34] and surface observations [35] [36] that the positions of the main storm tracks that cross North America, which are generally associated with the jet stream, have moved northwards. This may be impacting the wind speed pattern in nearby areas. Although, similar research is not found in sub-continent region, it can be predicted from their analysis, the global climate change has played a major role in this declining wind speed. The severe decrease of speed in cold season is also explained by increasing temperature in Mountainous region. The region which otherwise would be much colder, air with higher density would move to replace hotter air in lower belts. This pattern is affected.

3.2. Weibull and Rayleigh Distribution

The variation of wind speeds is often described using Weibull & Rayleigh density function. These are statistical tool which are widely accepted for evaluation of local wind probabilities and considered as a standard approach. *Methods of Moments* was used to calculate both weibull parameters. To calculate weibull parameters, yearly mean wind speed and standard deviation were calculate and shown in **Table 2**. **Table 2** shows yearly weibull parameters and average weibull parameters for whole 10 years. It is seen from table that, while scale factor varies between 5.66 and 8.13, the shape factor ranges from 2.83 to 3.84. The 10 year average value of scale factor and shape factor are 6.69 and 3.03 m/s respectively. As it can be seen from **Table 2**, the highest and the lowest of k parameter belongs to 2004 & 2009 respectively. From the result, it is obvious that shape parameter has small variation compared to scale parameter. It has been found that for most wind conditions value of k varies from 1.5 to 3, whereas c ranges from 3 to 8 [37]. Value of c is within the range specified but value of k is offset for this location. k being shape parameter shows how peaked the wind distribution is. For this belt, the wind distribution is peaked compared to general trend. Ration of k/c is a crucial factor as this will determine peak frequency. The high value of k/c will be useful for predicting most probable speed with greater accuracy. Apart from that, methods used for parameter evaluation is also responsible for this difference. Ahmed [11] shows value of k varied between 2.3029 and 3.2592 when four different methods were compared. MOMs as shown a better predictor by researchers, we stick to this value.

Table 2. Yearly mean wind speed for cold and warm seasons in Jumla.

Parameter	2004	2005	2006	2007	2008	2009	2010	2011	2013	2014	Whole Year
V_m (m/s)	7.35	6.52	6.82	6.42	5.65	5.9	5.45	5.07	5.34	5.13	5.98
σ (m/s)	2.13	2.5	2.07	2.01	1.85	2.26	1.8	1.76	1.85	1.81	2.15
k (-)	3.84	2.83	3.65	3.54	3.36	2.83	3.32	3.14	3.16	3.09	3.03
c (m/s)	8.13	7.32	7.56	7.13	6.29	6.63	6.07	5.66	5.97	5.73	6.69

Table 3 shows characteristics wind speed predicted from Weibull model & Rayleigh model. The most probable wind speed, wind speed which is carrying the maximum energy, predicted mean speed and root mean cube speed were calculated. The V_{mp} for weibull ranged from 5.01 to 7.52 m/s with an average of 5.86 m/s whereas for Rayleigh ranged from 4 to 5.75 m/s with an average of 4.73 m/s. Also, the highest value of V_{op} was at 2004. Weibull predicted it to be 9.07 m/s whereas Rayleigh predicted it as 11.5 m/s. Mean speed predicted by weibull was same as predicted main speed because same equation (6, 19) are used to calculate value of shape factor and mean speed. Rayleigh model predicted mean wind speed with maximum deviation of 0.37 m/s with 80% of difference not over 0.1 m/s. Root mean cube speed was calculated with an average value of 6.7 m/s. **Figure 2** shows histogram of the actual frequency distribution for all these years with the Weibull and Rayleigh function for fitting a wind data probability distribution. The difference between these two function is shape parameter k. Estimated average Weibull distribution shape parameter 3.03 which is different with Rayleigh distribution shape parameter 2. As it can be seen in figure, Weibull distribution fits the time series data more appropriately than Rayleigh distribution. Several statistical tools were used to analyze the error in fitting weibull and Rayleigh distribution. **Table 4** shows evaluation of Weibull and Rayleigh distribution and MPE, MAPE, RMSE and Chi-Square goodness-of-fit were used to evaluate them. **Table 4** shows χ^2 is 0.03 for Weibull and 0.46 for Rayleigh. χ^2 with lower value shows better goodness of fit. The MPE, MAPE & RMSE for Weibull distribution were 18%, 30% and 0.012 while these indices for Rayleigh distribution were 70%, 84% and 0.024 respectively. Literatures have found that the weibull model predict the actual value better than in comparison to the Rayleigh model which is supported by this study [8] [9] [12] [19]. Dhunny *et al.* [38] have analysed among 7 distribution to support Weibull distribution is better probability distribution.

3.3. Wind Power Density

The power density calculated from measured probability density distributions and those obtained from models are presented in **Table 5**. As it can be seen in **Table 5**, the average value of wind power calculated from mean wind speed was 131.31 W/m^2/year. Extreme values of wind power calculate using root mean cube speed (for 2004 and 2011) were 306.36 and 109.29 W/m^2/year with an average of 184.93 W/m^2/year. The values of wind power which have been calculated by applying root mean cube speed approach were higher than that of arithmetic mean wind speed values. The wind power density calculated from root mean cube speed and predicted by weibull model are similar. Average wind power density predicted using Weibull and Rayleigh model are 183.28 and 244.82 W/m^2/year. **Table 5** clearly shows wind power estimated by Rayleigh model is higher than that predicted by Weibull model. Betz limit gives maximum efficiency of wind power conversion system (wind turbine) to be 0.593. Best wind turbine have efficiency of around 0.4. So, maximum power generation capability is 183.28 (0.4) W/m^2.

Errors in calculating the power density using Weibull and Rayleigh models are presented in **Table 6**. Taking arithmetic mean wind speed as reference speed, maximum error of 45.36% in Weibull and 88.11% in Rayleigh distribution were estimated. When root mean cube speed was taken as reference, maximum error for Weibull of 3.86% with remaining of calculated errors below 0.9% and for Rayleigh of 43.52%. As was shown before, the weibull function describe the observed values of wind speed reasonably well, so it can be easily guessed that predicted and observed value of wind power density will be similar. This analogy shows root mean cube speed measures wind power density more accurately than arithmetic mean speed. Results of wind power density shows Jumla is in class 3 [39]. Class 1 are generally not suitable for wind turbine applications whereas class 2 areas are marginal. Areas which are classified as class 3 or greater are suitable for most wind turbine applications [40]. For initial years, Jumla was a class 6 region and the decreasing wind speed has put in class 3 today. It is a big

Table 3. Characteristics wind speed (m/s).

Year	Weibull			Rayleigh			V_{rmc}
	V_{mp}	V_{op}	V_m	V_{mp}	V_{op}	V_m	
2004	7.52	9.07	7.35	5.75	11.50	7.21	7.93
2005	6.28	8.85	6.52	5.18	10.36	6.49	7.48
2006	6.93	8.53	6.82	5.35	10.70	6.70	7.42
2007	6.50	8.09	6.42	5.04	10.09	6.32	7.00
2008	5.67	7.23	5.65	4.45	8.90	5.28	6.21
2009	5.68	8.00	5.90	4.69	9.37	5.88	6.67
2010	5.45	7.00	5.45	4.29	8.59	5.38	5.99
2011	5.01	6.62	5.07	4.00	8.01	5.02	5.62
2013	5.29	6.97	5.34	4.22	8.44	5.29	5.93
2014	5.05	6.74	5.13	4.05	8.11	5.08	5.70
Average	5.86	7.90	5.98	4.73	9.46	5.93	6.70

Table 4. Evaluation of Weibull and Rayleigh distribution.

Index	Weibull	Rayleigh
MPE	18%	70%
MAPE	30%	84%
RMSE	0.012	0.024
χ^2	0.03	0.46

Table 5. Wind power density (W/m^2/year) of Jumla (2004-2014).

Year	Measured		Predicted	
	Mean Speed	Root Mean Cube Speed	Weibull	Rayleigh
2004	244.61	306.36	306.33	439.70
2005	170.91	257.93	248.43	321.50
2006	195.34	251.18	249.72	354.19
2007	163.02	211.09	211.04	297.01
2008	111.05	147.65	147.18	204.00
2009	126.72	182.52	183.91	238.31
2010	99.56	132.52	132.64	183.20
2011	80.06	109.29	109.64	148.53
2013	93.75	128.46	128.11	173.83
2014	82.94	114.25	114.62	154.26
Average	131.31	184.93	183.28	244.82

challenge to forecasting power density with similar trend of decreasing speed and if similar trend is seen, it may be class 2 region some years later which is a subject of concern for investors. Similarly, road transportation puts a question mark for such projects is Himalayan region.

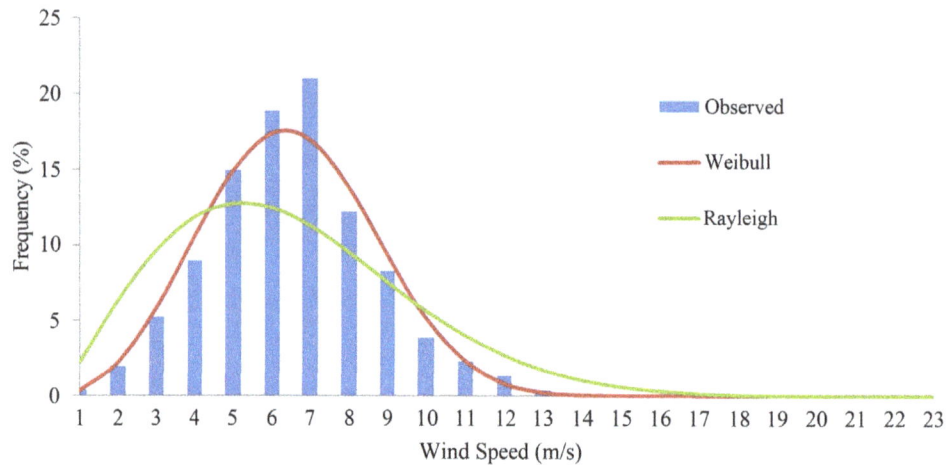

Figure 2. Comparison of observed and predicted wind speed frequencies of Jumla.

Table 6. Error (%) values in calculating wind power density (2004-2014).

Reference	Mean Speed		Root Mean Cube Speed	
Year	Weibull	Rayleigh	Weibull	Rayleigh
2004	25.23	79.76	−0.01	43.52
2005	45.36	88.11	−3.68	24.65
2006	27.84	81.32	−0.58	41.01
2007	29.46	82.19	−0.02	40.70
2008	32.53	83.70	−0.32	38.16
2009	45.13	88.06	0.76	30.57
2010	33.23	84.01	0.09	38.24
2011	36.95	85.52	0.32	35.90
2013	36.65	85.42	−0.27	35.32
2014	38.20	85.99	0.32	35.02
Average	39.58	86.44	−0.89	32.39

4. Conclusions

In the present study we discussed analysis whose objective was to investigate the potential of wind energy resource in Jumla. For this purpose, wind speed data of Jumla station (DHM) were analyzed over a 10 year period from 2004 to 2014 (2012 excluded). The probability density distributions and power density were derived from time series data. Weibull and Rayleigh probability density function have been fitted to the measured probability distributions. The wind power density has been evaluated. The most important outcomes of the study can be summarized as follows:

1) Jumla is shown to be a marginal site (Class III) for wind energy generation as the region possesses moderate wind characteristics. This is shown by average monthly & yearly wind speed along with wind power density. Himalayan region is supposed to have higher power generation capability in Nepal. Jumla, as one of perceived potential site, being a class III region shows Nepal has modest probability of wind energy generation capability in large scale.

2) There is a decreasing trend in yearly & monthly wind speed in Jumla which is a subject of concern. Decrease of wind speed is more than in warm season. We emphasize global climate change for this effect. This will be a subject of concern for Government while they're focusing on diverse source of energy after fuel crisis in Nepal and looking forward for wind energy development.

3) Warm season has higher mean wind speed compared to cold season and this will make sure, energy generation during warm and cold season don't have much difference as density varies with temperature.

4) The Weibull distribution is fitting the measured probability distribution better than Rayleigh distribution and supports the studies done in other parts of the world.

These results fulfill our four reasons of estimation of weibull parameter mentioned in introduction section. Meteorological data was calibrated to characterize the past data which was used to predict power generation capability. The results obtained are satisfying. Meanwhile, further investigations are to be done based on a more detailed and systematic analysis of wind speed patterns. Similarly, wind rose is not shown because of non-availability of information on wind direction from the station. Future works shall be guided in that path.

Acknowledgements

Supports by Department of Mechanical Engineering, Pulchowk Campus, Institute of Engineering, Tribhuwan University is gratefully acknowledged and also the author would like to thank Department of Hydrology and Meteorology, Government of Nepal for providing wind data of Jumla station.

References

[1] Albuhairi, M.H. (2006) Assessment and Analysis of Wind Power Density in Taiz—Republic of Yemen. *Assiut University Bulletin for Environmental Researches*, **9**, 13-21.

[2] Ramachandra, T. and Shruthi, B. (2005) Wind Energy Potential Mapping in Karnataka, India, Using GIS. *Energy Conversion and Management*, **46**, 1561-1578. http://dx.doi.org/10.1016/j.enconman.2004.07.009

[3] Hernandez-Escobedo, Q., Manzano-Agugliaro, F., Gazquez-Parra, J.A. and Zapata-Sierra, A. (2011) Is the Wind a Periodical Phenomenon? The Case of Mexico. *Renewable and Sustainable Energy Reviews*, **15**, 721-728. http://dx.doi.org/10.1016/j.rser.2010.09.023

[4] Wentink Jr., T.W. (1976) Study of Alaskan Wind Power and Its Possible Applications. Final Report, 1 May 1974-30 Jan 1976, Geophysical Institute, Alaska University, Fairbanks.

[5] Justus, C.G., Hargreaves, W.R. and Yalcin, A. (1976) Nationwide Assessment of Potential Output from Wind-Powered Generators. *Journal of Applied Meteorology*, **5**, 673-678.

[6] Baynes, C. and Davenport, A. (1975) Some Statistical Models for Wind Climate Prediction. *Preprints Fourth Conference Probability and Statistics in the Atmospheric Sciences*, Tallahassee, 18-21 November 1975, 1-7.

[7] Rehman, S., Halawani, T.O. and Husain, T. (1994) Weibull Parameters for Wind Speed Distribution in Saudi Arabia. *Solar Energy*, **53**, 473-479. http://dx.doi.org/10.1016/0038-092X(94)90126-M

[8] Fyrippis, I., Axaopoulos, P.J. and Panayiotou, G. (2010) Wind Energy Potential Assessment in Naxos Island, Greece. *Applied Energy*, **87**, 577-586. http://dx.doi.org/10.1016/j.apenergy.2009.05.031

[9] Oner, Y., Ozcira, S., Bekiroglu, N. and Senol, I. (2013) A Comparative Analysis of Wind Power Density Prediction Methods for Çanakkale, Intepe Region, Turkey. *Renewable and Sustainable Energy Reviews*, **23**, 491-502. http://dx.doi.org/10.1016/j.rser.2013.01.052

[10] Odo, F.C., Offiah, S.U. and Ugwuoke, P.E. (2012) Weibull Distribution-Based Model for Prediction of Wind Potential in Enugu, Nigeria. *Advances in Applied Science Research*, **3**, 1202-1208.

[11] Ahmed, S.A. (2013) Comparative Study of Four Methods for Estimating Weibull Parameters for Halabja, Iraq. *International Journal of Physical Sciences*, **8**, 186-192.

[12] Islam, M.R., Saidur, R. and Rahim, N.A. (2011) Assessment of Wind Energy Potentiality at Kudat and Labuan, Malaysia Using Weibull Distribution Function. *Energy*, **36**, 985-992. http://dx.doi.org/10.1016/j.energy.2010.12.011

[13] Safari, B. and Gasore, J. (2010) A Statistical Investigation of Wind Characteristics and Wind Energy Potential Based on the Weibull and Rayleigh Models in Rwanda. *Renewable Energy*, **35**, 2874-2880. http://dx.doi.org/10.1016/j.renene.2010.04.032

[14] Oyedepo, S.O., Adaramola, M.S. and Paul, S.S. (2012) Analysis of Wind Speed Data and Wind Energy Potential in Three Selected Locations in South-East Nigeria. *International Journal of Energy and Environmental Engineering*, **3**, 7. http://dx.doi.org/10.1186/2251-6832-3-7

[15] Abbas, K., Alamgir, K., Ali, A., Khan, D. and Khalil, U. (2012) Statistical Analysis of Wind Speed Data in Pakistan. *World Applied Sciences Journal*, **18**, 1533-1539.

[16] WECS (2008) Energy Sector Synopsis Report.

[17] Surendra, K.C., Khanal, S.K., Shrestha, P. and Lamsal, B. (2011) Current Status of Renewable Energy in Nepal: Op-

portunities and Challenges. *Renewable and Sustainable Energy Reviews*, **15**, 4107-4117. http://dx.doi.org/10.1016/j.rser.2011.07.022

[18] Ghimire, M., Poudel, R.C., Bhattarai, N. and Luintel, M.C. (n.d.) Wind Energy Resource Assessment and Feasibility Study of Wind Farm in Mustang. *Journal of Institute of Engineering*, **8**, 93-106. http://dx.doi.org/10.3126/jie.v8i1-2.5099

[19] Pishgar-Komleh, S.H., Keyhani, A. and Sefeedpari, P. (2015) Wind Speed and Power Density Analysis Based on Weibull and Rayleigh Distributions (A Case Study: Firouzkooh County of Iran). *Renewable and Sustainable Energy Reviews*, **42**, 313-322. http://dx.doi.org/10.1016/j.rser.2014.10.028

[20] Buenestado-Caballero, P., Jarauta-Bragulat, E. and Hervada-Sala, C. (2006) Weibull Parameters Distribution Fitting in the Surface Wind Layer. *International Association for Mathematical Geology* 11*th International Congress*, Liège, 3-8 September 2006, 6-9.

[21] Simiu, E. and Heckert, N.A. (1996) Extreme Wind Distribution Tails: A "Peaks over Threshold" Approach. 539-547.

[22] Ouarda, T.B.M.J., Charron, C., Shin, J.-Y., Marpu, P.R., Al-Mandoos, A.H., Al-Tamimi, M.H., *et al.* (2015) Probability Distributions of Wind Speed in the UAE. *Energy Conversion and Management*, **93**, 414-434. http://dx.doi.org/10.1016/j.enconman.2015.01.036

[23] Weibull, W. (1951) A Statistical Distribution Function of Wide Applicability. *Journal of Applied Mechanics*, **103**, 293-297.

[24] Akdağ, S.A. and Dinler, A. (2009) A New Method to Estimate Weibull Parameters for Wind Energy Applications. *Energy Conversion and Management*, **50**, 1761-1766. http://dx.doi.org/10.1016/j.enconman.2009.03.020

[25] Azad, A., Rasul, M. and Yusaf, T. (2014) Statistical Diagnosis of the Best Weibull Methods for Wind Power Assessment for Agricultural Applications. *Energies*, **7**, 3056-3085. http://dx.doi.org/10.3390/en7053056

[26] Chang, T.P. (2010) Wind Speed and Power Density Analyses Based on Mixture Weibull and Maximum Entropy Distributions. *International Journal of Applied Science and Engineering Research*, **8**, 39-46.

[27] Carlin, P.W. (1997) Analytical Expressions for Maximum Wind Turbine Average Power in a Rayleigh Wind Regime. *ASME Wind Energy Symposium*, Reno, 6-9 January 1997, 1-9.

[28] Patel, M. (2005) Wind and Solar Power Systems: Design, Analysis, and Operation. CRC Press, Boca Raton. http://dx.doi.org/10.1201/9781420039924

[29] Keyhani, A., Ghasemi-Varnamkhasti, M., Khanali, M. and Abbaszadeh, R. (2010) An Assessment of Wind Energy Potential as a Power Generation Source in the Capital of Iran, Tehran. *Energy*, **35**, 188-201. http://dx.doi.org/10.1016/j.energy.2009.09.009

[30] Caretto, L. (2010) Use of Probability Distribution Functions for Wind. California State University Northridge, Calif.

[31] Crutcher, H.L. (1957) On the Standard Vector-Deviation Wind Rose. *Journal of Meteorology*, **14**, 28-33. http://dx.doi.org/10.1175/0095-9634-14.1.28

[32] Dore, A.J., Vieno, M., Fournier, N., Weston, K.J. and Sutton, M.A. (2006) Development of a New Wind-Rose for the British Isles Using Radiosonde Data, and Application to an Atmospheric Transport Model. *Quarterly Journal of the Royal Meteorological Society*, **132**, 2769-2784. http://dx.doi.org/10.1256/qj.05.198

[33] Iacono, M.J. (2009) Why Is the Wind Speed Decreasing? *Journal of Geophysical Research: Atmospheres*, **114**, 1-3.

[34] Yin, J.H. (2005) A Consistent Poleward Shift of the Storm Tracks in Simulations of 21st Century Climate. *Geophysical Research Letters*, **32**, L18701. http://dx.doi.org/10.1029/2005GL023684

[35] Leibensperger, E.M., Mickley, L.J. and Jacob, D.J. (2008) Sensitivity of US Air Quality to Mid-Latitude Cyclone Frequency and Implications of 1980-2006 Climate Change. *Atmospheric Chemistry and Physics*, **8**, 7075-7086. http://dx.doi.org/10.5194/acpd-8-12253-2008

[36] Wang, X.L., Wan, H. and Swail, V.R. (2006) Observed Changes in Cyclone Activity in Canada and Their Relationships to Major Circulation Regimes. *Journal of Climate*, **19**, 896-915. http://dx.doi.org/10.1175/JCLI3664.1

[37] Kaldellis, J.K. (1999) Wind Energy Management. Stomoullis, Athens.

[38] Dhunny, A.Z., Lollchund, M.R., Boojhawon, R. and Rughooputh, S.D.D.V. (2014) Statistical Modelling of Wind Speed Data for Mauritius. *International Journal of Renewable Energy Research*, **4**, 1056-1064.

[39] Mohammadi, K. and Mostafaeipour, A. (2013) Using Different Methods for Comprehensive Study of Wind Turbine Utilization in Zarrineh, Iran. *Energy Conversion and Management*, **65**, 463-470. http://dx.doi.org/10.1016/j.enconman.2012.09.004

[40] Elliott, D. and Holladay, C. (1987) Wind Energy Resource Atlas of the United States. NASA STI/Recon Technical Report N, 87, 24819.

PERMISSIONS

All chapters in this book were first published in EPE, by Scientific Research Publishing; hereby published with permission under the Creative Commons Attribution License or equivalent. Every chapter published in this book has been scrutinized by our experts. Their significance has been extensively debated. The topics covered herein carry significant findings which will fuel the growth of the discipline. They may even be implemented as practical applications or may be referred to as a beginning point for another development.

The contributors of this book come from diverse backgrounds, making this book a truly international effort. This book will bring forth new frontiers with its revolutionizing research information and detailed analysis of the nascent developments around the world.

We would like to thank all the contributing authors for lending their expertise to make the book truly unique. They have played a crucial role in the development of this book. Without their invaluable contributions this book wouldn't have been possible. They have made vital efforts to compile up to date information on the varied aspects of this subject to make this book a valuable addition to the collection of many professionals and students.

This book was conceptualized with the vision of imparting up-to-date information and advanced data in this field. To ensure the same, a matchless editorial board was set up. Every individual on the board went through rigorous rounds of assessment to prove their worth. After which they invested a large part of their time researching and compiling the most relevant data for our readers.

The editorial board has been involved in producing this book since its inception. They have spent rigorous hours researching and exploring the diverse topics which have resulted in the successful publishing of this book. They have passed on their knowledge of decades through this book. To expedite this challenging task, the publisher supported the team at every step. A small team of assistant editors was also appointed to further simplify the editing procedure and attain best results for the readers.

Apart from the editorial board, the designing team has also invested a significant amount of their time in understanding the subject and creating the most relevant covers. They scrutinized every image to scout for the most suitable representation of the subject and create an appropriate cover for the book.

The publishing team has been an ardent support to the editorial, designing and production team. Their endless efforts to recruit the best for this project, has resulted in the accomplishment of this book. They are a veteran in the field of academics and their pool of knowledge is as vast as their experience in printing. Their expertise and guidance has proved useful at every step. Their uncompromising quality standards have made this book an exceptional effort. Their encouragement from time to time has been an inspiration for everyone.

The publisher and the editorial board hope that this book will prove to be a valuable piece of knowledge for researchers, students, practitioners and scholars across the globe.

LIST OF CONTRIBUTORS

Oleksandr Mokin, Borys Mokin and Vadym Bazalytskyy
Department of Renewable Energy and Transport Electrical Systems and Complexes, Vinnytsia National Technical University, Vinnytsia, Ukraine

Zaccheus O. Olaofe
Climate System and Analysis Group, University of Cape Town, Rondebosch, South Africa

Timo Wundsam and Sascha M. Henninger
Department of Physical Geography, University of Kaiserslautern, Kaiserslautern, Germany

Sagarkumar M. Agravat and T. Harinarayana
Gujarat Energy Research and Management Institute, Gandhinagar, India

N. V. S. Manyam
Department of Electrical Engineering, University of Petroleum and Energy Studies, Dehradun, India

Sanket Mankar
School of Mechanical and Building Sciences, Vellore Institute of Technology, Vellore, India

Seyedali Meghdadi and Tariq Iqbal
Department of Electrical Engineering, Memorial University of Newfoundland, St. John's, Canada

Kedharnath Sairam and Mark G. Turner
School of Aerospace Systems, University of Cincinnati, Cincinnati, USA

Luqman Muhammed Audu
Department of Mechanical Engineering, Auchi Polytechnic, Auchi, Nigeria

Olugbenga Olanrewaju Noah
Department of Mechanical Engineering, University of Lagos, Lagos, Nigeria

Ogaga Kenneth Ajaino
Department of Mechanical Engineering, Delta State Polytechnic, Ogwashuku, Nigeria

Friday Chukwuyem Igbesi
Department of Mechanical Engineering, Delta State Polytechnic, Otefe-Oghara, Nigeria

Miguel Edgar Morales Udaeta, Antonio Gomes dos Reis, José Aquiles Baesso Grimoni and Antonio Celso de Abreu Junior
GEPEA/EPUSP, Energy Group of the Department of the Electrical Energy and Automation Engineering/ Polytechnic School of the University of São Paulo, São Paulo, Brazil

Takaaki Kono and Takahiro Kiwata
Research Center for Sustainable Energy & Technology, Kanazawa University, Kanazawa, Japan

Akira Yamagishi, Shigeo Kimura and Nobuyoshi Komatsu
Division of Mechanical Science and Engineering, Kanazawa University, Kanazawa, Japan

Hua Ji
Waterloo CFD Engineering Consulting Inc., Waterloo, Ontario, Canada

Zhongxian Men and Fue-Sang Lien
Waterloo CFD Engineering Consulting Inc., Waterloo, Ontario, Canada
Department of Mechanical & Mechatronics Engineering, University of Waterloo, Waterloo, Onatrio, Canada

Eugene Yee
Department of Mechanical & Mechatronics Engineering, University of Waterloo, Waterloo, Onatrio, Canada
Defence Research and Development Canada, Suffield Research Centre, Medicine Hat, Alberta, Canada

Yongqian Liu
School of Renewable Energy, North China Electric Power University, Beijing, China

Shengzhi Chen
Department of Mechano-Micro Engineering, Tokyo Institute of Technology, Kanagawa, Japan

Chongho Youn and Toshiharu Kagawa
Interdisciplinary Graduate School of Science and Engineering, Tokyo Institute of Technology, Kanagawa, Japan

Maolin Cai
School of Automation Science and Electrical Engineering, Beihang University, Beijing, China

Robert W. Furness, Mark Trinder and David MacArthur
MacArthur Green, Glasgow, UK

Andrew Douse
Scottish Natural Heritage, Great Glen House, Inverness, UK

Manzar Ahmed, Uzma Amin, Suhail Aftab and Zaki Ahmed
Electrical Engineering Department, Faculty of Engineering, University of South Asia, Lahore, Pakistan

Francesco Ruggiero and Graziarosa Scaletta
Dipartimento di Scienze dell'Ingegneria Civile e dell'Architettura, Politecnico di Bari, Bari, Italy

Jose M. "Chema" Martínez-Val Piera
ETSI Minas, Universidad Politécnica de Madrid, Madrid, Spain

Sergei Soldatenko
A.M. Obukhov Institute of the Atmospheric Physics of the Russian Academy of Sciences, Moscow, Russia

Lev Karlin
Russian State Hydrometeorological University, St. Petersburg, Russia

Amitesh Pandey
Jaypee Cement Limited, Rewa, India

Ayush Parajuli
Department of Mechanical Engineering, Pulchowk Campus, Tribhuwan University, Lalitpur, Nepal

Index